New Zealand Society of Animal Production Conference (NZSAP 2021)

Postponed until 2022

Proceedings of the New Zealand Society of Animal Production
Volume 81

Online
1-3 February 2022

ISBN: 978-1-7138-5219-3

Printed from e-media with permission by:

Curran Associates, Inc.
57 Morehouse Lane
Red Hook, NY 12571

CURRAN ASSOCIATES INC.
proceedings
.com

Some format issues inherent in the e-media version may also appear in this print version.

New Zealand Society of Animal Production (Inc.) (NZSAP)
c/o Kate Crookston
PO Box 955
Cambridge, New Zealand

www.nzsap.org.nz

Additional copies of this publication are available from:

Curran Associates, Inc.
57 Morehouse Lane
Red Hook, NY 12571 USA
Phone: 845-758-0400
Fax: 845-758-2633
Email: curran@proceedings.com
Web: www.proceedings.com

TABLE OF CONTENTS

INVITED REVIEW: A review of the current sheep industry in New Zealand and opportunities for change to meet future challenges.

PL Johnson[a*], S-AN Newman[a], KM McRae[a], TJ van der Weerden[a], M Brown[b] and DR Scobie[c]

[a]*AgResearch, Invermay Agricultural Centre, Private Bag 50034, Mosgiel, New Zealand;* [b]*AgResearch, Grasslands Research Centre, Private Bag 11008, Palmerston North, New Zealand;* [c]*AgResearch, Lincoln Research Centre, Private Bag 4749, Christchurch, New Zealand*

**Corresponding author:* Email: tricia.johnson@agresearch.co.nz

Abstract

Like all farming systems, sheep operations face increasing economic, environmental, cultural, and social challenges, including climate change and consumer expectations about their food. Genetic improvement has the potential to help with these challenges. This review explores the importance of genetic improvement on welfare and disease traits in meeting these challenges. For all wool breeds, there is a need for traits such as being polled, having a short tail and having a bare breech and a bare belly to ensure an ethical outcome. Sheep that either seasonally lose their fleece, or do not grow wool, represent an opportunity. Predicted climate change poses significant animal-health challenges, including the internal parasite *Haemonchus contortus*, facial eczema, flystrike and pneumonia, that are likely to encroach on a wider geographic range. Given consumer expectations around reduced chemical inputs and improved animal welfare, genetic improvement will play a critical role in developing sheep that are more tolerant to such health challenges. While several key physical characteristics required for the future exist in minor breeds currently in New Zealand, their performance under a range of New Zealand commercial farming conditions has not been evaluated. Some overseas breeds, such as Katahdin and Barbados Blackbelly, may offer opportunities. Consideration needs to be given to the use of novel gene pools both within and outside the New Zealand sheep industry for it to be able to meet future challenges. To achieve this end, more information is needed about the overall merit of these novel breeds, to be able to identify the best strategy to take towards breeding sheep that address the future needs of the New Zealand sheep industry.

Keywords: sheep; genotypes; breeding; future

Introduction

Sheep farming is a significant contributor to New Zealand's economy, via employment throughout the value chain, and national and international revenue from product sales. However, like all farming systems, sheep operations face an increasing number of economic, environmental, cultural and social challenges. These include competition from alternative land uses (Journeaux et al. 2017), the impact of climate change on production systems (McRae et al. 2018b), consumer expectations about their food and food-production systems (Montossi et al. 2013), increasing concerns about the environmental impact of farming (Nielsen Company 2019), and competition from synthetic products (Antedote Group 2018).

Sheep are a domesticated, small ruminant (*Ovis aries*) that originated in the Middle East and Asia. With domestication, man has created a range of phenotypes associated with their requirements in a wide range of environments. This has resulted in the development of a range of breeds with variations in their genotype and associated phenotypic expression for a wide range of traits. While these different genotypes can interbreed, the phenotypic ranges within many genotypes do not overlap, thereby creating a wide variation in phenotypic characteristics within the *O. aries* species (Cottle 2010).

Production, health, and welfare of animals are influenced by a combination of genotype, environment and management. Ensuring that appropriate genotypes are utilised forms the basis of the sheep industry's ability to remain profitable whilst fulfilling consumer expectations. The appropriate genotypes may not necessarily be those that predominate today. This paper provides a historical perspective on New Zealand sheep genotypes, then discusses what changes the industry needs to make for the future.

Current Situation

Genetic diversity of the New Zealand sheep flock

As with all New Zealand's farmed livestock species, sheep are the result of the importation of exotic breeds. The foundations of the New Zealand sheep flock were imported from Australia and the United Kingdom in the early to mid-1800s. Initially, the Merino was the favoured breed because of its fine wool. When the ability to export meat was established in 1882, dual-purpose sheep, capable of producing both a saleable carcass and wool suitable for manufacturing, became increasingly popular. Through the early to mid-1900s new breeds were imported, predominantly from the United Kingdom. New breeds were also developed within New Zealand, with aspects of different introduced breeds integrated to produce animals better suited to the environment while growing products suited for export. The three examples that continue to be farmed today are the Corriedale (Lincoln or Leicester x Merino), Perendale (Cheviot x Romney) and Coopworth (Border Leicester x Romney).

Attempts to import genetically diverse breeds from Europe included the introduction of the Texel, Finnish Landrace, Gotland Pelt, Oxford Down and White-headed Marsh in the late 1980s. In the early 1990s Awassi, Karakul and East Friesian were also imported. Since 2016, despite revised biosecurity measures, an increasing number of genetically diverse animals have been imported, some for milk production (for example the Lacaune), others for their shedding attributes (for example the UltraWhite, EasyCare, Exlana) and others for their meat attributes (for example Australian White, Lambpro).

Within New Zealand, there has been the use of imported breeds alongside traditional Romney, Perendale and Coopworth breeds in composite maternal breeding programmes, particularly following the introduction of the Texel and Finnish Landrace. This ranges from introgression, with the Texel muscling gene now present in many other breeds in which the mutation was not historically found, through to the establishment of stabilised composites, such as the TEFRom (comprises Romney, Texel and East Friesian genotypes), Highlander (comprises Romney, Finn and Texel genotypes) and Kelso Maternal (contains Romney, Coopworth, Finn, Texel and East Friesian genotypes). There has also been a rise in the number of terminal (meat- production) composites.

The New Zealand sheep population peaked at 70 million sheep wintered in 1982, but has progressively declined to 26.2 million sheep wintered in 2020. Romney, or Romney-type sheep, as defined by the person completing the census document, remains the predominant breed, accounting for approximately 50% of the New Zealand breeding ewe population, but with "composite" being the second most populous ewe breed now, ahead of the Perendale and Coopworth (Beef + Lamb New Zealand Economic Service 2020).

Products derived from sheep

When the New Zealand sheep industry was established, the major product was wool, which could be shipped to the United Kingdom, plus meat for local consumption. As relative product values and new markets have emerged an increased range of products has been produced.

Meat. New Zealand currently has diverse international markets for sheep-meat products, with key markets in North Asia, the European Union and North America. The amount of sheep-meat, consisting of lamb (young sheep aged under one year of age) and mutton (sheep aged over one year of age) produced by New Zealand farmers, has remained relatively stable over the last twenty years despite the size of the national ewe flock halving during that time (Beef + Lamb New Zealand Economic Service 2020).

Pelts. Sheep skin or 'pelts' have been harvested as a by-product since the carcass industry began. Sheep skin can be used for two major products when they are tanned with the wool remaining on them or the wool removed and tanned as a leather (Scobie 2010). Pelts can quickly deteriorate unless they are cooled and processed soon after removal from the animal. However, once they are part processed or "pickled" they can be stored for some time. Stockpiling of pickled pelts can affect demand and supply relationships, but demand has been depressed by "faux furs" and synthetic alternatives which are cheaper and more uniform in much the same way as alternatives for yarns. Currently pelts are considered a by-product and farmers do not receive a nominated return reflecting their quality.

Wool. The New Zealand wool clip is broadly divided into three classes, coarse (>32.5 μm), medium (24.6-32.5 μm) and fine wool (<24.5 μm). Products that can be manufactured from each of these wool types vary, and so does the price the raw wool can attract (Petrie 1995). Fine wool suitable to make garments worn against the skin is more valuable than coarse wool made into interior furnishings. The price differential between fine and coarse wool has increased over the last two decades. As of 2019, fine wool has been worth over six times more than coarse wool, and two to three times more than medium wool (Beef + Lamb New Zealand Economic Service 2020).

In 1970 approximately 2% of the wool offered for sale by auction was finer than 24.5 μm (Merino) and 75% was coarser than 32.5 μm, whereas by 2020 approximately 8% of the wool offered for sale by auction was finer than 24.5 μm (Merino) and 65% was coarser than 32.5 μm (Beef + Lamb New Zealand Economic Service 2020).

While the costs of wool harvesting have increased over this period, the price rise is independent of the wool type. In contrast, the low price of coarse wool has meant that most sheep farmers now receive a negative net return from the production of coarse wool.

Milk. Compared to meat, skins and wool, the New Zealand dairy-sheep industry is in its infancy. Establishment of the industry was contingent on market development, and a suitable gene pool being available. Most traditionally farmed breeds in New Zealand do not produce adequate levels of milk on which to develop a financially viable industry (Peterson & Prichard 2016). Recently, three large-scale enterprises, milking thousands of ewes, has expanded the industry from small-scale family operations milking 50-200 ewes producing specialty cheeses and milk for local markets, to high-value infant formula, nutritional powders and luxury high-value sheep-milk foods for international markets.

Where sheep are farmed

The absolute area and class of land utilised for farming sheep in New Zealand has changed significantly over time.

Attributes of the land

The farmed areas of New Zealand are made up of a variety of land types, ranging from flat land to rugged steep hill country, which are generally categorised into land-use capability classification classes ranging from Class 1 (few limitations for cultivation) to Class 8 (unsuitable for grazing or production forestry and is best managed for catchment protection and/or conservation or biodiversity) (Beef + Lamb New Zealand Economic Service 2020).

New Zealand is situated between 35.6° and 47° latitude south of the equator. Historically, the climate has been described as temperate, however, there is great climatic variation through the country.

The combination of land class and climate influences pasture growth, both in terms of seasonality and quantity. This variation influenced the choice of breeds of sheep that were imported. The Merino, which is more suited to arid and extreme climates, became the common breed in the South Island high country, whilst for the rest of the country, breeds from the temperate United Kingdom predominated.

Drivers of land-use change

Initial land use was directed by product prices in association with climate and land class, and the requirements for settlement. As product prices have changed, pressure on land use for different enterprises has changed, in association with an increased population and the need for increased urbanisation.

A significant change between the 1990s and the early 2010s, was the conversion of large areas of better-quality land, where production could be increased through the application of fertiliser and/or irrigation, from sheep and beef units to dairy farms. This change resulted in a significant decline in both the national sheep population and the number of sheep farms.

While historically 'flood' irrigation has been used in some areas, centre-pivot irrigation has seen dramatic increases in recent decades. This has resulted in dry areas such as the Mackenzie Basin, which were extensive sheep and beef farms, being converted to dairy farms. The environmental sustainability of such decisions has been questioned, however, in the medium-term it is unlikely that such properties will revert to sheep.

Sheep are considered to be more environmentally friendly than dairy and beef cattle, with less direct damage to soil structure, a more-even spread of lower volumes of urine and lower water requirements. Hence, on the permeable volcanic soils surrounding Lake Taupo, sheep have displaced dairy cattle as a direct result of environmental considerations where restrictions have been placed on the number of dairy cattle that can be grazed to reduce the amount of nitrogen draining from the farmed land back into the lake.

Forestry. The Emissions Trading Scheme (ETS) was introduced in 2008, with tradeable carbon credits earned through the sequestration of carbon, largely in the form of fast-growing trees. Whilst in the first decade there was some planting of harvestable species, there has been a significant increase in forestry conversions since 2017, as a direct result of government initiatives to increase carbon sequestration. Previously, smaller areas within farms would be converted to pine plantations, however, there are now examples where entire farms are purchased and planted. Most farms considered for such purposes were sheep and beef farms close to ports, with adequate rainfall and suitable elevation.

Current New Zealand sheep industry. As a consequence of the factors described above, there has been a significant shift in land use during recent decades with a decline in sheep production in some areas. In addition to the reduction in area farmed, the number of sheep farms has reduced due to the amalgamation of neighbouring farms. Despite the decrease in numbers of sheep this has been offset by increased per-head productivity (Beef + Lamb New Zealand Economic Service 2020).

Future of the New Zealand sheep industry

In the future, the New Zealand sheep industry must become increasingly dynamic and responsive to a variety of factors that influence its social license and profitability relative to other production systems. Whilst there have always been expectations placed on farming systems about the ethical production of meat, the advent of social media has resulted in both negative and positive aspects of production systems being highlighted and rapidly propagated globally. There is also a growing expectation of transparency throughout the food chain, with the desire of many consumers to know where their food comes from. With the rise in popularity of alternative protein sources, consideration of consumer expectations must become a critical part of decision making by breeders and farmers. Furthermore, if a proportion of the industry can achieve the standards and expectations of the consumer through a mixture of breeding and management then all farmers will be required to achieve similar outcomes.

While many farming practices are common across all production species, some are unique to farming sheep. Additionally, the role that production animals have in contributing to climate change is of importance to consumers. Several profitability drivers stem from the ability of animals to remain productive in a changing climate, with the additional health challenges that this presents, as well as the relative value of different products and the relative emphasis placed on them in various production systems.

The following sections capture welfare expectations, attributes of the animal that need to be considered to address health outcomes in a changing climate, and what the suite of products from sheep could look like.

Meeting welfare expectations

Animal welfare has always been a paramount concern of consumers, however, some surgical practices that used to occur on-farm, remained 'under the radar' for many years before being phased out. There has also been a change in perception and previous notions of practices that involve short-term pain for a long-term beneficial advantage to the animal concerned are now being challenged as unacceptable.

Tail length. The docking of tails is almost ubiquitously used on New Zealand sheep farms to reduce dags and flystrike and to decrease the cost of crutching animals. Tail docking has been challenged by animal welfare and animal rights activists and is prohibited in some countries. A genetic solution is an attractive option as it would be permanent and cumulative.

Many breeds worldwide have short tails, and there are reports of short-tailed animals occurring in long-tailed breeds. In most breeds, the trait is thought to be polygenic, occurring as a result of multiple genes. When short-tail phenotypes occur in long-tailed breeds, it is thought to be a result of a mutation in one or two genes of major effect, but the phenotype has almost invariably been associated with other deleterious effects (James et al. 1990). James (2006) reviewed heritability estimates for tail length from all over the world with estimates ranging from 0.37 to 0.84.

In New Zealand, all the major commercial breeds have naturally long, thin tails but sources of short-tail genes are represented in some minor breeds; Finnish Landrace and Gotland Pelt have very short tails, and the Texel and Dorper both have "stubby" tails (Scobie et al. 2007).

Horns. While most domesticated sheep breeds are polled, hornedness is a traditional breed 'standard' for some breeds, such as the Merino, Dorset Horn and Wiltshire Horn. Horns pose both a welfare risk, with injury to other animals and humans, and a product quality risk, with pelt damage and bruising of the carcass before slaughter. Horns are exclusively under genetic control. Mutations that result in polled animals are known and can be selected for. If there is no genetic variation in a horned population of interest, the polled mutation can be introduced through repeated back-crossing as was done to create the Poll Dorset and Polled Wiltshire.

Non-body wool. There are areas of skin around the edges of the main fleece-growing area that produce little or no wool of commercial value, but the presence of wool growing on these areas can have welfare considerations. For example, the absence of breech wool can reduce the incidence of dags, and the bareness of wool on the breech or belly simplifies shearing and removes the requirement for crutching prior to main shearing, reducing costs and animal handling requirements.

Breech bareness is a heritable trait (0.33±0.06; Scobie et al. 2007) that is most extensive in East Friesian, Texel and Finnish Landrace sheep. Belly bareness is also a heritable trait (0.31±0.06; Scobie et al. 2011). It occurs sporadically in a variety of breeds including the Cheviot, Border Leicester, Texel and Finnish Landrace. In 1997 Scobie et al. (1997) proposed breeding a sheep with a genetically short tail and no wool on the head, legs, belly or breech. There has been a limited uptake of this approach by farmers.

Whether physically docked or genetically shortened, the tail will still have wool that can become soiled and flystruck. Sheep have been identified that naturally moult their tail wool but not their fleece. Scobie et al. (2016) reported this response is most likely due to genes unrelated to moulting of the fleece. Sheep that moulted wool from the tail exhibited fewer dags at weaning and again at hogget shearing.

'Wool blindness' was a major problem in Romney, Corriedale and Merino sheep in the 1950s and 1960s. Cockrem et al. (1956) showed woolly heads to be associated with reduced fecundity. Within-breed selection of bare-headed rams, as well as the use of Perendale and Coopworth rams in coarse-wool flocks, overcame the antagonist relationship.

Mulesing. Merino sheep historically had skin wrinkles across their body, particularly in the perianal area, predisposing the sheep to flystrike. Mulesing is a procedure in which a large area of skin is removed from the perianal area, historically using shearing blades without anaesthetic. Healing results in the skin around the tail becoming tight and growing little wool, thereby reducing the susceptibility of sheep to flystrike. New Zealand began phasing out mulesing in 2007 and an official ban came into effect in 2018. Genetic variability in body wrinkle and breech cover has been demonstrated in Merinos and their genetic correlation with other production traits to allow for selection against body wrinkle and breech cover (Hatcher & Preston 2018).

Farming healthy animals

Healthy animals are the most productive animals, while poor animal health has welfare implications. Management of animal health currently relies on the use of chemicals such as antibiotics, anthelmintics and zinc, which is against market trends that seek low chemical inputs in farming. One of the key concerns associated with the use of chemicals in animal production systems is their potential contribution to antibiotic resistance in human diseases (Magnusson 2020). Additionally, the effect of treated animals entering the human food chain is questioned. Although these chemicals have withholding periods during which the animal cannot be milked or enter the food chain, there are premiums offered for antibiotic-free products.

Whilst there are many health challenges confronting sheep, the challenges are often geographically influenced due to climate variability across the country. Projected future climate change of an increase in CO_2 concentrations to $\approx$ 540 ppm with a 0.8°C increase in temperature by 2040 will potentially affect sheep health in a variety of ways. Winter and spring rainfall and westerly winds are very likely to increase in the west of the North and South Islands, while conditions are very likely to be drier in the east and north. Summer will be wetter in the east of both islands, while the frequency of extreme rainfall and drought will increase in most areas of New Zealand (Ministry for the Environment 2018; Tait et al. 2016). Modelling has been carried out to consider changes in livestock classes given different climate change scenarios (McRae et al. 2018b). Using current populations of sheep, it is predicted that very few, if any, sheep will be farmed in the upper North Island, with a decreasing proportion being farmed in the lower North Island and upper South Island as a result of animal health challenges.

Health challenges that affect New Zealand sheep, with specific emphasis on the potential that genotypes can play in ensuring animals can remain productive, and withstand disease challenges with reduced need for chemical inputs are discussed below.

Facial eczema (FE). FE is a disease of major concern. The disease results from ingestion of the mycotoxin sporidesmin, found in spores from the fungus *Pithomyces chartarum* (Smith 2000). Climate-change projections indicate an increase in the geographical spread of the causative fungus, resulting in outbreaks of FE in currently unaffected regions (Dennis et al. 2014).

Current strategies for preventing severe FE include protecting animals through ingestion of zinc compounds (Smith et al. 1977), spraying pasture with fungicide (Parle & di Menna 1972), pasture management or alternative feeds (Towers 1986), and breeding animals for tolerance (Morris et al. 1989). There are both between- and within-breed differences in facial-eczema tolerance (Morris et al. 1990; Morris et al. 1994; Morris et al. 1995). With a moderate to high heritability (0.42-0.45), and no significant negative genetic correlations with any production traits (Campbell et al. 1981; McRae et al. 2016b; Morris et al. 1995), including FE tolerance in a genetic selection program will generally be beneficial to sheep production.

Ryegrass staggers. Ryegrass staggers is a disease of the nervous system that occurs in sheep grazing ryegrass-dominant pastures. It results from the ingestion of neurotoxins produced by a lolium endophyte (*Acremonium loliae*) growing in the leaf sheath (West et al. 2009). It is unlikely that an animal will die from staggers, however, production consequences arising from a loss of neuromuscular coordination in affected animals can be significant. Large investments have been made in the production and breeding of non-toxic endophyte strains and grass cultivar associations. An alternative approach is to breed sheep resistant to ryegrass staggers (Morris et al. 2007), though this has not achieved the same industry uptake as has selection for FE resistance.

Gastrointestinal parasites. Gastrointestinal nematode parasites are one of the most serious causes of disease in domestic ruminants in New Zealand (Perry & Randolph 1999). The most common species of parasite harboured by sheep are *Haemonchus, Ostertagia, Trichostrongylus, Nematodirus* and *Cooperia*. Economic losses arise from decreased production, and the cost of treatment. Climate has a direct impact on the geographic incidence of specific parasite species. The most economically-damaging species of parasites tend to be found nationwide, although there are regional differences. *Haemonchus contortus* (Barbers Pole worm) is the least tolerant of cold, and as a result, occurs with greater frequency and intensity in the North Island (Hervé et al. 2003; Lawrence et al. 2007; McKenna et al. 1973). Warming of regions further south due to climate change may bring this challenge to the South Island and reduce productivity in flocks that are unaccustomed to it.

Differences in resistance to parasites within and between breeds are well documented (Bishop & Morris 2007). The most common approach to estimating host resistance is utilising faecal egg count (FEC), which is moderately heritable (0.2-0.3), with a wide variability among individuals (Bishop & Morris 2007; Safari et al. 2005). Using FEC, rapid genetic progress has been demonstrated in both experimental and commercial flocks in New Zealand and Australia (McEwan et al. 1995; Woolaston et al. 1991).

Flystrike and dags. The prevalence of flystrike in New Zealand has been estimated at 3–5% of the national flock (Heath & Bishop 1995, 2006). Studies over the last few decades in research and commercial flocks, have estimated that incidence rates over the summer months range from 0.4-33% (Brandsma & Blair 1997; Litherland et al. 1992; Pickering et al. 2015a; Scobie & O'Connell 2002, 2010). It has been estimated that flystrike costs the sheep industry \$40 to \$50 million annually (Beef + Lamb New Zealand 1999), primarily through decreased wool and body growth, and the cost of control strategies, including insecticides and shearing of wool from the breech area (crutching). While insecticides have been a relatively effective control measure, increasing consumer concern about insecticide use (Sneddon & Rollin 2010) has led to greater interest in utilising host genetic resistance to control flystrike. Differences among breeds have been observed, with the New Zealand Wiltshire having a lower incidence of flystrike compared to Merinos and Romneys (Litherland et al. 1992), and among other composites and crosses (Scobie & O'Connell, 2002, 2010).

The accumulation of faecal material around the breech area of sheep (dags), is a major attractant for blowflies (*Calliphora stygia*, *Lucilia cuprina* and *L. sericata*) to lay their eggs to initiate flystrike. This is of interest to the New Zealand sheep industry for a variety of reasons, including financial penalties for presenting animals with dags at slaughter (due to contamination of carcasses), reduced wool value, and the labour cost associated with crutching all ages and classes of sheep on a farm. Both juvenile and adult dag score are moderately heritable (0.40-0.44; McRae et al. 2016c; Pickering et al. 2012; Scobie et al. 2008). Dag score is also a useful indirect predictor of flystrike (genetic correlation 0.71), especially for breech strike (Pickering et al. 2012). Dag score is a trait that is available within the Sheep Improvement Ltd. (SIL) genetic evaluation service to facilitate selection against dags in flocks recording this trait. It is currently not included within the national selection indices but can be added when breeders are interested in selecting to reduce dags.

Footrot. Footrot is caused by infection of the hoof with *Dichelobacter nodosus*, after the proliferation of the more common *Fusiformis necrophorum* in the hoof tissues. It can cause debilitating lameness and dramatic loss of productivity in sheep flocks, including lower weight gain and fecundity. Environmental conditions that allow transmission and development of the disease include moisture and warm temperatures which may prevail with predicted climate change.

The development of genetic resistance to footrot has been demonstrated in Romney (Skerman et al. 1988) and Corriedale sheep (Skerman & Moorhouse 1987). A recognised problem in selecting Merino sheep against footrot is that they are frequently run in environments that rarely experience outbreaks (Skerman 1986) and breeders tend to concentrate on control rather than genetic progress when outbreaks occur. Despite this, a breeding value for footrot has been available for New Zealand Merinos since 2020 through MERINOSELECT in association with the New Zealand Merino Company (MNZ) (Walkom et al. 2018).

Pneumonia. Respiratory disease in sheep is etiologically complex, resulting from an interaction of bacterial, mycobacterial and viral infectious agents, the defence mechanisms of the host, and environmental factors (Alley 2002). Enzootic pneumonia is common, particularly in hoggets. The economic cost of pneumonia is high due to direct losses of stock on-farm; lower weight gains and lower wool production associated with sub-clinical infections (Alley 1987; Goodwin et al. 2004; Goodwin-Ray 2006; McRae et al. 2016a); and condemnation and downgrading of carcasses at slaughter (West et al. 2009).

Temperature, humidity and inclement weather have all been implicated as risk factors for pneumonia (West et al. 2009). Studies investigating the susceptibility or resistance to pneumonia of different breeds of sheep are lacking, although anecdotal evidence suggests that Merinos may be less susceptible. Given that temperature is a risk factor for pneumonia (Knowles et al. 1995), this may be due to the greater heat tolerance of Merinos (Carles 1983). Within-breed variation in susceptibility to chronic, non-progressive pneumonia in New Zealand lambs has been examined, with heritability estimates for pneumonic lesions ranging between 0.07 and 0.16 (McRae et al. 2016a; McRae et al. 2018a). This suggests that genetic selection to lower the prevalence of pneumonia is feasible.

Heat stress. Extreme heat or cold are not experienced in New Zealand to the degree that they are in other countries where sheep are run. Nevertheless, there is a growing interest in the ability of New Zealand livestock to adapt to climate change and to tolerate heat stress. Ausseil et al. (2019) mapped the change in the number of days with mild, moderate and severe heat stress for dairy cows in New Zealand. For all the climate-change scenarios, there was an increase in the risk of moderate heat stress, with up to two more weeks per year of moderate heat stress, spread across the country. For severe heat stress, the risk also increases, with a more pronounced risk towards the end of this century.

Although the majority of reports to date on heat stress have related to cattle, they may have applicability to sheep, particularly hair sheep. Genetic variation in the response to heat stress has been identified in cattle, both between-breeds (Paula-Lopes et al. 2003; Romero et al. 2013) and within a breed (Mackinnon et al. 1991; Ravagnolo & Misztal 2000). The value of genomic selection as a tool for breeding for heat tolerance has been reported in dairy cattle (Garner et al. 2016). Additionally, in conjunction with existing genetic and genomic tools, new tools such as methylation assays can estimate the level of stress an individual has been exposed to during its lifetime (Murgatroyd et al. 2009). These tools can be used to predict longevity and to identify individuals that are genetically better adapted to their environment (Joy et al. 2020).

Minimising environmental impact

Climate change is of increasing concern for consumers, again magnified by the reach of social media. The role that agriculture, industry and population growth has in climate change is a hotly debated topic. Ruminants are a significant source of methane emissions globally (Schaefer et al. 2016), and one of the main contributors to New Zealand's greenhouse gas (GHG) inventory (MfE 2017). Agricultural production systems need to establish a path towards reduced greenhouse gas emissions.

Over the last 30 years, the amount of sheep meat produced has been relatively stable, despite a 50% reduction in ewe numbers, meaning absolute emissions for the sheep sector have been steadily declining, and are currently 31% below 1990 levels (Beef + Lamb New Zealand Economic Service 2020). Methane emissions are, in part, under genetic control in sheep, with heritability estimates ranging between 0.2 and 0.3 (Jonker et al. 2018). Given it is a difficult trait to measure, there are limited data to address whether specific breeds of sheep have significantly different methane emissions, but genetic variation exists within all populations investigated to date (Jonker et al. 2018). Currently, there is no direct relative economic value associated with breeding to reduce methane emissions, however, it is an important trait that should be factored into future breeding objectives.

Nitrous oxide (N_2O) is New Zealand's second most important agricultural greenhouse gas, representing 22% of agricultural emissions, or 11% of total GHG emissions. N_2O emissions are primarily derived from dung and urine excreted by animals. While dung emission factors, which represents direct N_2O emissions from excreta deposited during grazing, are the same for sheep and cattle, N_2O emissions from sheep urine are 50% or less than those of cattle, depending on the hill slope (van der Weerden et al. 2020). It is unclear why sheep have a lower emission factor than cattle, but it has been proposed that it may relate to the volume of urine deposited onto the soil and/or the effect of diet on subsequent urine characteristics. Genetic variability in milk-urea nitrogen in dairy cattle has been demonstrated and is being selected against (Beatson et al. 2019). No research has been undertaken on genetic variability in urinary nitrogen concentration or N_2O emissions in sheep.

Products

Meat. As the size of the national flock has decreased, the amount of meat produced has been maintained, in part through increased fecundity. When compared to cattle and deer, most New Zealand sheep breeds have longer anoestrus, thereby limiting their range of potential mating and lambing dates to best match pasture growth curves. If climate changes make seasons less extreme, there may be renewed farmer interest in year-round lambing to maintain or further increase production (Lewis et al. 1998; Morel et al. 2004).

One of the requirements of a year-round lambing system, is that ewes need to be successfully bred outside of their normal breeding season. This has proven challenging for most New Zealand breeds, however, some are reported to have extended breeding seasons that are under genetic control, such as the Merino, Poll Dorset (McQueen & Reid 1988) and Dorset Horn (Kelly et al. 1976).

Meat, and in particular fat, continues to be of concern to consumers. The evidence is conflicting, but generally, it is perceived that the consumption of excess red meat, and in particular meat high in fat, is associated with negative health outcomes. This is, in part, driving the rise in alternative protein sources which are perceived to be healthier. To maintain the role of sheep meat in consumer choices for protein intake, consideration needs to be given to making sheep meat meet consumer expectations as a healthy product.

New Zealand sheep meat offers a point of difference in many overseas markets in that the sheep are raised and finished on pasture prior to slaughter, resulting in an improved fatty acid composition compared to grain-feed meat, with reduced levels of saturated fats (Howes et al. 2015). Niche products have been developed by groups of entrepreneurs using specialised pasture-based finishing systems such as Te Mana Lamb (Alliance Group), Provenance Lamb (Crutchley family) and Coastal Lamb (Redmayne and 16 other families).

Some attributes of meat can be influenced by processing methods, such as electrical stimulation and specific carcass cooling and chilled-storage protocols to increase tenderness. However, several attributes of meat are fixed at the time the lamb leaves the farm on which it was finished. These include the lean-to-fat ratio and the composition of the fat. Traditionally, there has been a drive to reduced carcass fat due to perceived negative human-health outcomes associated with its consumption. However, there is an increasing recognition of the need for more intramuscular fat for improved taste and flavour (Realini et al. 2021).

To ensure the eating quality of New Zealand lamb, a minimum of 3% intramuscular fat is required, but many carcasses do not reach that threshold (Realini et al. 2021). Intramuscular fat is a highly heritable trait (0.83±0.11; Johnson et al. 2018) and progress can be made through genetic selection, as demonstrated by Te Mana Lamb and FarmIQ Systems Ltd. To date, collection of large-scale data sets across the industry to identify outliers has been limited due to the cost and complexity of generating intramuscular fat estimates. Automated approaches are being investigated (Hitchman et al. 2021).

The composition of fatty acids, whether in subcutaneous or intramuscular depots, is influenced by genotype and management, with two impacts on meat quality. First, certain fatty acids are negatively associated with human-health outcomes, particularly some saturated fatty acids (Astrup et al. 2019), whilst others such as the long-chain omega type may be positively associated with human-health outcomes (New Zealand Ministry of Health 2016). Second, the eating experience of lamb can be changed depending on the fatty acid composition (Enser & Wood 1993). Significant genetic variation exists for fatty acid composition in New Zealand sheep (PL Johnson, Unpublished data). There are anecdotal reports of large differences in the melting point of fat in some fat-tailed/fat-rumped breeds from Kazakhstan and surrounding countries. There is also evidence of different fatty acid profiles in fat-tailed Damara rams, which have been shown to have very high proportions of non-terminal branched-chain fatty acids in the tail adipose tissue (Alves et al. 2013), particularly those with lower melting points.

Pelts. Pelts are considered a by-product in all sheep breeds worldwide, aside from the Karakul and Gotland breeds (Näsholm & Eythórsdóttir 2011). The physical properties of sheep pelts are important in determining their end-use. Hair sheep are reported to have better quality pelts than are wool sheep (Carneiro et al. 2010; Cloete et al. 2000; Jacinto et al. 2004; Oliveira et al. 2007), although results were variable. Faults of genetic, parasitic or management origin in pelts, and their impact on processing and end-product use have been reviewed by Scobie (2010). Pelts are currently of very low commercial value in New Zealand with farmers not receiving a nominated return reflecting their quality.

Wool. As outlined earlier in this review, fine wool and coarse wool have two distinctly different end-uses resulting in vastly different net wool returns to fine- and coarse-wool growers. While there are ongoing attempts to revive 'old' markets and create 'new' uses for coarse wool in the hope of lifting prices, there has been little change in the coarse wool price.

There is a physiological difference between sheep growing fine and coarse wool. Sheep growing fine wool produce slower-growing lambs less suited for meat production whereas sheep growing coarse wool tend to be dual purpose producing both wool, albeit of low value, and lambs well-suited for meat production and/or milk for sheep-milk production. It is, therefore, likely that a proportion of the sheep population producing fine and medium wool will be retained for wool production, with a proportion of coarse-wool growers attempting to remove the fleece through selection and breeding. Even though the future for coarse wool is uncertain, some breeders are likely to continue to farm dual-purpose breeds.

There is genetic variation in wool characteristics, such as fleece weight, fibre diameter, fibre crimping and wool colour, that affect wool returns (Wuliji et al. 2001; Sumner et al. 2007; Scobie et al. 2012). These affect manufacturing performance and, hence, wool returns reflected in the relative economic value of each characteristic (Sumner et al. 2008). Although there have been recent efforts to encourage the production of finer wool, uptake has been limited. A transition to a finer-wool breed or a breed without wool would take time and incur costs that are not well understood (Farrell et al. 2020).

There is also a need to consider the management costs associated with dagging, crutching and shearing, relative to fleece weight and wool price, and whether fleece weight and/or the area on the body growing wool, should be genetically reduced. Irrespective of whether the main fleece is retained, there is merit in the removal of low-value wool from the head, belly and breech areas, and in investigating genetic variants that moult their tail wool (Scobie et al. 2007; Scobie et al. 2012).

An option is to consider hair or shedding sheep. Most wool breeds have been selected for continuous wool growth, but certain breeds and composites have retained the ability to shed their wool, with considerable variation in the speed and extent of shedding. The ability to shed wool is under significant genetic control, with evidence of a single putative dominant gene (Pollott 2011; Matika et al. 2013). Reported heritability estimates for shedding range from 0.26 to 0.80 (Johnson et al. 2007; Matika et al. 2013; Vargas Juardo et al. 2016; Guy et al. 2021). Sheep that shed their wool, or hair sheep that do not grow wool at all, have become an increasing proportion of the Australian sheep-meat industry. There are genotypes of sheep in New Zealand that do not require shearing, including the shedding New Zealand Wiltshire, Damara, Dorper and Meat master (developed in South Africa from Damara x White Dorper x Wiltshire x Van Rooy). Like wool breeds, there are considerable phenotypic differences among hair-sheep breeds, depending on their origin. Some breeds have short, slick, hair coats, absent of wool, while others grow thicker coats that contain a mixture of hair and wool fibres that shed annually. Hair sheep offer a better solution to flystrike than those which moult annually, when farmed for meat production. There is evidence that hairy birth coats are associated with improved lamb survival (Allain et al. 2014) due to the greater insulative properties of hair versus wool.

A consideration for wool is fleece pigmentation. The expression of colour and the distribution of pigment over the body of sheep is controlled by single genes at several loci with several alleles operating at each loci (Sponenberg et al. 1996). Pigmented sheep are less susceptible to photosensitisation (Baazaoui et al. 2020). In farming systems where wool is not a source of income, sheep that are pigmented may dominate. They will be less susceptible to skin cancers, FE lesions, and may have a selective advantage in the New Zealand climate, particularly with climate change. However, wool colour will remain an issue for wool and meat processors where white wool contaminated with pigmented wool will incur a price discount.

Milk. Developing a viable sheep milk industry in New Zealand requires genetic improvement to achieve higher production levels than are currently achieved in dairy-sheep herds (Scholtens et al. 2017). High-producing French Lacaune and Manech breeds were imported in 2016, in addition to the previously imported East Friesian and Awassi. Although there have been marked improvements in milk yield with these breeds, they originate from intensively farmed environments where they are often housed and fed total mixed rations, which is very different from the pastoral system in New Zealand. In addition to challenges faced by farming of sheep in a pastoral system, farmers milking sheep have the added challenges of milking ability and udder morphology, mastitis resistance, out-of-season breeding, animal welfare and disease management.

International marketing of sheep-milk products, including high-value infant formula and nutritional powders, is based on New Zealand's image as a 'Clean and Green' nation with a global reputation for using safe, sustainable and trusted methods of production. New Zealand must uphold this image, especially as public concern mounts against the impact of intensive farming practices on the environment. The dairy-cow industry faces significant management challenges relating to its effect on the environment. This provides both a challenge, with intensive sheep-dairy farming facing similar issues, but also an opportunity as milking sheep are less detrimental to the environment. This potentially makes milking sheep in New Zealand a more sustainable option than milking cows. Crossing exotic purebred animals over local breeds would enable selection for increased dairy production while maintaining environmental adaptation of the local breeds. This breeding strategy has been used to develop the DairyMeade breed from East Friesians in New Zealand.

Summary of attributes for future New Zealand sheep

The discussion above highlights the fact that sheep farmed in New Zealand have been chosen to fit the climate and bred to deliver products that ensured economic viability under a historical context. Markets, climate, optimal production systems, and consumer expectations around ethical and low-input systems are evolving. Further alterations will be required for the sheep industry to remain relevant and viable.

Key attributes of sheep to better meet current and anticipated market changes as indicated by consumers and the general population will include polled animals with short tails and bare points (head, leg, belly and breech), that exhibit increased disease tolerance to FE, internal parasites and footrot, increased heat tolerance, that either grow no wool or grow wool fetching a higher price in the market, and which have a low environmental footprint. Depending on the on-farm production system there will also need to be emphasis on meat, pelts, wool and milk in terms of both quantity and quality.

Options for changing the attributes of the New Zealand flock

Whilst management technology will play a part in addressing the future needs of the New Zealand sheep industry, breeding will have a very significant role due to the moderate-to-high heritabilities of many traits required to achieve the outcomes required to future-proof sheep-production systems. To date, New Zealand sheep genetic-improvement programmes have had a strong emphasis on growth and reproduction traits, with some emphasis on disease traits such as resistance to internal parasites or facial eczema in Romney or Coopworth sheep. In the future, there will need to be an increased emphasis on disease traits and characteristics that address welfare considerations.

What genotypes already in New Zealand could contribute

Many genotypes/breeds already present in New Zealand have attributes of interest, including Australian White, Awassi, Damara, Dorper, East Friesian, Ezicare, New Zealand Finn, Hampshire Down, Lacaune, Manech, Meatmaster and New Zealand Wiltshire. However, these tend to be minority breeds and their genetic merit for the full range of traits required by the sheep of the future is unknown, in particular, disease tolerance or reproductive merit in commercial-farming scenarios.

Additional international genotypes that exhibit traits of interest

There are reported to be over 1,000 breeds of sheep globally. Many are native to isolated geographical regions with low population sizes and geographical ranges. International sheep that are of importance to New Zealand are likely to include breeds where emphasis has been placed on lack or loss of wool, together with heat and parasite tolerance. Note that for the important disease trait of FE tolerance, there are no other countries in the world where selection for this trait is undertaken.

Potential breeds of interest include Afrikaner (South Africa), Assaf (Israel), Barbados Blackbelly (Caribbean), Black Headed Persian (Africa) or Van Rooy (South Africa), as well as composites such as Exlana (England), Katahdin (USA), Cashmore Nudies, Low Footprint Lamb, Ultrawhites and Wiltipoll (Australia). However, the comparative performance of these breeds in a common environment is largely unknown, and even information on the size of the animals and reproductive characteristics are often confounded with production systems that are significantly different to those in New Zealand.

Regulations and risks surrounding the importation of novel genetic material

The introduction of any novel genetic material into New Zealand has associated biosecurity risks. Being an island nation, New Zealand has remained free of many diseases associated with livestock species across the world, in a large part due to the strict controls implemented since the middle of last century. Biosecurity risks include Maedi-visna virus, which results in respiratory and nervous clinical symptoms, Jaagsiekte sheep retrovirus, the causative agent of contagious lung cancer in sheep, called ovine pulmonary adenocarcinoma (OPA), and scrapie.

In 2016, a significant change of importation rules occurred. There are now a limited number of countries (United Kingdom, France, Canada and Australia) from which New Zealand can currently import novel genetic material, without the need for further protocol development in the exporting country. Importation from many of the countries in which hair breeds are available, such as South Africa and America, is not acceptable. However, the majority of the South African hair-sheep breeds have already been imported into Australia, and importations are possible from Canada.

Due to the importation rules, choices for which animals are imported are not based on animals with the best genetic merit, rather, those that are available. As few animals are imported this increases the risks that they may be carrying undesirable recessive mutations, particularly if the genetic base remains small. This has already occurred in breeds introduced into New Zealand, including microphthalmia in Texels (Becker et al. 2010; Roe et al. 2003), chondrodysplasia in Texels (Thompson et al. 2005), cerebellar cortical abiotrophy in Wiltshires (Johnstone et al. 2005) and dermatosparaxis in white Dorpers (Vaatstra et al. 2011).

Breeding the sheep of the future

For decades, New Zealand sheep breeders have bred sheep that are suited to the environment and produced what has been desired by existing markets. Genetic improvement has been the result of selection applied using tools such as the Beef + Lamb New Zealand Genetics evaluation system, managed by SIL, and the MERINOSELECT service from Australia. SIL offers recording for the majority of traits of interest.

For genetic progress to be possible, there must first be genetic variation in the trait of interest within the population that is to be selected. For the likes of tail length, where there is variation within long-tailed Romney sheep, the ability to ultimately achieve the short tail of the Finnish Landrace is unlikely, even with full selection pressure, unless *de novo* mutations occur which are identified and selected for, or introgressed into a breed. Other traits, such as shedding, have been heavily selected against within modern wool breeds so there is little genetic variation left in the population. In the case of disease traits, genetic variation in tolerance/susceptibility is likely to occur within most breeds. With significant selection pressure applied, as is the case for facial eczema, it can take decades for good tolerance to be achieved, and even those animals are not resistant to a severe challenge. With the introduction of genomic selection as part of the breeding toolbox, there is potential for faster genetic gain, provided adequate numbers of animals are measured and genotyped for the traits of interest.

Many breeds possess the attributes identified as important for the future. As many of these attributes occur within minority breeds, it is unlikely that a one-for-one breed substitution with any of the current breeds will occur. Instead, crossbreeding programmes will be required. Introgression of traits of interest, or generation of composites involving the traits of interest, will not be a single-generation outcome and will require dedicated recording and selection programmes. First crosses, and their resulting intermediate phenotypes, are often considered 'ugly' as they exhibit considerable variation. Stabilisation is a long-term process. For example, modelling showed it would take 12-15 years of crossbreeding to achieve a fully-shedding flock when transitioning from a purebred Romney flock to a fully shedding Wiltshire-Romney crossbred flock (Farrell et al. 2020).

The ability to fast-track introgression of traits of interest, or stabilise a new composite breed using genomic technologies, will be dependent on specific mutations being identified soon after birth, rather than waiting until the trait is fully expressed in later life.

A question remains as to which breeds should be targeted within such breeding programmes. Given there have been no studies that have directly compared the full suite of production and health characteristics under a common environment, the relative attributes of different breeds cannot be properly compared.

Psychology of changing breeds

Farmers work in environments where multiple variables with different risk profiles and complex interactions, affect their businesses. Within these complex systems, farmers have to make numerous, interlinked decisions (Cornish et al. 2015). Most studies about how humans make decisions emphasise rationality as the major function in the process, with decision making influenced by many factors, such as a person's values, goals, biases, personalities, emotions and stage of life (Kahneman 2011). In addition to these influences, farmer decision making may be further influenced by the decision maker's stage of life in the farming lifecycle, tolerance to risk, stress levels, and commonly held beliefs that have been passed down from one generation to the next.

Further research is needed to determine the beliefs, drivers and motivations behind the current choice of sheep breed and whether breeders are open to changing breed or to a step change in a trait in the future.

Conclusions

There are currently 26.2 million sheep in New Zealand that are genetically diverse with a wide array of characteristics. For several reasons, at least a proportion of the national flock will need to change from the traditionally farmed breeds to ensure sheep farming remains a profitable enterprise that meets consumer and public expectations.

Whilst Merino and Merino-based sheep will continue to be farmed for their high-value wool, there is a need to reduce the fibre diameter of wool grown by coarse wool sheep for wool production to remain profitable for these sheep. For wool-growing sheep, there is a need for consideration around the physical traits of short tails, bare bellies and bare breeches for ethical outcomes. It may even be appropriate for a proportion of the national flock to shift to farming sheep that grow no wool in some areas of New Zealand, such as the upper half of the North Island. With warmer temperatures there will also need to be increased emphasis on breeding for disease tolerance to maintain productivity and meet welfare expectations. Increased attention to product quality is also likely to be needed to meet stricter consumer expectations.

A number of the key physical characteristics required for the future sheep are already present in minor breeds already in New Zealand but their attributes across the full range of relevant productive traits are not known under New Zealand farming conditions. For overseas breeds, excluding those in Australia, the Katahdin (USA) and Barbados Blackbelly (Caribbean) warrant further investigation. However, the importation of overseas breeds does bring with it the risk of importing diseases like Maedi-visna, which could result in disease incursions similar to *Mycoplasma bovis* in cattle.

Resistance to endemic diseases is unlikely to be better in imported breeds, though the tolerance of the Finnish Landrace to FE proves that will not always be the case (Morris et al. 1994). Resistance to *H. contortus* and other gastrointestinal parasites is worthy of consideration for imported genotypes. Resistance to extreme heat also has merit as there are some tropical and desert-adapted breeds that may be valuable in a changing climate.

Whichever genotypes are identified, the most-appropriate breeding programmes to fast-track incorporation of important traits into the national flock will be required. In addition to identifying the right genotypes, there is also a need for social science to understand breeders'and commercial farmers' attitudes to changing breeds, as many are traditionalists and exhibit reluctance to changing key attributes of their breed. Climate change and societal pressure are setting the thresholds for acceptable farming practices and sheep breeders need to evolve their sheep flocks to meet future challenges.

Acknowledgements

Thanks to Roland Sumner for his input into this manuscript.

References

Allain D, Foulquié D, Autran P, Francois D, Bouix J 2014. Importance of birthcoat for lamb survival and growth in the Romane sheep breed extensively managed on rangelands. Journal of Animal Science 92: 54-63.

Alley MR 1987. The effect of chronic non-progressive pneumonia on weight gain of pasture-fed lambs. New Zealand Veterinary Journal 35: 163-166.

Alley MR 2002. Pneumonia in sheep in New Zealand: an overview. New Zealand Veterinary Journal 50: 99-101.

Alves SP, Bessa RJB, Quaresma MAG, Kilminster T, Scanlon T, Oldham C, Milton J, Greeff J, Almeid AM 2013. Does the fat tailed Damara ovine breed have a distinct lipid metabolism leading to a high concentration of branched chain fatty acids in tissues? PLoS One 8: e77313.

Antedote Group 2018. Future of meat - how should New Zealand's red meat sector respond to alternative protein advancements? A report prepared for Beef + Lamb New Zealand. http://beeflambnz.com/sites/default/files/levies/files/Alternative%20Proteins%20summary%20report.pdf [accessed 20 June 2020].

Astrup A, Bertram HCS, Bonjour J-P, de Groot LCP, de Oliveira Otto MC, Feeney EL, Garg ML, Givens I, Kok FJ, Krauss RM, Lamarche B, Lecerf J-M, Legrand P, McKinley M, Micha R, Michlaski M-C, Mozaffarian D, Soedamah-Muthu SS 2019. WHO draft guidelines on dietary saturated and trans fatty acids: time for a new approach? British Medical Journal 366: l4137.

Ausseil A-G, van der Weerden T, Beare M, Teixeira E, Baisden T, Lieffering M, Guo J, Keller L, Law R, Noble A 2019. Climate change impacts on land use suitability. A report for Deep South and Our Land and Water National Science Challenges. Manaaki Whenua Landcare Research. 78 p.

Baazaoui I, McEwan J, Anderson R, Brauning R, McCulloch A, Van Stijn T, Bedhiaf-Romdhani S 2020. GBS data identify pigmentation-specific genes of potential role in skin-photosensitization in two Tunisian sheep breeds. Animals 10: 5.

Beatson P, Meier S, Cullen N, Eding, H 2019. Genetic variation in milk urea nitrogen concentration of dairy cattle and its implications for reducing urinary nitrogen excretion. Animal 13: 1-8.

Becker D, Tetens J, Brunner A, Bürstel D, Ganter M, Kijas J for the International Sheep Genetics Consortium. 2010. Microphthalmia in Texel sheep is associated with a missense mutation in the paired-like homeodomain 3 (PITX3) gene. PLoS ONE 5: e8689.

Beef + Lamb New Zealand 1999. Combating flystrike and lice through using a model approach. Research and Development Brief No.54, Beef + Lamb New Zealand.

Beef + Lamb New Zealand Economic Service 2020. Compendium of New Zealand Farm Facts 44th Edition. Beef + Lamb New Zealand Economic Service reports.

Bishop SC, Morris CA 2007. Genetics of disease resistance in sheep and goats. Small Ruminant Research 70: 48-59.

Brandsma JH, Blair HT 1997. Heritability of resistance to flystrike in New Zealand Perendale sheep. Proceedings of the New Zealand Society of Animal Production 57: 41-42.

Campbell AG, Meyer HH, Henderson HV, Wesselink C 1981. Breeding for facial eczema resistance - a progress report. Proceedings of the New Zealand Society of Animal Production 41: 273-278.

Carles AB 1983. Sheep production in the tropics. New York: Oxford University Press. 213 p.

Carneiro H, Louvandini H, Paiva SR, Macedo F, Mernies B, McManus C 2010. Morphological characterization of sheep breeds in Brazil, Uruguay and Colombia. Small Ruminant Research 94: 58-65.

Cloete SWP, Snyman MA, Herselman MJ 2000. Productive performance of Dorper sheep. Small Ruminant Research 36: 119-135.

Cockrem FR, Barton RA, Rae AL 1956. Face cover in Romney sheep. Proceedings of the New Zealand Society of Animal Production 16: 59-65.

Cornish D, Creelman Z, England D, Long J, Long B, Mudge B, Nicholson C 2015. Farm decision making: the interaction of personality, farm business and risk to make more informed decisions. http://www.grainandgraze3.com.au/resources/Farm_Decision_Making.pdf [accessed 20 June 2020].

Dennis NA, Amer PR, Meier S 2014. Predicting the impact of climate change on the risk of facial eczema outbreaks throughout New Zealand. Proceedings of the New Zealand Society of Animal Production 74: 1-6.

Enser M, Wood JD 1993. Effect of time of year on fatty acid composition and melting point of UK lamb. Proceedings of the 39th International Congress of Meat Science and Technology 2: 74.

Farrell LJ, Morris ST, Kenyon PR, Tozer PR 2020. Modelling a transition from purebred Romney to fully shedding Wiltshire-Romney crossbred. Animals 10: 2066.

Garner JB, Douglas ML, Williams SRO, Wales WJ, Marett LC, Nguyen TTT, Reich CM, Hayes BJ 2016. Genomic selection improves heat tolerance in dairy cattle. Scientific Reports 6: 34114.

Goodwin KA, Jackson R, Brown C, Davies PR, Morris RS, Perkins NR 2004. Pneumonic lesions in lambs in New Zealand: patterns of prevalence and effects on production. New Zealand Veterinary Journal 52: 175-179.

Goodwin-Ray KA 2006. Pneumonia and pleurisy in sheep: Studies of prevalence, risk factors, vaccine efficacy and economic impact. Unpublished PhD thesis. Massey University, Palmerston North, New Zealand.

Guy SZY, Kellier J, Brown DJ 2021. Preliminary genetic analysis of wool-shedding ability in sheep. Proceedings of Australian Association of Animal Sciences 33: 184.

Hatcher S, Preston JWV 2018. Genetic relationships of breech cover, wrinkle and wool coverage scores with key production traits in Australian Merino sheep. Small Ruminant Research 164: 48-57.

Heath ACG, Bishop DM 1995. Flystrike in New Zealand. Surveillance 22: 11-13.

Heath ACG, Bishop DM 2006. Flystrike in New Zealand: An overview based on a 16-year study, following the introduction and dispersal of the Australian sheep blowfly, *Lucilia cuprina* Wiedemann (Diptera: Calliphoridae). Veterinary Parasitology 137: 333-344.

Herve M, McAnulty RW, Logan CM, Sykes AR 2003. Regional variations in the nematode worm populations of breeding ewes in New Zealand. New Zealand Veterinary Journal 51: 159-164.

Hitchman S, Loeffen M, Reis MM, Craigie CR 2021. Robustness of Hyperspectral imaging and PLSR model predictions of intramuscular fat in lamb *M. longissimus lumborum* across several flocks and years. Meat Science 176: 108492.

Howes NL, Bekhit AEA, Burritt DJ, Campbell AW 2015. Opportunities and implications of pasture-based lamb fattening to enhance the long-chain fatty acid composition in meat. Comprehensive Reviews in Food Science and Food Safety 14: 22-36.

Jacinto MAC, Sobrinho AGS, Costa RG 2004. Anatomic and structural characteristics of wool and non-wool sheep (*Ovis aries* L.) in regard to the physico-mechanical aspects of the leather. Brazilian Journal of Animal Science 4: 1001-1008.

James PJ 2006. Genetic alternatives to mulesing and tail docking in sheep: a review. Australian Journal of Experimental Agriculture 46: 1-18.

James PJ, Gare DR, Singh AW, Clark JP, Ponzoni RW, Ancell PM 1990. Studies of the potential for breeding short tail Merinos. Wool Technology and Sheep Breeding 38: 106-111.

Johnson PL, O'Connell D, Dodds KG, Sumner RMW, McEwan JC, Pearson AJ 2007. Wool shedding as a trait for genetic improvement using marker assisted selection? Proceedings of the Association for the Advancement of Animal Breeding and Genetics 17: 541-544.

Johnson PL, Agnew M, Shorten PR, Reis MM, Cullen N, Craigie CR 2018. Preliminary genetic parameter estimates for intramuscular fat percentage in New Zealand lamb. Proceedings of the World Congress on Genetics Applied to Livestock Production 11: 611.

Johnstone AC, Jolly RD, Johnson CB 2005. Cerebellar cortical abiotrophy in Wiltshire sheep. New Zealand Veterinary Journal 53: 93.

Jonker A, Hickey SM, Rowe SJ, Janssen PH, Shackell GH, Elmes S, Bain W, Wing J, Greer GJ, Bryson B, MacLean S, Dodds KG, Pinares-Patiño CS, Young EA, Knowler K, Pickering NK, McEwan JC 2018. Genetic parameters of methane emissions determined using portable accumulation chambers in lambs and ewes grazing pasture and genetic correlations with emissions determined in respiration chambers. Journal of Animal Science 96: 3031-3042.

Journeaux P, van Reenen E, Manjala T, Pike S, Hanmore I 2017. Analysis of drivers and barriers to land use change – A report prepared for the Ministry for Primary Industries. https://www.mpi.govt.nz/dmsdocument/23056-ANALYSIS-OF-DRIVERS-AND-BARRIERS-TO-LAND-USE-CHANGE [accessed 20 June 2020].

Joy A, Dunshea FR, Leury BJ, Clarke IJ, DiGiacomo K, Chauhan SS 2020. Resilience of small ruminants to climate change and increased environmental temperature: a review. Animals 10: 867.

Kahneman D 2011. Thinking, fast and slow. London: Penguin Books. 499 p.

Kelly RW, Allison AJ, Shackell GH 1976. Seasonal variation in oestrous and ovarian activity of five breeds of ewes in Otago. New Zealand Journal of Experimental Agriculture 4: 209-214.

Knowles TG, Brown SN, Warriss PD, Phillips AJ, Dolan SK, Hunt P, Ford JE, Edwards JE, Watkins PE 1995. Effects on sheep of transport by road for up to 24 hours. Veterinary Record 136: 431-438.

Lawrence KE, Lethwick DM, Rhodes AP, Jackson R, Heuer C, Pomroy WE, West DM, Waghorn TS, Moffat JR 2007. Management of gastrointestinal nematode parasites on sheep farms in New Zealand. New Zealand Veterinary Journal 55: 228-234.

Lewis RM, Notter DR, Hogue DE, Magee BH, Bergmann JAG 1998. Lambing frequency in the STAR accelerated lambing system. Proceedings of the 6th World Congress on Genetics Applied to Livestock Production 27: 51-54.

Litherland AJ, Sorenson E, Niezen J, Bishop D 1992. A pilot evaluation of shedding breeds compared with non-shedding breeds for susceptibility to nematodes and flystrike. Proceedings of the New Zealand Society of Animal Production 52: 233-235.

Mackinnon MJ, Meyer K, Hetzel DJS 1991. Genetic variation and covariation for growth, parasite resistance and heat tolerance in tropical cattle. Livestock Production Science 27: 105-122.

Magnusson U 2020. Prudent and effective antimicrobial use in a diverse livestock and consumer's world. Journal of Animal Science 98: S4-S8.

Matika O, Bishop SC, Pong-Wong R, Riggio V, Headon DJ 2013. Genetic factors controlling wool shedding in a composite Easycare sheep flock. Animal Genetics 44: 742-749.

McEwan JC, Dodds KG, Watson TG, Greer GJ, Hosking BC, Douch PGC 1995. Selection for host resistance to roundworms by the New Zealand sheep breeding industry: The WORMFEC Service. Proceedings of the Association for the Advancement of Animal Breeding and Genetics 11: 70-71.

McKenna PB, Campbell E, Riffkin GG, Rutherford DM, Shirley DK 1973. The incidence of *Haemonchus contortus* infections in sheep abomasa received at animal health laboratories in New Zealand. New Zealand Veterinary Journal 21: 11-16.

McQueen JPM, Reid TC 1988. The development of an autumn lambing flock of Dorset x Romney ewes without the use of hormones. Proceedings of the New Zealand Society of Animal Production 48: 87-90.

McRae KM, Baird HJ, Dodds KG, Bixley MJ, Clarke SM 2016a. Incidence and heritability of ovine pneumonia, and the relationship with production traits in New Zealand sheep. Small Ruminant Research 145: 136-141.

McRae KM, Cullen NG, Amyes NC, Johnson PL 2016b. An update on genetic parameters for facial eczema tolerance in sheep. Proceedings of the New Zealand Society of Animal Production 76: 43-45.

McRae KM, Dodds KG, Newman S-AN, Brito LF, Pickering NK, McEwan JC 2016c. Preliminary estimates of genetic parameters for adult dag scores in New Zealand sheep and their relationship with juvenile measurements. Proceedings of the New Zealand Society of Animal Production 76: 144–145.

McRae KM, Rowe SJ, Baird HJ, Bixley MJ, Clarke SM 2018a. Genome-wide association study of lung lesions and pleurisy in New Zealand lambs. Journal of Animal Science 96: 4512-4520.

McRae KM, Rowe S, McEwan JC 2018b. Potential alterations in New Zealand sheep, beef cattle and deer numbers due to climate change: what can genetics offer. Proceedings of the New Zealand Society of Animal Production 78: 146-150.

MfE 2017. New Zealand's greenhouse gas inventory 1990–2015. http://www.mfe.govt.nz/node/23304/ [Accessed 1 June 2020].

Ministry for the Environment 2018. Climate change projections for New Zealand: Atmospheric projections based on simulations undertaken for the IPCC 5th Assessment, 2nd Edition. Wellington: Ministry for the Environment.131 p.

Montossi F, Font-i-Furnols M, del Campo M, San Julián R, Brito G, Sañudo C 2013. Sustainable sheep production and consumer preference trends: Compatibilities, contradictions, and unresolved dilemmas. Meat Science 95: 772-789.

Morel PCH, Kenyon PR, Morris ST 2004. Economical analysis of year round lamb production. Proceedings of the New Zealand Society of Animal Production 64: 179-182.

Morris CA, Amyes NC, Orchard RL 2007. Responses to selection in ryegrass staggers lines of sheep. Proceedings of the New Zealand Society of Animal Production 67: 1-5.

Morris CA, Towers NR, Campbell AG, Meyer HH, Wesselink C, Wheeler M 1989. Responses achieved in Romney flocks selected for or against susceptibility to facial eczema, 1975-87. New Zealand Journal of Agricultural Research 32: 379-388.

Morris CA, Towers NR, Tempero HJ, Cox NR, Henderson HV 1990. Facial eczema in Jersey cattle: heritability and correlation with production. Proceedings of the New Zealand Society of Animal Production 50: 255-259.

Morris CA, Towers NR, Wesselink C, Amyes NC 1994. Susceptibility of Finnish Landrace, Romney, and Finn × Romney lambs to a sporidesmin challenge. New Zealand Journal of Agricultural Research 37: 547-552.

Morris CA, Towers NR, Wheeler M, Wesselink C 1995. Selection for or against facial eczema susceptibility in Romney sheep, as monitored by serum concentrations of a liver enzyme. New Zealand Journal of Agricultural Research 38: 211-219.

Murgatroyd C, Patchev AV, Wu Y, Micale V, Bockmühl Y, Fischer D, Holsboer F, Wotjak CT, Almeida OFX, Spengler D 2009. Dynamic DNA methylation programs persistent adverse effects of early-life stress. Nature Neuroscience 12: 1559-1566.

Nasholm A, Eythórsdóttir E 2011. Characteristics and utilization of sheep pelts. Small Ruminant Research 101: 182-187.

New Zealand Ministry of Health 2016. Nutrient reference values for Australia and New Zealand: including recommended dietary intakes. Version 1.2, updated September 2017 https://www.health.govt.nz/publication/nutrient-reference-values-australia-and-new-zealand [accessed 20 June 2020].

Nielsen Company 2019. Climate issues facing famers: sustainable land management and climate change research programme – A report prepared for the Ministry of Primary Industries https://www.mpi.govt.nz/dmsdocument/33747/direct [accessed 20 June 2020].

Oliveira RJF, Costa RG, Sousa WH, Medeiros AN, Dal Monte MAB, Aquino D, Oliveira CJB 2007. Influence of genotype on physico-mechanical characteristics of goat and sheep leather. Small Ruminant Research 73: 181-185.

Parle JN, di Menna ME 1972. Fungicides and the control of *Pithomyces chartarum*. New Zealand Journal of Agricultural Research 15: 54-63.

Paula-Lopes FF, Chase CC Jr, Al-Katanani YM, Krininger CE 3rd, Rivera RM, Tekin S, Majewski AC, Ocon OM, Olson TA, Hansen PJ 2003. Genetic divergence in cellular resistance to heat shock in cattle: differences between breeds developed in temperate versus hot climates in responses of preimplantation embryos, reproductive tract tissues and lymphocytes to increased culture temperatures. Reproduction 125: 285-294.

Perry BD, Randolph TF 1999. Improving the assessment of the economic impact of parasitic diseases and of their control in production animals. Veterinary Parasitology 84: 145-168.

Peterson S, Prichard C 2016. Imagine there's no 'dairy': From milk to 'milks'. In: Massey C, editor. The New Zealand & Food Annual, Why Waste a good crisis? Palmerston North: Massey University Press. pp. 268-281.

Petrie OJ 1995. Harvesting of Textile Animal Fibres. FAO Agricultural Services Bulletin 122. Rome, Italy: Food and Agriculture Organization of the United Nations. 101 p.

Pickering NK, Blair HT, Hickson RE, Johnson PL, Dodds KG, McEwan JC 2015. Estimates of genetic parameters for breech strike and potential indirect indicators in sheep. New Zealand Veterinary Journal 63: 98-103.

Pickering NK, Dodds KG, Blair HT, Hickson RE, Johnson PL, McEwan JC 2012. Genetic parameters for production traits in New Zealand dual-purpose sheep, with an emphasis on dagginess. Journal of Animal Science 90: 1411-1420.

Pollett GE 2011. A suggested mode of inheritance for wool shedding in sheep. Journal of Animal Science 89: 2316-2325.

Ravagnolo O, Misztal I 2000. Genetic component of heat stress in dairy cattle, parameter estimation. Journal of Dairy Science 83: 2126-2130.

Realini CE, Pavana E, Johnson PL, Font-i-Furnols M, Jacob N, Moon C, Agnew M, Bermingham EN, Craigie CR 2021. Role of intramuscular fat level and composition on consumer linking scores of M. longissimus lumborum from pasture finished lambs. Meat Science 173: 108380.

Roe WD, West DM, Walshe MT, Jolly RD 2003. Microphthalmia in Texel lambs. New Zealand Veterinary Journal. 51: 194-195.

Romero RD, Montero Pardo A, Montaldo HH, Rodríguez AD, Hernández Cerón J 2013. Differences in body temperature, cell viability, and HSP-70 concentrations between Pelibuey and Suffolk sheep under heat stress. Tropical Animal Health and Production 45: 1691-1696.

Safari E, Fogarty NM, Gilmour AR 2005. A review of genetic parameter estimates for wool, growth, meat and reproduction traits in sheep. Livestock Production Science 92: 271-289.

Schaefer H, Fletcher SEM, Veidt C, Lassey KR, Brailsford GW, Bromley TM, Dlugokencky EJ, Michel SE, Miller JB, Levin I, Lowe DC, Martin RJ, Vaughn BH, White JWC 2016. A 21st-century shift from fossil-fuel to biogenic methane emissions indicated by $^{13}CH_4$. Science 352: 80-84.

Scholtens MR, Peterson SW, Gunson K, Gunson A, Lopez-Villalobos N 2017. Prototype genetic evaluation in a New Zealand dairy sheep flock. Proceedings of the New Zealand Society of Animal Production 77: 121-124.

Scobie DR, Bray AR, O'Connell D 1997. The ethically improved sheep concept. Proceedings of the New Zealand Society of Animal Production 57: 84-87.

Scobie DR, O'Connell D 2002. Genetic reduction of tail length in New Zealand sheep. Proceedings of the New Zealand Society of Animal Production 62: 195-198.

Scobie DR, O'Connell D, Morris CA, Hickey SM 2007. A preliminary genetic analysis of breech and tail traits with the aim of improving the welfare of sheep. Australian Journal of Agricultural Research 58: 161-167.

Scobie DR, O'Connell D, Morris CA, Hickey SM 2008. Dag score is negatively correlated with breech bareness score of sheep. Australian Journal of Experimental Agriculture 48: 999-1003.

Scobie DR 2010. Skin. In: Cottle DJ editor. International sheep and wool handbook. Nottingham: Nottingham University Press. pp 661-675.

Scobie DR, O'Connell D 2010. Breech bareness reduces flystrike in New Zealand crossbred sheep. Animal Production Science 50: 599-602.

Scobie DR, O'Connell D, Morris CA, Hickey SM 2011. Inherited bareness of the belly reduces the time taken to shear ewes and hoggets. Animal Production Science 51: 176-182.

Scobie DR, Bray AR, Smith MC, Woods JL, Morris CA, Hickey SM 2012. Wool staple tenacity in New Zealand Romney sheep: heritability estimates, correlated traits, and direct response to selection. Animal Production Science 52: 448-455.

Scobie DR, Dodds KG, O'Connell D 2016. Moulting of tail wool reduced dag accumulation in yearling sheep. Australian Society of Animal Production. 31: 77-78.

Skerman TM 1986. Genetic variation and inheritance of susceptibility to footrot in sheep. Footrot in ruminants. Proceedings of a workshop, Melbourne 1985. CSIRO Division of Animal Health pp 73-75.

Skerman TM, Johnson DL, Kane DW, Clarke JN 1988. Clinical footscald and footrot in a New Zealand Romney Flock, phenotypic and genetic parameters. Australian Journal of Agricultural Research 39: 907-916.

Skerman TM, Moorhouse SR 1987. Broomfield Corriedales: a strain of sheep selectively bred for resistance to footrot. New Zealand Veterinary Journal 35: 101-106.

Smith BL, Embling PP, Towers NR, Wright DE, Payne E 1977. The protective effect of zinc sulphate in experimental sporidesmin poisoning of sheep. New Zealand Veterinary Journal 25: 124-127.

Smith BL 2000. Effects of low dose rates of sporidesmin given orally to sheep. New Zealand Veterinary Journal 48: 176-181.

Sneddon J, Rollin B 2010. Mulesing and animal ethics. Journal of Agricultural and Environmental Ethics 23: 371-386.

Sponenberg DP, Dolling CHS, Lundie RS, Rae AL, Renieri C, Lauvergne JJ 1996. Coat colour loci. In: Lavergne JJ, Dolling CHS, Renieri C editors. Mendelian inheritance in sheep 1996 (MIS 96). Clamart: COGOVICA/COGNOSAG, France. CaMerino: University of CaMerino, Italy. p. 13-57.

Sumner RMW, Clarke JN, Cullen NG 2007. Effect of divergent selection for wool bulk on live weight and wool characteristics in Perendale sheep. Proceedings of the New Zealand Society of Animal Production 67: 180-186.

Sumner RMW, McDermott AJ, Cox NR 2008. Relative economic value of wool processing parameters for New Zealand strong wool between 2003 and 2007. Proceedings of the New Zealand Society of Animal Production 68: 53-56.

Tait A, Sood A, Mullan B, Stuart S, Bodeker G 2016. Updated climate change projections for New Zealand for use in impact studies. Synthesis Report RA1. Climate Change Impacts and Implications (CCII) for New Zealand to 2100. MBIE contract C01X1225.

Thompson KG, Blair HT, Linney LE, West DM, Byrne T 2005. Inherited chondrodysplasia in Texel sheep. New Zealand Veterinary Journal 53: 208-212.

Towers NR 1986. Facial eczema - Problems and successes in control. Proceedings of the New Zealand Grassland Association 47: 121-127.

Vaatstra BL, Halliday WD, Waropastrakul S 2011. Dermatosparaxis in two White Dorper lambs. New Zealand Veterinary Journal 59: 258-260.

van der Weerden TJ, Noble AN, Luo J, de Klein CAM, Saggar S, Giltrap D, Gibbs J, Rys G 2020. Meta-analysis of New Zealand's nitrous oxide emission factors for ruminant excreta supports disaggregation based on excreta form, livestock type and slope class. Science of the Total Environment 732: 139235.

Vargas Jurado N, Leymaster KA, Kuehn LA, Lewis RM 2016. Estimating heritability of wool shedding in a cross-bred ewe population. Journal of Animal Breeding and Genetics 133: 396-403.

Walkom SF, Bunter KL, Raadsma H, Brown DJ, Gibson W, Swan AA, Börner V, Ferguson MB 2018. Estimation of breeding values for footrot in New Zealand Merino sheep. Proceedings of the World Congress on Genetics Applied to Livestock Production 11: 261.

West DM, Bruère AN, Ridler AL 2009. The sheep: Health, disease and production. Palmerston North: Massey University Press. 477 p.

Woolaston RR, Windon RG, Gray GD 1991. Genetic variation in resistance to internal parasites in Armidale experimental flocks. In: Breeding for disease resistance in sheep. Gray GD, Woolaston RR editors. Melbourne: Australian Wool Corporation. pp. 1-9.

Wuliji T, Dodds K, Land J, Andrews R, Turner P 2001. Selection for ultrafine Merino sheep in New Zealand: Heritability, phenotypic and genetic correlations of live weight, fleece weight and wool characteristics in yearlings. Animal Science 72: 241-250.

Seasonal variation in pasture mass offered was not associated with changes in live weight, or body condition score, in three cohorts of horses.

BR Adams[a], EK Gee[a], PJ Back[ab] and CW Rogers[ab]

[a]*School of Veterinary Science, Massey University, New Zealand;* [b]*School of Agriculture and Environment, Massey University, New Zealand*

**Corresponding author:* Email: b.adams@massey.ac.nz

Abstract

Many New Zealand equine owners/stud managers rely on body condition to adjust feed supply for weight management, rather than objective weigh scales. This cohort observational study examined the inter-relationship of pasture supply with monthly BCS and live weight (LW) in three groups (n = 7, 8, 9) of grazing-horses of median 6/9 BCS (range 4-8). Dry-matter (DM) intake was estimated from weekly pasture mass (using rising-plate meter), monthly pasture cuts, and fortnightly LW records. Total pasture DM at the start of grazing ranged from 559 to 3511 kg DM/ha, and grazed lawns (nutritive areas) ranged from 364 to 2601 kg DM/ha. Total estimated DM offered (including hay when offered) ranged from 1.46 to 26.37 kg DM/horse/day (0.3% to 5.3% of horses LW). Within-horse maximum monthly gains and losses were 8% LW and -13% LW respectively. Median monthly BCS change was 0/9 (range -2/9 to 1/9). There was no significant association between monthly changes in LW and BCS, indicating BCS may lack the sensitivity to manage weight. During periods of reduced DMI (below estimated maintenance) there were no significant decreases in LW or BCS. These data implied that the horses may have altered transit time to improve the relative digestibility of the limited feed on offer.

Keywords: horse; weight; body condition score; pasture; intake

Introduction

In New Zealand, horses are predominately maintained at pasture year-round which often makes it difficult to precisely quantify feed supply and feed demand. Both managers of commercial properties and recreational owners tend to assess nutritional requirements using body condition score (BCS) rather than live weight (LW) (Fernandes et al. 2014; Goh et al. 2020; Rogers et al. 2017b). However, there may be a lack of sensitivity to weight loss when using BCS as it is only an estimate of subcutaneous fat stores. Studies have shown some BCS assessment areas may have a better representation of overall weight changes compared with other areas, and short-term changes in weight may not be reflected in a change in BCS as visceral fat may be mobilised more rapidly than subcutaneous fat (Dugdale et al. 2010; Dugdale et al. 2012; Suagee et al. 2008). Owners can also be subjective, and lack precision in their estimation of weight loss and BCS, with 37% of owners of obese horses (as judged by researchers) reporting they were not overweight (Wyse et al. 2008), and only 11% of professional and non-professional horse owners able to recognise overweight horses from photos (Morrison et al. 2017).

Catalano et al. (2019) reported that Thoroughbred horses had an average difference of 26 kg (4.9% LW) between BCS when using the Henneke et al. (1983) nine-point scale (1/9 = emaciated, 9/9 = obese), compared to 8 kg (6.8% LW) for miniature-type and 15 kg (3.2% LW) for saddle-type horses. Other authors have reported a similar range of 2.8-4.6% LW difference between BCS amongst various horse-types, equating to 10-39 kg LW (Martinson et al. 2014; Catalano et al. 2016). Using a 0-5 BCS scale Carroll & Huntington (1988) estimated a similar 25-30 kg LW difference between 0.5 BCS (equating to 1 BCS in the scale of Henneke et al. (1983)) for horse-types.

New Zealand Thoroughbred and Standardbred broodmares are frequently maintained on pasture and conserved forage alone, with minimal concentrate-feed supplementation (Rogers et al. 2007). For these type of horses (500-600 kg LW) maintenance requirements should be met with a daily dry matter intake (DMI) of 1.5-2% LW, and DE of 69-84 MJ/day (NRC 2007). Voluntary DMI can be increased in individuals and in response to environmental conditions but is usually within the range of 2-3% LW (Chavez et al. 2014; Grace et al. 2002; Ringmark et al. 2013). Standardbred horses are able to maintain body condition under restricted pasture levels with minimal supplementation, implying higher feed conversion efficiency (Rogers et al. 2017a; Rogers et al. 2020), and may be termed a thrifty phenotype. Even under commercial management conditions horses exhibit selective grazing behaviour, establishing lawns (nutritive areas) and avoiding roughs (eliminative or latrine areas with high density of faeces). On New Zealand commercial farms, establishment of lawns and roughs has been reported, with Standardbred mares having lawns with post-grazing mass of 1000 kg DM/ha. Thoroughbred farms typically do not graze the pasture as low as Standardbred farms with lawns of 1300-1800 kg DM/ha (Bengtsson et al. 2018; Rogers et al. 2007; Rogers et al. 2020). Once the pasture mass of the lawns drops under 1500 kg DM/ha there appears to be less reluctance from weanling horses to consume some areas in the roughs (Rogers et al. 2018). Under higher stocking densities (5.2-5.6 horses/ha) Standardbreds have been observed to graze these areas equally despite hay offered *ad lib.*, thereby maintaining a low, even pasture (Medica et al. 1996).

This two-year observational study aimed to examine if there was a significant relationship between BCS, LW and estimated DM intake in a cohort of non-commercial horses kept intensively at pasture, observing and accounting for their selective grazing behaviour.

Materials and methods

This observational study was conducted from August 2017 until August 2019 at the Massey University Veterinary Large Animal Teaching Unit, a farm comprising 57.92 ha (effective 42.1 ha) of flat to rolling river terrace utilised for keeping herds of grazing animals for the purposes of teaching. The soil type of the farm was predominately Tokomaru silt loam. The paddocks utilised by horses were a mix of grass species (predominately perennial ryegrass (*Lolium perenne*), but also cocksfoot (*Dactylis glomerate*), fescue (*Festuca arundinacea*), browntop (*Agrostis tenuis*), meadow grass (*Poa trivialis*)), white clover (*Trifolium repens*), red clover (*Trifolium pratense*) and weeds (typically dock (*Rumex obtusifolius)* and buttercup *(Ranunculus repens)*). There was a standard annual fertiliser application to all accessible paddocks (30% potash super, Ravensdown) applied at a rate of 450 kg/ha in April 2018 and January 2019.

Twenty-four horses (19 Standardbred, three Thoroughbred, one Kaimanawa, and one Stationbred), comprising of 14 mares and 10 geldings, were managed in three consistent groups based on their routine use for different teaching activities (n=7, 8, and 9 for G1, G2 and G3 respectively) (Guinnefollau et al. 2020). The horses were 13±4 years at the start of observation, kept at pasture all year, and had been on the farm for >4 years. Movement of stock and supplementary feeding was determined by farm management based on a number of criteria including body condition/weight, visual assessment of pasture levels and availability of paddocks.

BCS and weight

Body condition was monitored by the same technician monthly using the Henneke (1983) nine-point scale, and live weight (LW) in kilograms (kg) recorded approximately fortnightly using a Tru-Test weighing system (MP 800 load bars, XRP2 chip panel reader, XR5000 weigh scale indicator). One horse was not consistently assessed for BCS and was removed from analysis. Use and care of the horses was monitored by the Massey University Animal Ethics Committee (protocols 17/94, 18/97, 19/76).

Intake

Farm and paddock pasture growth (DM kg/ha) was estimated using protected (caged) areas of pasture representative of total pasture cover, at two locations on-farm (Year 1) or within paddocks during each set-stocked period (SSP) per horse-group (Year 2). Each cage was placed in areas representative of average pasture cover, with a 30 x 60 cm quadrat sample of pasture collected monthly from each cage, cut to ground level to mimic grazing which created lawns, and dried in an oven to estimate pasture growth (kg DM/ha). Mass of total pasture, lawns and roughs of horse-grazed paddocks was estimated using a rising-plate meter (RPM) weekly by the same technician using an L- or W-shaped transect, and converted into kg DM/ha using the equation below (Jenquip 2017).

Average rising-plate meter reading x 158 + 200 = pasture cover (kg DM/ha)

The daily pasture DMI for each SSP was estimated using the difference between pre- and post-grazing total pasture mass, divided by days grazed, with the addition of average daily pasture growth rate for that period, and multiplying by 0.7 (utilisation factor based on the companion paper Adams et al. (2021) and Rogers et al. (2017a)). Typical pasture energy content during the year was based on average pasture ME of 10 MJ/kg DM found in similar pasture (Rogers et al. 2020).

Horses were supplemented with hay from May to October each year, with an estimated quantity offered recorded as part-bale/horse/day with a 20% wastage factor (due to being fed on the ground) and average ME of 9 MJ/kg DM (Fernandes 2016). Random bales were weighed each year to generate an estimated average bale weight and core samples were collected for DM calculation.

To compare these estimated feed-on-offer (pasture ± hay) data with required intake for maintenance, average monthly group LW was inserted into the maintenance equation 0.6 MJ × $kgLW^{0.75}$ (NRC 2007).

This was then divided by number of days in the month, to estimate daily DM required for maintenance.

Statistical analysis

Data were analysed for normality using density plots and histograms. Simple descriptive statistics are presented as mean ± standard deviation for parametric and median and inter-quartile range for non-parametric data. Maximum and minimum values are reported when emphasis is on the spread of the data. Relationship between intake and weight change was examined using linear regression. Analysis was conducted using R version 4.0.3.

Results

Pasture management

During the two-year observation period, the horses grazed 21 different paddocks (0.046 ha to 2.9 ha in size) in a semi-set-stocked pattern. Occasionally farm management split groups into smaller groups for limited periods of time. The total area utilised by the horses was 26.67 ha (63% of effective farm area). Within individual paddocks, there was considerable variation in stocking density ranging from 1.9-43.5 horses/ha (1,016.5-23,272.5 kg LW/ha) due to variation in paddock sizes. The median SSP was 15 days (IQR 8-28.5 days) with considerable range of 2-123 days. The paddock with the longest set-stock period of 123 days (October 2017-February 2018) was grazed by horses most frequently (480/730 days of the study), as presented in the companion paper Adams et al. (2021).

Intake

Estimated paddock pasture growth ranged from 3.1 kg DM/ha/day to 131.8 kg DM/ha/day, with the monthly farm average ranging from 6.5 kg DM/ha/day to 84.7 kg DM/ha/day. Year one (Y1) and year two (Y2) had similar average monthly pasture growth rates from March to August. In Y1, there was a drought during the summer which resulted in lower-than-average pasture growth. Total annual pasture growth (DM/ha) was 11.6 t in Y1 and 12.8 t in Y2.

The median total pasture pre-grazing mass was 1,472 (range 582-3,512) kg DM/ha. Grazed lawns were a median 729 (IQR 542-1,047), 756 (IQR 590-1,340), and 850 (IQR 657-996) kg DM/ha for G1, G2 and G3 respectively. Total pasture post-grazing mass was median 890 (range 427 to 2367) kg DM/ha across all groups. The estimated daily pasture DMI ranged from 1.46 to 26.37 kg/horse (approximately 2.72-49.11 g/kg LW) and total DMI including hay when offered ranged from 1.8 to 26.37 kg DM/horse/day (3.35-49.11 g/kg LW or 0.3% to 5.3% LW) per SSP.

Based on average group LW at the beginning and end of each month, the required pasture DMI for maintenance ranged from 1.2% to 1.3% kg LW. The horses had an estimated DMI at or above that required for maintenance for 79%, 64% and 89% of observation days for G1, G2 and G3 respectively. The maximum consecutive period when estimated DMI was below maintenance was 61 days for G1 (five SSP during autumn 2019), 65 days for G2 (during two SSP in summer 2018-2019), and 35 days for G3 (one SSP during summer 2017-2018).

BCS and weight

The mean horse LW was 537±56 kg and the mean group LW was 3,818±94 kg, 4,511±102 kg, 4,444±179 kg for G1, G2, and G3 respectively. Each group gained weight from their first to their final monthly average weight (P<0.05, Table 1). Two horses lost weight (< 4% of LW), one as a result of a period of illness towards the end of the 2-year period (the weight previous to illness was used for group weight); the remaining horses gained between 2% and 10% LW. Individual horse's maximum monthly LW gain and loss was 8% and -13% of average monthly LW (median 0%, IQR -2%-2%). Despite the variation in DMI and periods of estimated DMI below maintenance, there was no clear association of this variation in estimated DMI with recorded weight change during each SSP (R^2 = 0.0367, P>0.05, Fig. 1).

Figure 1 The estimated dry-matter intake presented as the difference between available feed and estimated maintenance requirements (as percentage of live weight) *vs*. weight changes (percentage live weight) during each set-stocked period of grazing for three horse groups kept at pasture over two years.

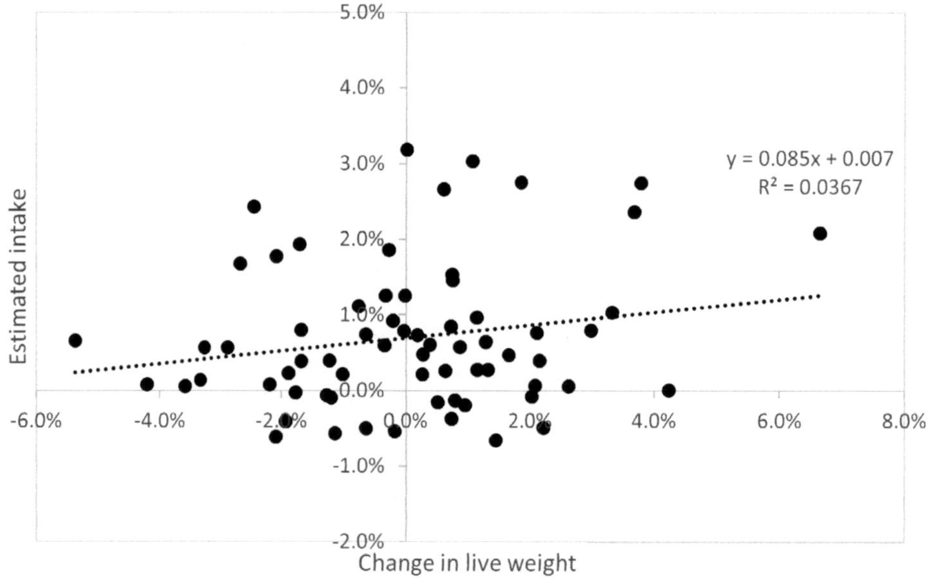

Table 1 Live weight (LW) and body condition score (BCS, nine-point scale) figures comparing initial and final monthly measures of three horse groups (G1, G2, G3) kept at pasture for two years. Also represented is the mean (± SD) of the individual changes in LW over the study period.

	First measure		Final measure		Individual change
	LW kg (mean±SD)	BCS (median (IQR))	LW kg (mean ± SD)	BCS (median (IQR))	LW kg (mean ± SD)
G1	3589±48[a]	6 (5-6)	3778±42[b]	6 (5-7)	42±13
G2[1]	4360±71[a]	5 (5-6)	4547±61[b]	5 (5-6)	26±18
G3	4174±34[a]	5 (4-6)	4383±40[b]	6 (5-6)	25±16
All horses	11991±60[a]	5 (5-6)	12679±55[b]	6 (5-7)	29±20

Different superscript within a row are significantly different P<0.05, [1]23 horses assessed for BCS (G2 had one horse removed from analysis).

From initial to final observation, the median BCS of all horses increased from 5/9 to 6/9. Examination of BCS within groups identified that for G3 the median BCS increased from 5/9 to 6/9. There was no change in the median BCS of the horses in G1 (6/9) or G2 (5/9).

Median monthly change in BCS per horse was 0/9 (range -2 to 2). When all horses were categorized by their median BCS during observations, five horses had a median BCS of 5/9 (IQR 5-7), fourteen horses a median of 6/9 (IQR 5-6), and four horses a median 7/9 (IQR 6-8). Within each median BCS category, there was little variability in monthly weight, with the majority of monthly weight changes falling within ± 4% LW (93%, 93%, and 95% for median BCS 5/9, 6/9, and 7/9 respectively). Based on the mean LW of 537 kg, a 4% change in LW represented approximately 21 kg. For individuals with no change in monthly BCS there was a median weight fluctuation of 0% (IQR -2%-1%) and 99% of monthly changes in BCS varied by a single BCS or less (Fig.2).

Deterministic modeling of the periods of below maintenance, and the cumulative negative deficit and weight loss calculated using estimated 30 MJ of ME for every 1 kg of body fat mobilised (Nicol & Brookes 2017) indicated a disparity between the theoretical loss (expected over 200 kg LW/group) and the observed estimated gain in LW (38-74 kg LW/group, Table 2).

Discussion

New Zealand's temperate climate allows the majority of horses to be maintained at pasture all year, often only supplemented with conserved forage to offset the seasonal pasture growth. The horses in this study had similar or lower available total pasture mass on offer as previously reported for horses kept for recreation, and lower than empty broodmares on Standardbred and Thoroughbred farms (Rogers et al. 2020, Rogers et al. 2007). The latrine behaviour of horses effectively means that 70% of the pasture, identified as lawns (grazing area), is available for grazing and influences the estimated DM on offer (Adams et al. 2021, Rogers et al. 2017a). The RPM measurements identified limited pasture mass on offer, resulting in horses consuming the new pasture growth as the daily DMI, thus, utilisation of pasture was high. This management system did provide the ability to have relatively good precision when estimating the daily DMI. Based on the data collected, it was estimated that for these cohorts of horses, maintenance requirements were achieved when they were consuming ~1.3% of LW as pasture and/or conserved forage DM. Despite the low level of pasture mass on offer in the lawns, for many of the estimated DMI recordings the horses were above maintenance (64-89% of the observation period) due to daily pasture growth and hay supplementation. It should be noted that RPM estimates do not separate out desired and non-palatable species and as such these periods of above maintenance may be over-estimated.

There were periods of sustained limited feed supply when the DMI was estimated to be below maintenance. The expectation was that these periods would have been associated with fluctuations in either LW and/or BCS.

Figure 2 Monthly body condition score (BCS) change and associated change in live weight (LW) of 23 grazing horses as a percentage (%) of initial LW each month. The width of the box represents the number of observations for each BCS category.

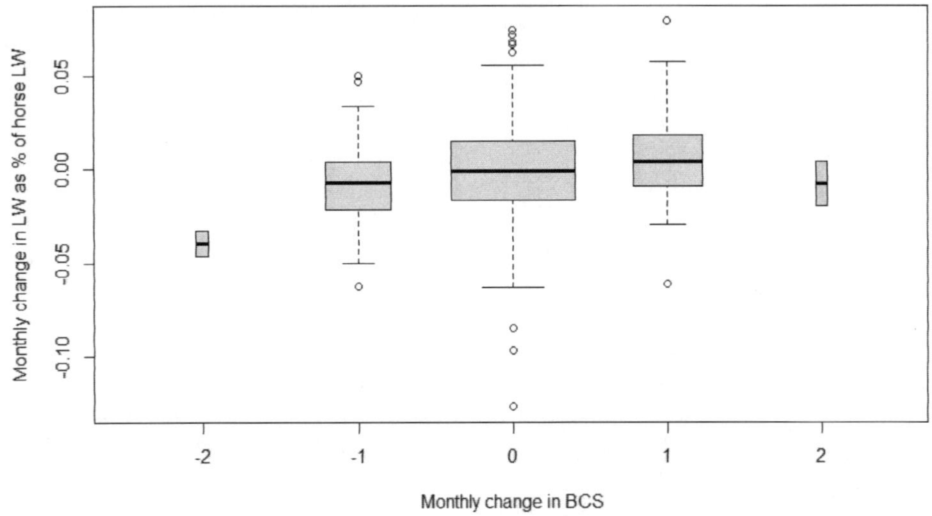

Table 2 Deterministic modelling of periods of negative energy balance (MJ) in grazing horses and the expected weight loss (kg adipose tissue) and actual LW change (kg gained) seen in horses associated with this.

Group	n=	Period (days)	Energy deficit per horse (MJ ME)	Expected adipose loss (kg) per horse	Expected group weight loss (kg)	Actual estimated group weight gain (kg)
G1	7	61	1101.19	34.41	240.88	59
G2	8	65	1032.53	32.27	258.13	38
G3	9	35	826.48	25.83	232.45	74

The variations in LW and BCS recorded with these horses indicated that low intakes were not sufficient to stimulate significant mobilisation of subcutaneous fat and/or loss of weight. It is estimated that mobilisation of 1 kg of body fat yields ~30 MJ of ME in grazing animals (Nicol & Brookes 2017). When these data were used to deterministically model changes in LW, it implied an over-estimation of fluctuations, with groups appearing to maintain or gain weight during these periods. This highlights that the limited changes in BCS observed in this study were not solely due to lack of sensitivity in the scoring system related to mobilisation of visceral fat over subcutaneous fat.

The lack of change in either LW or BCS within the sustained periods of mild energy-supply deficit implies that the horses may have some compensatory mechanism to alter the relative feed value of the limited feed supply. Horses, as monogastric hindgut fermenters, can respond to reduced DMI by decreasing the transit time of feed (Clauss et al. 2014; Edouard et al. 2008) along with subtle differences in the hindgut microbiota (Fernandes 2016). Within a multiple regression analysis there was an increase in the digestibility coefficient for DE with a reduction in transit time (Morel 2010). This provides some insight into how these variables may contribute to provide an apparent increase in feed value.

Much of the LW change recorded each month were below the 5% LW threshold for normal variation associated with intake (feed and water) and output (faeces and urine) (Webb & Weaver 1979), and generally below those reported in the literature associated with a change in BCS (Catalano et al. 2019; Martinson et al. 2014). As this was an observational study and data was captured opportunistically when the horses were in the central yards for teaching purposes, there is some inherent variability in the weight measurements. However, the length of the observation period (two years) and the high frequency of recording should, to a certain extent, have minimised the error associated with such an experimental design. There was also limited variation in LW during periods of very high estimated DMI (up to 5.3% LW). Others have reported high pasture DMI (Longland et al. 2011) and concentrate feed DMI (Argo et al. 2002) but this was associated with increased LW. A high percentage of weeds in the botanical composition of the pasture may have elevated the estimate of pasture mass (companion paper Adams et al. (2021)).

These data collected during the observational study support the hypothesis that Standardbred horses, which provided the major breed composition of the herds observed, appear to be efficient feed converters. Estimation of DMI and true feed value for horses at pasture is difficult, however, these data indicate that reduced DMI may be associated with compensatory mechanisms such as increased transit time and changes in hindgut microbiota, potentially increasing the apparent feed value of the restricted pasture available.

Acknowledgements

The authors would like to thank Mike Reilly, Andrew Smith, Jalissa Gage, and Daniel Sneddon for their assistance in gathering data.

References

Adams BA, Gee EK, Back PJ, Rogers CW 2021. Defecation behaviour reduces horses's utilisation of pasture. New Zealand Journal of Animal Science and Production 81: 74-80.

Argo CM, Cox JE, Lockyer C, Fuller Z 2002. Adaptive changes in the appetite, growth and feeding behaviour of pony mares offered ad libitum access to a complete diet in either a pelleted or chaff-based form. Animal Science 74: 517-528.

Bengtsson J, Rogers CW, Back PJ, Emanuelson U, Roca J, Gee EK, Bolwell CF 2018. Characteristics of the grazing and farm management of broodmares on commercial Thoroughbred stud farms during spring. New Zealand Journal of Animal Science and Production 78: 88-91.

Carroll CL, Huntington PJ 1988. Body condition scoring and weight estimation of horses. Equine Veterinary Journal 20: 41-45.

Catalano DN, Coleman RJ, Hathaway MR, Ne AE, Wagner E, Tyler PJ, McCue ME, Martinson KL 2019. Estimation of actual and ideal bodyweight using morphometric measurements of miniature, saddle-type, and Thoroughbred horses. Journal of Equine Veterinary Science 78: 117-122.

Catalano DN, Hathaway MR, Martinson KL, Coleman RJ, McCue ME, Rendahl AK 2016. Estimation of actual and ideal bodyweight using morphometric measurements and owner guessed bodyweight of adult draft and warmblood horses. Journal of Equine Veterinary Science 39: 38-43.

Chavez S, Siciliano PD, Huntington GB 2014. Intake estimation of horses grazing tall fescue (Lolium arundinaceum) or fed tall fescue hay. Journal of Animal Science 92: 2304-2308.

Clauss M, Schiele, K, Ortmann S, Fritz J, Codron D, Hummel J, Kienzle E 2014. The effect of very low food intake on digestive physiology and forage digestibility in horses. Journal of Animal Physiology and Animal Nutrition 98: 107-118.

Dugdale AHA, Curtis GC, Cripps P, Harris PA, Argo CM 2010. Effect of dietary restriction on body condition, composition and welfare of overweight and obese pony mares. Equine Veterinary Journal 42: 600-610.

Dugdale AHA, Grove-White D, Curtis GC, Harris PA, Argo CM 2012. Body condition scoring as a predictor of body fat in horses and ponies. The Veterinary Journal 194: 173-178.

Edouard N, Fleurance G, Martin-Rosset W, Duncan P, Dulphy R, Grange S, Baumont R, Dubroeucq H, Perez-Barberia FJ, Gordon IJ 2008. Voluntary intake and digestibility in horses: effect of forage quality with emphasis on individual animal variability. Animal 2: 1526-1533.

Fernandes K, Rogers C, Gee E, Bolwell C, Thomas D 2014. A cross-sectional survey of rider and horse demographics, and the feeding, health and management of Pony Club horses in New Zealand. Proceedings of the New Zealand Society of Animal Production 74: 11-16.

Fernandes KA 2016. Aspects of dietary management and dynamics of the faecal microbiota of horses and ponies (Equus caballus) in New Zealand : a thesis presented in partial fulfilment of the requirements for the degree of Doctor of Philosophy in Veterinary Science, Massey University, Palmerston North, New Zealand. Doctor of Philosophy (PhD) Doctoral, Massey University.

Goh SY, Gee EK, Thomas DG, Back PJ, Rogers CW 2020. Miniature horses in New Zealand: demographics and feeding management. New Zealand Journal of Animal Science and Production 80: 1-7.

Grace N, Shaw HL, Gee EK, Firth EC 2002. Determination of the digestible energy intake and apparent absorption of macroelements in pasture-fed lactating Thoroughbred mares. New Zealand Veterinary Journal 50: 182-185.

Guinnefollau L, Gee EK, Norman EJ, Rogers CW, Bolwell CF 2020. Horses used for educational purposes in New Zealand: A descriptive analysis of their use for teaching. Animals 10: 1547-1547.

Henneke DR, Potter GD, Kreider, JL, Yeates BF 1983. Relationship between condition score, physical measurements and body fat percentage in mares. Equine Veterinary Journal 15: 371-372. https://doi.org/10.1111/j.2042-3306.1983.tb01826.x

Jenquip, 2017. Handbook Manual Platemeter. Retrieved 14 August 2017. https://0281517e-9ece-4353-ac75-db19f265c1f8.filesusr.com/ugd/391576_bea7ba10a7ed4394b6517a655d37cfd3.pdf

Longland A, Ince J, Harris PA 2011. Estimation of pasture intake by ponies from liveweight change during six weeks at pasture. Journal of Equine Veterinary Science 31: 275-276.

Martinson KL, Coleman RC, Rendahl AK, Fang Z, McCue ME 2014. Estimation of body weight and development of a body weight score for adult equids using morphometric measurements. Journal of Animal Science 92: 2230-2238.

Medica DL, Hanaway MJ, Ralston SL, Sukhdeo MVK 1996. Grazing behavior of horses on pasture: Predisposition to strongylid infection? Journal of Equine Veterinary Science 16: 421-427.

Morel P 2010. Effect of age, live weight and dry matter intake on total tract nutrient digestibility in adult mares. Proceedings of the New Zealand Society of Animal Production 70: 152-156.

Morrison PK, Harris PA, Maltin CA, Grove-White D, Barfoot CF, Argo CM 2017. Perceptions of obesity and management practices in a UK population of leisure-horse owners and managers. Journal of Equine Veterinary Science 53: 19-29.

NRC 2007. Nutrient requirements of horses. Sixth revised edition. Washington DC: National Academies Press (Animal Nutrition Series). 341 p.

Nicol AM, Brookes IM 2017. The metabolisable energy requirements of grazing livestock. In: Rattray PV, Brookes IM, Nicol AM (Eds). Pasture and supplements for grazing animals. New Zealand Society of Animal Production. Pg. 151-172.

Ringmark S, Roepstorff L, Essen-Gustavsson B, Revold T, Lindholm A, Hedenstrom U, Rundgren M, Ogren G, Jansson A 2013. Growth, training response and health in Standardbred yearlings fed a forage-only diet. Animal 7: 746-753.

Rogers C, Gee E, Back P 2017a. Pastures and supplements in New Zealand commercial equine production systems. In: Rattray PV, Brookes IM, Nicol AM. (Eds.). Pasture and supplements for grazing animals. New Zealand Society of Animal Production. Pg. 303-315.

Rogers CW, Gee EK, Back PJ, van Zon S, Hirst RL, Pomroy WE 2018. Pasture use and management in commercial equine production systems. New Zealand Journal of Animal Science and Production 78: 40-44.

Rogers CW, Gee EK, Bolwell CF, Rosanowski SM 2017b. Commercial equine production in New Zealand. 2. Growth and development of the equine athlete. Animal Production Science. http://dx.doi.org/10.1071/AN16752.

Rogers CW, Gee EK, Firth EC 2007. A cross-sectional survey of thoroughbred stud farm management in the North Island of New Zealand. New Zealand Veterinary Journal 55: 302-307.

Rogers CW, Herbst A, Langhans K, Stone R, Hirst R, Gee EK, Back, PJ 2020. The management of Standardbred breeding stock on commercial properties. New Zealand Journal of Animal Science and Production 80: 84-89.

Suagee JK, Burk AO, Quinn RW, Petersen ED, Hartsock TG, Douglass LW 2008. Effects of diet and weight gain on body condition scoring in Thoroughbred geldings. Journal of Equine Veterinary Science 28: 156-166.

Webb AI, Weaver BM 1979. Body composition of the horse. Equine Veterinary Journal 11: 39-47.

Wyse CA, McNie KA, Tannahil VJ, Love S, Murray JK 2008. Prevalence of obesity in riding horses in Scotland. Veterinary Record 162: 590-591.

Impact of casein- versus whey-based lamb milk replacers on growth and health of artificially reared lambs

SA McCoard[a]*, J Ryrie[b], T MacDonald[c], S-Y Hea[a], D Pacheco[a] and D Stevens[b]

[a]*AgResearch Limited, Grasslands Research Centre, Tennent Drive, Palmerston North, New Zealand 4442;* [b]*Spring Sheep Dairy NZ LP, Auckland, New Zealand;* [c]*AgResearch Limited, Invermay Research Centre, Puddle Alley, Mosgiel, New Zealand*

**Corresponding author*: Email: sue.mccoard@agresearch.co.nz

Abstract

High cost of milk proteins has driven development of cheaper whey-based milk replacer (WBMR) as an alternative to skim-milk (casein)-based milk-replacer (CBMR) for artificial-rearing of young ruminants. This study compared growth, antibiotic/anti-inflammatory use and mortality in lambs reared on either WBMR (n=138) or CBMR (n=151) in the first three weeks of rearing at commercial scale. There was a diet x rank interaction (P=0.001) whereby CBMR-fed single/twin and triplet/quad lambs had the highest average daily gain (ADG) independent of birth-rank (362 *vs.* 358 g/d respectively, SED=16.7) compared to WBMR-fed lambs where higher ADG was observed in triplet/quad compared to single/twin lambs (275 *vs.* 213 g/d respectively, SED=16.7). Males tended to have higher ADG than females (311 *vs.* 294, SED=9.7 g/d, P=0.064). Overall, there was a tendency for greater mortality (11% *vs.* 4%; P=0.055), and overall incidence rate (*N*/100 animal weeks) for use of antibiotics (4.43 *vs.* 1.27; P=0.073) and anti-inflammatories (4.59 *vs.* 1.59; P=0.099) in WBMR- than CBMR-fed lambs. These results highlight that CBMR under an *ad-libitum* milk-feeding regime supports greater ADG independent of birth-rank and reduces the incidence of disease in the first three weeks of life compared to WBMR. Furthermore, WBMR under an *ad-libitum* milk-feeding regime results in lower ADG of single/twin-born lambs compared to their triplet/quad counterparts.

Keywords: lambs; artificial rearing; milk replacer; composition; growth; health; mortality

Introduction

In nature, lambs are naturally suckled by their dams and receive numerous small feeds of whole milk. One of the most important nutrients in milk replacers is protein. Therefore, choosing an appropriate protein source influences feed quality and animal production costs (Erickson et al. 1989). Milk replacers generally contain protein from two sources: milk protein and non-milk protein. Milk protein has balanced amino acid constituents, is highly digestible and is, therefore, considered an excellent protein source. However, the high cost of milk proteins in the US and Europe has resulted in skim-based milk replacers being uneconomic for calf-rearing systems (Davis & Drackley 1998). As a result, non-curding whey milk replacers have been developed and tested under United States and European conditions and are being used in artificial-rearing systems for ruminants including lambs and calves.

Recent studies indicate lower weight gain in artificially reared lambs (McCoard et al. 2021) and calves (Thomson et al. 2018) fed whey-based versus casein-based milk replacers while disease and mortality were similar. Artificial rearing of lambs is an essential part of some dairy sheep systems for generating replacements, and for for rearing surplus progeny and orphan/multiple-born lambs on traditional meat- and wool-producing sheep farms. As with calf-rearing systems, the cost of lamb rearing is a key consideration for producers. While the impact of whey-based milk replacers in calves has received attention in the literature (Huang et al. 2015; Thomson et al. 2018), the production performance of artificially reared lambs on casein- or whey-based milk replacers has not been described.

The objective was to evaluate the effect of a casein-based versus whey-based lamb milk replacer on the growth, health and survival of artificially reared East Friesian cross-bred lambs in the first three weeks of rearing using an *ad-libitum* milk-feeding system at commercial scale.

Materials and methods

All procedures in this study were approved by the AgResearch Grasslands Animal Ethics Committee (AE13960).

East Friesian cross-bred lambs (n=294) were sourced from naturally mated ewes lambed outdoors on pasture on a commercial dairy sheep farm in the central North Island. Lambs were separated from their dams at 2-3 days of age to provide sufficient time for colostral intake and were transferred immediately to the rearing facility between the 21st and 26th of August 2016. Upon entry to the dedicated commercial rearing facility, all lambs were weighed, tagged for identification and their navels dipped in iodine to reduce the risk of infection. The lambs were randomly allocated to one of two treatment groups balanced for sex, birth rank (single-twin or triplet-quadruplet) and date of birth to reflect commercial rearing practices. The lambs were group housed in 3 × 3 metre pens with eight pens per treatment group (16-20 lambs/pen with two teats per pen). Treatments started upon entry to the rearing shed and consisted of either a casein-based milk replacer formulated from 100% bovine milk proteins and fat (CBMR; n=151: 93 female, 58 male, 61 single-twin, 90 triplet-quad) or whey-based milk replacer formulated from bovine milk proteins, hydrolyzed wheat protein and vegetable oil (WBMR; n=143: 82 female, 61 male, 74 single-twin, 69 triplet-quad). All lambs were fed using *ad-libitum* warm reconstituted milk replacer (MR) at a mixing rate of 200 g/L using automatic feeders (DeLaval LKF1200, DeLaval, Hamilton, NZ). Lambs within each pen received the same treatment. Lamb starter meal and fresh water were available in the pens *ad-libitum.* Lambs within a treatment group were fed MR using a single automatic feeder. Therefore, individual intake of MR was not recorded. However, total amount of MR per group was recorded. The total amount of starter diet offered per pen was recorded and total intake per group calculated. All pens were bedded with wood chip and the rearing facility had good ventilation and dedicated rearing staff.

All CBMR and WBMR was purchased as a single batch. Whole sheep milk powder (WSMP) was also sourced from a commercial dairy sheep operation and used to enable the comparison of the nutritional composition profile between the two milk replacers evaluated in this study compared to whole sheep milk. Three samples of each batch of MR were pooled and composition analysed (Table 1) for ash (Furnace 550oC AOAC 942.05), crude protein (AOAC 968.06 Dumas method, N-P = 6.38), fat (Mojonnier, dairy AOAC989.05), lactose (Enzymatic method), minerals (Vacuum oven, AOAC 990.19, 990.21), and carbohydrate (by difference) by the Massey University Nutrition Laboratory (Palmerston North, New Zealand). Metabolisable energy and lactose content of the MR and WSMP was calculated using NRC (2002) equations. The milk protein composition of each diet was determined using high-throughput liquid chromatography (Day et al. 2015) using a Phenomenex Aeris™ 3.6 µm WIDEPORE XB-C18 200 Å, LC Column 250 x 4.6 mm Part Number:00G-4482-E0 (Phenomenex, Torrance CA, USA). Bovine milk protein standards were used for quantitation as sheep protein standards are not available (Sigma Aldrich, Auckland, New Zealand: κ-CN (C-0406), α-CN (C-6780), β-CN (C-6905), α-LA (L-5385), β-LG B (L-8005), and β-LG A (L-7880), β-LG (L3908). Amino acid profiling was performed as described by McCoard et al. (2021).

Dry matter (DM), crude protein (CP), acid detergent fibre (ADF), neutral detergent fibre (NDF), soluble sugars (SS), starch and ash in the starter diet were determined using the following procedures: DM (Method 945.15, AOAC, 2010), CP (Method 992.15, AOAC, 2010), ADF and NDF (Method 7.074, AOAC, 1990), soluble sugars (Method 942.05, AOAC, 2010), starch (Method 996.11, AOAC, 2010), and ash (Method 942.05, AOAC, 2012) by a commercial laboratory (Massey Nutrition Laboratory, Palmerston North, New Zealand).

Lamb live weight was measured upon entry to the rearing shed and exit from the rear shed, approximately three weeks later. After this time, the lambs were managed as a single mob and, therefore, there was no replication which prevented statistical analysis of the performance data. Therefore, our focus was on the first three weeks of rearing. All animal health issues that required animal health treatment intervention were recorded as part of the daily health-check regime for all individual animals. All animal health treatments were administered by trained staff according to the manufacturer's instructions and under guidance from the farm veterinarian as previously described (McCoard et al. 2021). All lamb deaths were recorded.

Lamb growth data were analysed using a linear mixed model (LMM). Lamb sex, date of birth, rank (single-twin vs triplet-quad), diet (MR type), and an interaction term between diet and rank were included in the model as fixed effects. Pen was used as a random effect (random intercept) to account for between-pen heterogeneity.

Generalised linear mixed models (GLMM) were used to model the effects of diet on lamb mortality, health events (pneumonia and lameness) and drug use (antibiotics and anti-inflammatories). A logit link function with a binomial distribution was used for modelling lamb mortality. A log link function with a negative binomial distribution and an offset using the natural logarithm of time (rearing period) was used for modelling health events and drug use. Diet was used as a fixed effect, and a random intercept was used for the pen to account for between-pen heterogeneity.

Statistical analysis was performed using R (R Core Team. 2021). GLMM modelling was performed using glmmTMB, and model fit was assessed using simulated quantile residuals from the DHARMa package. LMM modelling was performed using lme4 and model fit was assessed using residual plots (QQ-plots, residual vs fitted plots). Treatment contrasts, P values and confidence intervals were generated from the emmeans package.

Table 1 Mortality rate (%) and incidence rate and ratio[1] of drug use and selected animal health issues in lambs reared on casein-based MR (CBMR) or whey-based milk replacer (WBMR) in the first three weeks of life.

	WBMR	CBMR	Ratio (CI)	P value
Mortality (%)	11	4	-	0.055
Total antibiotics	4.43	1.27	3.5 (0.90, 13.80)	0.073
Total anti-inflammatories	4.59	1.59	2.89 (0.80, 10.2)	0.099
Pneumonia	1.89	0.70	2.70 (0.54, 13.50)	0.225
Lameness	1.50	0.26	5.82 (0.40, 85.10)	0.197

[1] Incidence rate and incidence ratio per 100 animal weeks. In this study, incidence rate is a measure of new cases of disease per unit of time at risk (i.e., animal week at risk), e.g., incidence rate of 4.43 per 100 animal weeks means we expect to see an average of 4.43 cases for every 100 animals observed during a one-week period within a treatment group. The incidence ratio is a measure of the incidence rates between two treatment groups, e.g., an incidence ratio of 3.5 means that the rate of total antibiotic use is 3.5 times greater in lambs exposed to WBMR compared with lambs exposed to CBMR.

Figure 1 Effect of milk replacer protein source (casein-based milk replacer, CBMR; whey-based milk replacer, WBMR), birth rank (single-twin vs. triplet-quad) and sex, and their interaction, on average daily gain. Data are presented as means ± standard error of the mean.

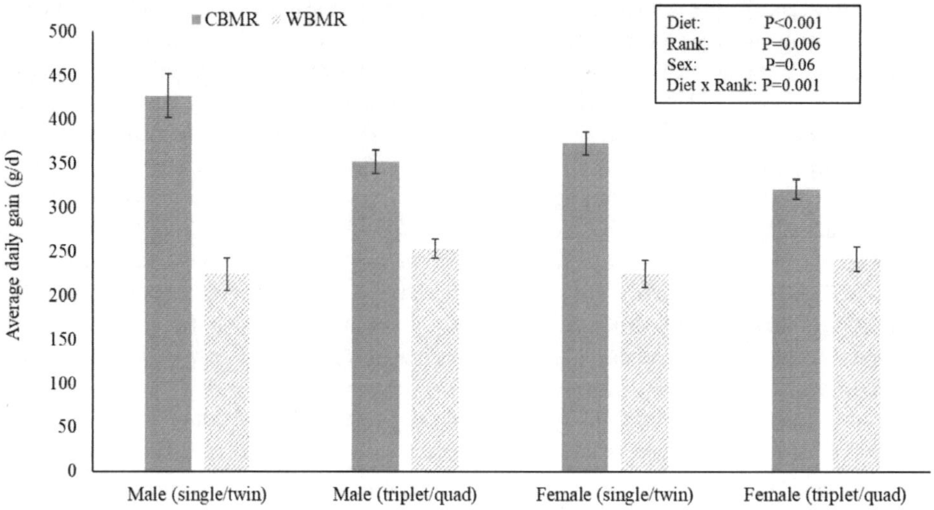

Results

Growth and DMI

Overall, lambs reared with *ad-libitum* CBMR grew faster than those reared on WBMR (116g/d per lamb difference). A diet-by-rank interaction was evident for average daily gain (ADG) whereby single/twin-born lambs reared on CBMR grew faster than triplet/quad-born lambs, while the opposite was observed for lambs reared on WBMR (Figure 1).

On average, the CBMR group consumed 6.8 kg compared to 5.3 kg total MR dry matter intake (DMI) per lamb in the WBMR. Similar intake of starter meal per lamb (0.8-0.9 kg/lamb) was recorded during the rearing period in both groups.

Health

During the three-week rearing period, mortality rate tended to be greater in lambs reared on *ad-libitum* WBMR compared to CBMR (11% vs. 4%, P=0.055). Lambs reared with WBMR tended to be 3.5 times (P=0.073) more likely to be treated with antibiotics and 2.9 times (P=0.099) more likely to be treated with anti-inflammatories than lambs reared with CBMR (Table 1). The main health issues observed were lameness and pneumonia with numerically higher incidence rates in the WBMR than CBMR group.

Milk and starter composition

The gross milk composition of both the CMBR and WBMR differed from WSMP (Table 2). Compared to WSMP, CBMR had greater carbohydrate and lactose content and lower protein, and fat content, while WBMR had a greater carbohydrate and lower protein, fat and lactose. The CBMR had greater protein, fat and lactose content, and lower carbohydrate content relative to the WBMR.

Compared to WSMP, the CBMR had lower levels of all the individual milk proteins except α-casein which was similar with the largest differences observed for β-casein (Table 2). Overall, CBMR had 12% lower total casein, 51% lower whey protein and 16% lower total milk protein compared to WSMP. Compared to WSMP, the WBMR had no casein milk proteins, almost four-fold more α-lactalbumin and almost two-fold more β-lactoglobulin, leading to just over two-fold more total whey protein but 78% less total milk protein. The CBMR contained 22% less total whey proteins compared to the WBMR, and WBMR contained no casein-protein with the CBMR having almost four times as much total milk protein compared to the WBMR. Overall, based on milk protein content, CBMR was more similar to WSMP than WBMR.

The amino acid content of the WSMP and the MR evaluated in this study are presented in Table 2. Compared to WSMP, CBMR had 9-23% lower EAA, 14-38% lower conditionally essential AA (CEAA) levels and 12-27% lower NEAA content leading to 18%, 20% and 18% lower overall EAA, CEAA and non-essential AA (NEAA) levels respectively and 19% lower total AA content. Compared to WSMP, WBMR had 7-50% lower EAA, 10-56% lower CEAA except cystine which was 77% greater, and 23-35% lower NEAA leading to 40%, 26% and 32% lower overall EAA, CEAA and NEAA levels respectively and 32% lower total AA content. Compared to WBMR, CBMR had 21% greater EAA levels driven by 8-62% higher levels of all EAA except threonine which was 13% lower and similar tryptophan content. CBMR had 32% greater CEAA levels relative to WBMR driven by 47-94% greater levels of arginine, proline and tyrosine while CBMR had lower levels of cystine and tyrosine compared to WMBR. Total AA content of CBMR was 19% greater than that of WBMR. Composition (%DM) of the starter diet was 20% CP, 6.4 % ADF, 15.7% NDF, 5.8% SS, 26% starch.

Table 2 Gross milk composition, individual milk proteins and amino acid composition of whole sheep milk powder (WSMP), casein-based milk replacer (CBMR), whey-based milk replacer (WBMR) and the percentage differences in amino acid composition between each component.

	WSMP	CBMR	WBMR
Composition			
Moisture % DM	1.3	2.6	2.6
Ash % DM	5.4	5.4	9.1
Protein % DM	32.1	25.9	22.8
Fat % DM	35.0	27.1	23.5
Carbohydrate % DM	28.8	41.6	44.5
Lactose g/100g DM	24.4	35.4	19.2
Calcium g/kg DM	10.7	9.2	10.2
Sodium g/kg DM	2.6	2.9	7.2
ME MJ/kg DM	23.8	21.8	20.3
Milk proteins (mg/g)			
κ-casein	41.8	35.2	0
α-casein	100.8	102.1	0
β-casein	248.6	207.5	0
α-lactalbumin	7.2	2.8	27.7
β-lactoglobulin	33.9	17.5	65.8
Total casein protein	391.2	344.8	0
Total whey protein	41.1	20.3	93.5
Total milk protein	432.2	365.1	93.5
Amino acids (mg/ml)			
Essential AA			
Histidine	8.4	6.8	4.2
Isoleucine	15.5	12.6	10.6
Leucine	30.2	24.1	20.0
Lysine	21.9	19.9	18.4
Methionine	6.8	5.5	3.5
Phenylalanine	14.5	12.1	8.2
Threonine	13.9	11.2	12.9
Tryptophan	4.4	3.4	3.4
Valine	20.9	16.0	11.3
Conditionally essential AA			
Arginine	10.2	8.7	5.9
Cystine	2.6	1.6	4.6
Glycine	6.2	4.8	5.6
Proline	31.0	24.0	16.3
Tyrosine	14.0	12.0	6.2
Non-essential AA			
Alanine	11.8	8.6	8.9
Aspartic acid	24.7	19.5	18.1
Glutamic Acid	66.5	55.5	51.3
Serine	16.2	14.2	10.5
Total EAA	136.5	111.6	92.5
Total CEAA	64.0	51.1	28.6
Total NEAA	119.2	97.8	88.8
Total AA	319.7	260.5	219.9

Note: Glutamine and Asparagine were not detected

Discussion

For the first few weeks of life, the diet of pre-ruminant lambs consists primarily of milk, and the gastrointestinal tract functions in a similar manner to that of monogastric species. This development period is characterised by rapid growth which is supported by high rates of skeletal muscle protein synthesis (Chien et al. 1993; Denne & Kalhan 1987). The neonatal period is also one of the most vulnerable for the establishment of pathogenic gastrointestinal infections which can impact growth and health, especially in intensive indoor-lambing systems where pathogen loads are higher than in outdoor environments (Tzipori et al. 1981). Therefore, the neonatal period is critical for the growth and health of pre-ruminants.

Antimicrobials, particularly in intensive livestock farming systems (ESVAC, 2013) are widely used to improve animal health, welfare and productivity. Such use has driven the growing concern around increased risk of development of resistance in animal and human pathogens (Van De Sande-Bruinsma et al. 2008). In the present study antibiotics were only used therapeutically. While neonates are particularly vulnerable to infectious pathogens and other animal health issues due to the immaturity of their immune system (Jones et al. 2010), this study illustrated that MR formulation can influence drug use in artificially-reared lambs consistent with our prior observations (McCoard et al. 2021). The results from this study suggest that use of WBMR is likely to increase the use of therapeutic drugs and negatively affect growth and health of lambs compared to CBMR.

Rapid pre-weaning growth rates in artificial lamb-rearing systems are desirable to reduce the time to reach weaning weights to reduce the cost of rearing. In the present study, one of the biggest changes in animal performance was the difference in growth rates and lamb mortality. While overall growth rates were lower in lambs reared in WBMR than CBMR, single- and twin-born lambs reared on WBMR were particularly compromised. Capacity for skeletal muscle protein synthesis is reduced in *in-utero* growth-restricted lambs such as multiple-born lambs (Sales et al. 2014) therefore, lower growth rates of triplets/quadruplets compared to singles and twins is expected, as was observed in the CBMR group. However, the lack of difference in growth rates between birth ranks, and lower overall growth rates compared to the CBMR group, indicates that the WBMR was unable to meet the requirements for growth, especially in single/twin-born lambs, which are heavier at birth and, therefore, have greater energy requirements for maintenance (Greenwood et al. 1998). The authors are not aware of prior studies that compare the growth rates of single-twin and triplet-quadruplet artificially reared lambs so comparisons are unable to be made. We have recently reported that growth and health is compromised in lambs artificially reared on a CBMR containing a greater proportion of whey protein, hydrolysed wheat and vegetable oil compared to a CBMR formulated from 100% milk proteins and fats (the same formulation used in the present study), while mortality was similar between groups (McCoard et al. 2021). In the present study, growth rates were substantially lower and mortality was higher than those in the aforementioned study.

Lower growth rates of the lambs fed the WBMR versus CBMR was likely due, at least in part, to lower average DMI of MR. Average DMI of MR per lamb for the CBMR group (322 g/lamb/d) were similar to other studies in which lambs were artificially reared indoors with *ad-libitum* milk allowance on automatic feeders (McCoard et al. 2021, ~300 g DM/d; Belanche et al. 2019, 300g DM/d; David et al. 2014, 329 g DM/d; Bimzcok et al. 2005, ~319 g DM/d) while MR intake per lamb in the WBMR group was lower (253 g/hd/d). Similar DMI were observed in our previous study independent of MR formulation (McCoard et al. 2021) despite lower inclusion rates of whey and hydrolysed wheat protein. Therefore, the lower DMI of WBMR than CBMR in the present study suggests the WBMR formulation may have negatively affected intake. In both the present study, and that of McCoard et al. (2021), the incidences of drug use and health issues were higher in lambs reared with a MR containing vegetable proteins and fats, suggesting health issues may also be a contributor to differences in DMI. However, incidence of health issues that did not require therapeutic drug use (e.g., scours) were not recorded in the present study and increased scour incidence was associated with lower ADG in lambs reared on a MR containing vegetable proteins and oils (McCoard et al. 2021). Other factors that can influence DMI include, competition at the feeder or adaptation to automatic feeders (David et al. 2014) and palatability of the diet (Miller-Cushon et al. 2014). Direct evaluation of these factors was not undertaken in the current study but warrant further investigation. Starter intake can also influence MR intake in calves (Khan et al. 2016) but to the authors' knowledge has not been evaluated in lambs. However, starter intake was very low and similar between groups (0.8-0.9 kg/lamb total intake during the study), which is consistent with low intake of starter observed in young lambs (McCoard et al. 2019; Belanche et al. 2019) and calves (Gerrits, 2019). It is important to note that individual intakes were not measured in this study due to group housing. Therefore, the reported overall differences in MR and starter intake cannot be evaluated statistically. However, that the size and shape of the pens, stocking density, number of teats per pen, and animal husbandry was consistent across all pens and treatments suggests that the differences in growth rate are more likely due to differences in the composition of the MR.

Neonates and growing animals have a high demand for amino acids for synthesis of protein, glucose and other cellular metabolites and have been demonstrated to be involved in functional roles that influence growth, development, health and survival (McCoard et al. 2016). The use of vegetable proteins as an alternative cheaper protein source to milk proteins for formulation of MR is growing in popularity (Gerrits 2019). In general, the digestive system of the young ruminant is poorly developed and can only digest a limited range of carbohydrates, fats and proteins (Davis & Drackley 1998). Substitution of skim milk powder with a mix of solubilised wheat protein concentrate and whey powder increases abomasal emptying of fat and protein and decreases digestibility of AA and total nitrogen, but not fat (Toullec & Formal, 1998). In the present study, the source of both milk protein and fats differed considerably so it is not possible to determine whether substitution of milk protein or milk fat contributed to the observed differences in animal performance. However, CP, milk protein type (casein *vs.* whey) and AA content differed considerably between WBMR and CBMR, with CBMR more closely reflecting the profile of WSMP. In the present study, the CBMR contained almost four times as much total milk protein compared to the WBMR. The similar protein content of the MR (25.2 *vs.* 22.2 % DM for CBMR and WBMR respectively) implies that the WBMR contained a much greater proportion of non-milk proteins than the CBMR. According to the product label, the WBMR contained hydrolysed wheat protein. In calves, the use of wheat gluten and rice protein concentration (80% CP content) as a source of protein for calf diets to substitute for milk protein has been associated with a reduction in growth rate as the proportion of vegetable protein increased (Hill et al. 2008). The type of vegetable protein is also important in calves as soybean concentrate and rice-protein isolate delivered similar growth performance to that of full milk protein, while lower growth rates were observed when hydrolysed wheat-protein or peanut-protein concentrate were used at the same nutrient (AA) levels (Li et al. 2008). Thus, the high inclusion rate of hydrolysed wheat protein at the expense of milk protein, coupled with lower DMI and supply and/or balance of AA including specific functional AA (e.g., arginine, leucine, methionine; McCoard et al. 2016), are likely major contributing factors to the reduction in growth rate of the lambs in the current study, as previously observed in lambs (McCoard et al. 2021) and calves (Huang et al. 2015). Future studies to establish the optimal AA profile to support the growth and health of artificially-reared lambs, compared to naturally-reared lambs, is warranted to provide insights into strategies to improve the formulation of MR to more closely mimic whole sheep milk. The impact of these MR formulations beyond three weeks of rearing is also warranted.

This study illustrates that young lambs can be reared on both casein- and whey-based milk replacers, albeit at a cost in terms of liveweight gain, health and welfare of lambs reared using the WBMR. The key result from this study was that CBMR, under an *ad-libitum* milk-feeding regime using automatic feeders, supports greater ADG independent of birth-rank, and tends to reduce mortality and incidence of disease requiring therapeutic drug use in the first three weeks of life compared to WBMR. Furthermore, compared to CBMR, WBMR under an *ad-libitum* milk feeding regime results in lower ADG of single-twin-born lambs compared to their triplet-quad counterparts. These effects are likely driven by a combination of lower ME and CP intake from MR and lower nutritional quality of the WBMR resulting from a lower inclusion rate of milk proteins which are highly digestible and have a balanced

AA profile for growth. We conclude that selection of MR for lamb rearing should consider not just feed cost, but also the impact on growth, health and mortality rates. Based on the observations from this study, feeding CBMR formulated from 100% milk proteins and fats is recommended for artificial rearing of lambs in the first three weeks of life.

Acknowledgements

The authors sincerely thank the Spring Sheep Milk Co. farm staff for their technical support for this study, Mike Agnew (AgResearch Grasslands) for the milk protein analyses, and John Rounce and Bryan Treloar for the amino acid profiling of the CBMR, WBMR and WSMP.

References

AOAC. 1990. Official methods of analysis. 15th ed. Washington, DC: AOAC.

AOAC. 2010. Official methods of analysis. 18th ed. AOAC International, Gaithersburg, MD.

AOAC. 2012. Official methods of analysis. 19th ed. AOAC International, Gaithersburg, MD.

Belanche A, Cooke J, Jones E, Worgan HJ, Newbold CJ 2019. Short- and long-term effects of conventional and artificial rearing strategies on the health and performance of growing lambs. Animal. https://www.sciencedirect.com/science/article/pii/S1751731118002100?via%3Dihub

Bimczok D, Rohl FW, Ganter M 2005. Evaluation of lamb performance and costs in motherless rearing of German Grey Heath sheep under field conditions using automatic feeding systems. Small Ruminant Research 60: 255265. https://doi.org/10.1016/j.smallrumres.2004.12.008

Chien PFW, Smith K, Watt PW, Scrimgeour CM, Taylor DJ, Rennie MJ 1993. Protein turnover in the human fetus studied at term using stable isotope tracer amino acids. American Journal of Physiology, Endocrinology and Metabolism 265: E31-E35.

David I, Bouvier F, Ricard E, Ruiesche J, Weisbecker J-L 2014. Feeding behaviour of artificially reared Romane lambs. Animal 8(6): 982-990. https://dx.doi.org/10.1017%2FS1751731114000603

Davis CL, Drackley JK 1998. The Development, Nutrition and Management of the Young Calf. Iowa State Univ. Press, Ames, IA.

Denne SC, Kalhan SC 1987. Leucine metabolism in human newborns. American Journal of Physiology, Endocrinology and Metabolism 253: E608-E615.

Erickson PS, Schauff DJ, Murphy MR 1989. Diet digestibility and growth of Holstein calves fed acidified milk replacers containing soy protein concentrate. Journal of Dairy Science 72: 1528-1533.

ESVAC (European Surveillance of Veterinary Antimicrobial Consumption). Third ESVAC report: Sales of veterinary antimicrobial agents in 25 EU/EEA countries in 2011. European Medicines Agency, London, UK. Accessed Jul. 17, 2013. www.ema.europa.eu/docs/enGB/documentlibrary/Report/2013/10/WC500152311.pdf.

Gerritse WJ 2019 Symposium review: Macronutrient metabolism in the growing calf. Journal of Dairy Science 102: 3684-91.

Greenwood PL, Hunt AS, Hermanson JW, Bell AW 1998. Effects of birth weight and postnatal nutrition on neonatal sheep: I. Body growth and composition and some aspects of energetic efficiency. Journal of Animal Science 76: 2354-2367.

Hill TM, Bateman HG, Aldrich JM, Schlotterbeck RL 2008. Effects of using wheat gluten and rice protein concentrate in dairy calf milk replacers. Professional Animal Scientist 24: 465-472.

Huang K, Tu Y, Si B, Xu G, Guo J, Guo F, Yang C, Diao Q 2015. Effects of protein sources for milk replacers on growth performance and serum biochemical indexes of suckling calves. Animal Nutrition 1: 349355. https://doi.org/10.1016/j.aninu.2015.11.012

Jones KD, Berkley JA, Warner JO 2010. Perinatal nutrition and immunity to infection. Paediatric Allergy Immunology 21: 564-576. https://dx.doi.org/10.1111%2Fj.1399-3038.2010.01002.x

Khan MA, Bach A, Weary DM, von Keyserlingk MAG 2016. Invited review: Transitioning from milk to solid feed in dairy heifers. Journal of Dairy Science 99: 885902. https://doi.org/10.3168/jds.2015-9975

Li H, Diao QY, Zhang NF, Fan ZY 2008. Growth, nutrient utilization and amino acid digestibility of dairy calves fed milk replacers containing different amounts of protein in the preruminant period. Asian-Australisian Journal of Animal Science 21: 1151-1158.

McCoard SA, Sales FA, Sciascia Q 2016. Amino acids in sheep production. Frontiers in Bioscience, Elite 8: 264-288.

McCoard SA, Hea S-Y, Karatiana D, Triggs J, Macdonald T 2021. Comparison of milk replacer composition and effects on growth and health of preruminant lambs, and health-associated costs of artificial rearing. Applied Animal Science 37: 176-185.

Miller-Cushon EK, Terré M, DeVries TJ, Bach A 2014. Palatability of protein source on dietary selection in dairy calves. Journal of Dairy Science 97: 4444-4454. http://dx.doi.org/10.3168/jds.2013-7816

R Core Team 2021. R: A language and environment for statistical computing. R Foundation for Statistical Computing, Vienna, Austria.

Sales FA, Pacheco D, Blair HT, Kenyon PR, Nicholas G, Senna Salerno M, McCoard SA 2014. Identification of amino acids associated with skeletal muscle growth in late gestation and at weaning in lambs of well-nourished sheep. Journal of Animal Science 92: 5041-5052. https://doi.org/10.2527/jas2014-7689

Thomson BC, Muir PD, Smith NB, Nieuwenhuis A 2018. The role of curding and non-curding calf milk replacers in NZ calf-rearing systems. New Zealand Journal of Animal Science and Production 78: 157-160.

Toullec R, Formal M 1998. Digestion of wheat protein in the preruminant calf: Ileal digestibility and blood concentrations of nutrients. Animal Feed Science Technology 73: 115–130. https://doi.org/10.1016/S0377-8401(98)00126-6

Tzipori S, Sherwood D, Angus K, Campbell I, Gordon M 1981. Diarrhea in lambs: experimental infections with enterotoxigenic Escherichia coli, rotavirus, and Cryptosporidium sp. Infectious Immunology 33(2): 401-406.

Van De Sande-Bruinsma N, Grundmann H, Verloo D, Tiemersma E, Monen J, Goossens H 2008. Antimicrobial drug use and resistance in Europe. Emerging Infection Diseases 14: 1722-1730.

Investigating the potential for global positioning satellite data to provide information on ewe behaviour around the time of lambing

PL Johnson[1]*, A Welsh[2], K Knowler[1] and P Pletnykov[3]

[1]*Invermay Agricultural Centre, AgResearch Ltd., Private Bag 50034, Mosgiel, New Zealand;* [2]*TwinFarm 808 Waimea Valley Rd, RD7, Gore, New Zealand;* [3]*Lincoln Research Centre, AgResearch Ltd., Christchurch, New Zealand*

**Corresponding author:* Email: tricia.johnson@agresearch.co.nz

Abstract

Global positioning satellite (GPS) devices allow spatial-temporal data sets to be generated which can describe ewe movement around the time of lambing and can potentially be linked to traits of interest, such as the timing and location of lambing events and maternal outcomes including lamb weaning weight. Collar-mounted GPS units were fitted to 25 ewes during their anticipated lambing period. Lambing timing and site were estimated based on a minimum of twice-daily checks of the ewes. Lambing and GPS data were ultimately available for 15 ewes. From the resulting data set, several traits were calculated including the average distance deviation from the central point of the three-hour time block. During the birthing event (at a resolution of hours), the ewes changed their movement behaviour and remained close to the centralised point of the three-hour block. There were significant associations among the weaning weight of the ewes' offspring (adjusted for rearing rank) and the GPS-derived traits: ewes that were comparatively inactive produced lambs with low weaning weights. Overall, this study demonstrates the potential of GPS-derived traits to be of relevance in studies investigating ewe behaviour around and after-lambing and its impact on production outcomes.

Keywords: lambing; global positioning satellites; weaning weights

Introduction

Within sheep-production systems the number of lambs a ewe weans relative to the number of lambs detected by pregnancy scanning, together with the weaning weight of surviving lambs, are important production metrics. These traits are influenced by lamb survival and the maternal ability of the ewe. A link between ewe behaviour around the time of lambing and lamb survival, and the growth potential of lambs has been demonstrated (O'Connor et al. 1985). In lamb-survival research, maternal behaviour was traditionally assed by how close the ewe comes to a human handling her lambs (often tagging) within 24 hours of birth. Although demonstrated to be a good proxy for maternal behaviour, it does not fully reflect how a ewe behaves without the presence of a human.

With the advent of wearable digital devices, there is an opportunity to capture information whilst ewes are undisturbed. The types of devices available to determine lambing events include global positioning satellite (GPS) devices (Dobos et al. 2014; 2015; Fogarty et al. 2020) and accelerometers (Smith et al. 2020), or a combination (Forgarty et al. 2021). A GPS device can capture not only spatial-temporal data on the location of the ewes, but can also be used to calculate several metrics, with examples in the literature including distances travelled, speed, area utilisation, and distance to other ewes. Accelerometer data can be used to infer what a ewe is doing at a given point in time, with algorithms developed to infer standing, walking, grazing and, in the case of lambing, the position/orientation of the ewe – if she is on her side or sitting upright.

In this field of research, to date, studies have mainly concentrated on determining attributes that can be derived from digital devices that indicate the onset of lambing enabling ewes to be monitored so that assistance can be provided if the ewe is undergoing a difficult birth (Dobos et al. 2014; Forgarty et al. 2021). Whilst this is an important opportunity, it requires commercial ewes to be wearing the appropriate devices and for real-time data acquisition, processing, and reporting to be available which, although possible, is cost-prohibitive for most commercial sheep operators. Alternatively, such data has value within research programmes to be retrospectively used to determine the date of lambing and her behaviour. It can also be used to explore the behaviour of the ewe around the time of lambing, including where she lambs and how she modifies her behaviour with lambs at foot. This resulting behavioural data can be analysed to investigate if they relate to lamb survival or the lamb-weaning-weight outcomes.

This paper reports on the potential for GPS devices to determine the timing and location of lambing, and whether GPS devices can demonstrate variation in ewe behaviour post-lambing that can be linked with weight of lambs weaned. Lamb survival, although an important trait, could not be investigated due to small numbers in the trial.

Materials and methods

The choice of ewes and location for this trial were based on ethics committee requirements that the ewes were viewed twice daily to eliminate risks associated with the ewes becoming entangled in fences or other objects. As such, the best resource for this first-stage research was a flat-land property with easy visual inspection of all fences within the paddocks. All data collection was approved by the AgResearch Animal Ethics Committee.

Data collection

Global positioning satellite devices mounted on collars were fitted on 25 maternal ewes lambing on a property near Gore, Southland, New Zealand (Latitude -45.970367, Longitude168.745957). Most ewes remained with a paddock 4.04 ha in size, with *ad libitum* access to a ryegrass-dominant pasture for the duration of the data collection. Some ewes were removed from the paddock for assistance with lambing.

The collars used were custom-made (DataCarter) for use in sheep, with GPS componentry and an 18Ah D cell lithium thionyl battery within a 3D-printed housing made of thick plastic, with a webbing collar and a plastic clasp used to secure the collar around the neck. The weight of the housing and battery (approximately 290 g) was sufficient for the housing to be held on the bottom side of the neck of the ewe with minimal movement. The GPS componentry was based on a passive antenna and a UBlox NeoM8 chip. The chips were programmed to record a position every four seconds if the ewe was actively moving; when stationary a position was recorded every ten minutes. The position was recorded as latitude and longitude coordinates, which were loaded into ArcGIS 9 (ArcMap Version 9.3, USA) for further processing and analysis. The batteries were estimated to last for 18 days, with a draw of 1Ah per day.

The collars were fitted on September 21st 2018 and removed after 18 days. The ewes were monitored at least twice daily and if they had lambed since the last monitoring, an estimate of lambing time was manually recorded within a 9- or 15-hour window (corresponding to lambed during the day or night). The location of the ewes at the time they were observed was recorded.

The management system involved small or weak lambs being removed from ewes within twenty-four hours of birth, and fostered onto ewes outside of this study group, or artificially reared.

Data sets were generated from 20 of the collars, but only 15 could be used in the study, with two ewes lambing outside of the period the collars were active, and three ewes removed from the lambing paddock for assistance with lambing. The 15 data sets that remained, represented ewes that gave birth to 1, 2, 3 or 4 lambs and reared 1, 2 or 3 lambs (due to lamb deaths or removal of small lambs as described above).

Analysis

The data sets generated were edited to remove GPS records that occurred during turning on and fitting the collars, together with ewe movement between the yards and the paddocks and *vice versa.* Data during the study period that did not map within the paddock boundaries were also removed. Less than 1% of data points needed to be removed for this reason.

The cleaned GPS data sets were analysed to generate five variables. The data were assigned to three-hour blocks starting from midnight. The distance and time between successive GPS records were calculated, and speed was calculated as the distance divided by time. The distances were also summed within the three-hour blocks to provide the total distance travelled for each time block. The final trait, which has not been described in similar literature, involved the calculation of the central point of the three-hour block for each ewe and then calculating a distance deviation from this central point for each record within the time block. This is similar to the minimum convex polygon box concept as used by Fogarty et al. (2021) but provides a more relatable measure of movement, or lack thereof. If the value is small, it indicates that either a ewe is resting/asleep or that, although moving, is remaining within a very restricted area. Alternatively if the value is larger, it indicates that the ewe is actively moving through its environment.

Data on the number of lambs detected at scanning, together with the number of lambs reared, and the weaning weight of those lambs, were extracted from the Sheep Improvement Limited (SIL) database. The rearing rank does not always reflect all lambs that survived, as some lambs were removed within a day of birth and fostered or artificially reared, but rather reflects the number of lambs that the ewe successfully reared through to weaning. Given a mix of single- twin- and triplet- reared lambs, the weaning weight data was adjusted (scaled) to reflect rearing rank. This was achieved by dividing the weight of the lamb by the average weight for that rearing rank (31.8 kg for singletons, 26.6 kg for twins and 23.9 kg for triplets) and multiplying it by the overall average weaning weight (27 kg).

The calculated GPS variables were plotted against time to allow visualisation of the data. The data for the GPS variables were averaged from the time the ewes gave birth to the end of the recording period.

Analysis of the data was undertaken using SAS (SAS, 2005). The relationships among the different GPS variables were investigated using simple correlation analysis. The averaged GPS data was used in a simple correlation analysis with adjusted weaning weight to determine the relationships among the GPS variables and adjusted weaning weight.

Results

The approximate timing of lambing could be inferred for all ewes through the plotting of the data and observing changes in the GPS variables. Ewes exhibited a significant decrease in movement and spatial utilisation at the time they were manually recorded to have given birth, followed by changes in patterns of behaviour compared with the pre-lambing period. For some ewes, this was preceded by a rise in activity and spatial utilisation as is often described when ewes are searching for a lambing site, however, this behaviour alone was not predictive. Examples of the average distance from three-hour central point across time and in relation to lambing events are shown in Fig. 1. For one individual, there was a false event that fitted the criteria of a significant decrease in movement and spatial utilisation, however, this was associated with the ewe being cast and not able to move (Fig. 1). Consistent with

Figure 1 The average distance from three-hour central point as predicted from global position satellite devices for two ewes. Lambing events are characterised by a reduction in distance from the central point of the three-hour block after lambing for several time periods. Grey-line ewe recorded as lambing at 6 am on September 30. Black-line ewe recorded as lambing at 11.30 am on September 24, noting that she was recorded as cast on the morning of September 23.

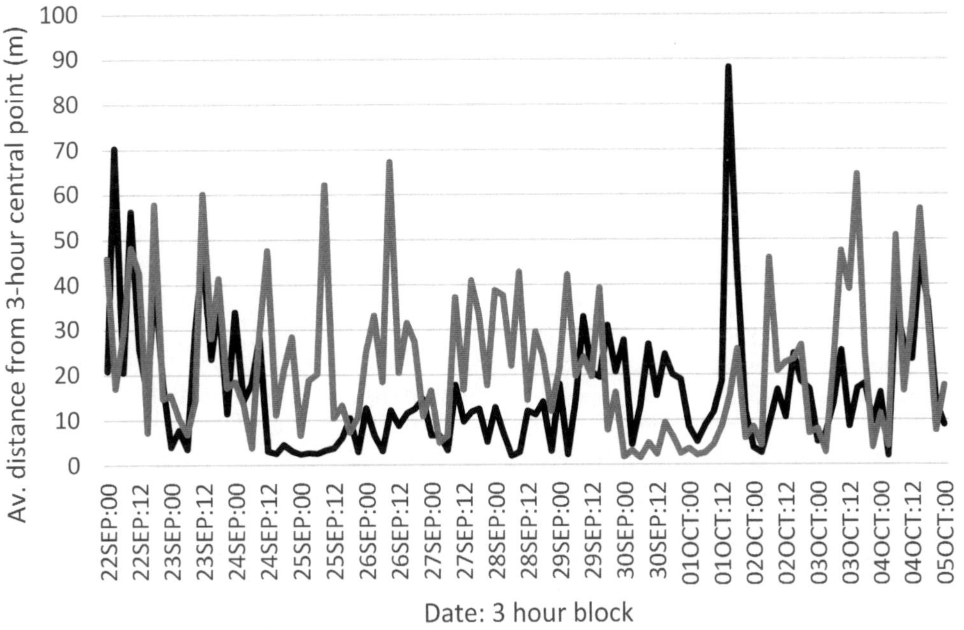

Figure 2 Location of lamb sites for 15 ewes on which global positioning satellite data were available.

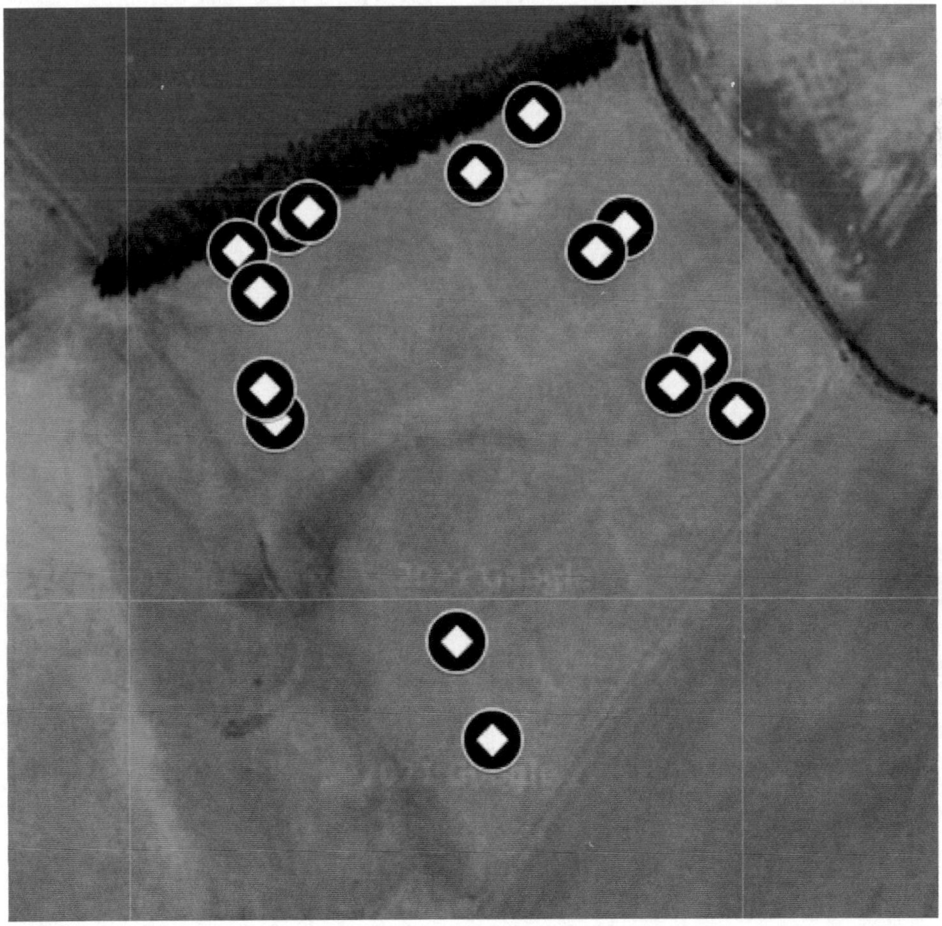

the literature, it was not possible to predict that a ewe was lambing until hours after lambing, when changes in data patterns were obvious.

The location of the lambing site, as predicted by the time of the behaviour change, corresponded to the manually recorded locations for all ewes. Lambing sites were not uniformly distributed throughout the paddock (Fig. 2), with selected areas of the paddocks "lambing sites", one being along the tree line, although other ewes lambed in the middle of the paddock with no obvious geographical features that would have predicted this to be a preferred lambing site versus other areas of the paddock that were not selected.

When comparing the different GPS variables calculated, the correlations were high and ranged from 0.69 to 0.99, with an expected negative correlation between the time and distance traits, that is the longer between records the less distance the ewes moved.

The correlation among unadjusted weaning weight and the GPS traits was not significant for any combination of traits (P>0.1). When the weaning weight data were adjusted for rearing rank, significant correlations were observed among the adjusted weaning weights and the GPS traits, the most significant being with the average distance between GPS fixes and the time between GPS fixes (P=0.04-0.08). Ewes with lower average distances between GPS fixes (or increased time between GPS fixes) having lambs with lower adjusted weaning weights. Visual observation of the correlation indicated that there was a significant outlier, with the ewe wearing Collar 2 having extreme levels of movement (and shorter time between GPS fixes) compared with other ewes, but only weaning small lambs. If this ewe was excluded, the correlations all rose to be highly significant (r>0.65, P<0.05 with most P<0.01). The data and correlations are presented in Table 1. The averaged data for each ewe is included, given the novelty of this type of data. The temporal profile for the length of time between GPS fixes for a typical ewe, a ewe with greater times between fixes (whose lambs had low adjusted weaning weights) and the outlier ewe is given in Fig. 3.

Discussion

Much of the literature relating to the use of GPS devices fitted to ewes around the time of lambing has focussed on attempting to predict when a ewe is going to, or has, lambed, so that the ewe can be checked to ensure a successful birth has taken place. The approach taken in this study had different objectives in investigated the potential of GPS data to estimate when lambing had taken place

Table 1 Lambing data and ewe global positioning satellite (GPS)-derived traits based on data collected during the lambing period and the correlation between the GPS-derived traits and adjusted weaning weight. Correlation data were calculated including and excluding data from Collar ID 2 which was identified as an outlier ewe in terms of behaviour.

	Collar ID															Correlation with Adj WWT			
																All Ewes		Excl. ID 2	
	2	1	3	5	6	7	9	10	12	16	17	18	20	21	23	r	P value	r	P value
Number lambs born	3	3	4	2	3	3	2	3	3	3	1	3	3	3	4				
Number of lambs weaned	2	2	2	2	3	1	1	2	0	2	1	3	2	1	1				
Average weaning weight (kg)	22.8	26.1	20.4	29.1	22.2	35.8	33.4	30.1		32.8	30.4	25.7	24.9	35.2	24.2				
Average weaning weight adjusted for NLB (kg)	23.1	26.5	20.7	29.5	25	30.4	28.4	30.6		33.3	25.8	29	25.3	29.9	20.5				
Average distance from 3-hour central point (m)	30	18	9	20	14	19	26	27	21	21	22	18	17	14	9	0.42	0.13	0.71	0.007
Total distance within 3-hour period (m)	349	210	84	219	140	149	254	275	250	259	275	216	214	189	117	0.36	0.2	0.66	0.01
Average distance between fixes (m)	4.8	4	2.5	4.5	4	3.6	4.2	4.4	4.5	4.6	4.5	4.6	4.4	4.4	3.6	0.53	0.05	0.68	0.01
Time between fixes (s)	187	317	468	275	379	362	231	245	252	231	238	275	308	299	396	-0.51	0.06	-0.73	0.004
Speed (m/s)	0.04	0.03	0.01	0.03	0.02	0.02	0.04	0.04	0.03	0.04	0.04	0.03	0.03	0.03	0.02	0.46	0.1	0.71	0.006

Figure 3 The average time between global-positioning-satellite fixes from devices fitted to ewes during lambing. The dotted-line ewe exhibited a more-typical profile of periods of time between GPS fixes and weaned two heavy lambs. Light-grey-line ewe exhibited consistently short periods of time between GPS fixes, indicating that it was very active; this ewe was extremely active and weaned two very light lambs. Dark-grey-line ewe exhibited consistently long periods of time between GPS fixes indicating that it was relatively inactive; this ewe weaned two very light lambs.

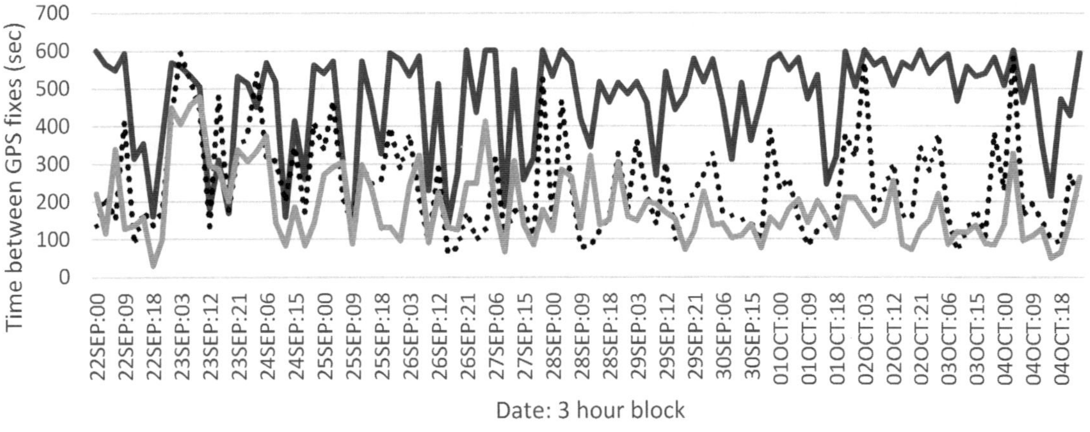

retrospectively and to determine whether any variables generated from the GPS data were associated with lamb weaning weights.

There are a wide range of other ways that the GPS data set could have been analysed, including considering how long ewes exhibited very restricted movement post-lambing and the type of diurnal pattern exhibited post-lambing, however, such analysis was outside of the scope of this preliminary study. Another trait often referred to in GPS/lambing analysis is the distance to closest peers (Dobos et al. 2014; Forgarty et al. 2021), however, this trait could not be accurately calculated as only approximately half of the ewes in the mob were fitted with GPS devices and, as such, any between-animal proximity data would not have been accurate.

The reasons for the relationship among the GPS traits and adjusted weaning weight cannot be inferred from this data set, however, for the two ewes with the lightest lamb weaning weights, both ewes had given birth to quadruplets and only successfully reared one or two lambs. Whether the reduction in distances covered was the result of restricted movement in the ewes or the lambs cannot be determined. That said, among the remaining ewes, there was also evidence that greater ewe movement (but not to the extreme) was associated with higher lamb weaning weights. This would suggest that more active ewes can harvest more and better feed and, as such, have greater milk production.

These results demonstrate variation and links between ewe movement and lamb weaning weight, a result that should be validated through the collection of larger data sets. Larger data sets would also allow the development of machine-learning models that can estimate the lambing event, however, as shown by Smith et al. (2020) absolute values will not inform lambing outcomes. Rather, individual ewes exhibit unique patterns and changes in patterns of behaviour that are associated with their lambing. Ultimately, sufficiently large data sets should be targeted to understand whether the patterns of behaviour at lambing and post-lambing are under genetic control or are the result of non-genetic variation. Additionally, that the data is collected from resources where lambs are not artificially removed to understand whether any patterns of behaviour, including lambing site, are associated with lamb-survival outcomes.

Overall, this study has demonstrated the potential for GPS data sets to provide novel insights into ewe behaviour at the time of, and post-lambing, and the potential to link them to productive outcomes.

Acknowledgements

Funding for this study was provided by the AgResearch Strategic Science Investment Fund. The authors thank the extended team at Twin Farm for supporting the study and collecting the on-farm data.

References

Dobos RC, Dickson S, Bailey DW, Trotter MG 2014. The use of GNSS technology to identify lambing behaviour in pregnant grazing Merino ewes. Animal Production Science 54: 1722-1727.

Dobos RC, Taylor DB, Trotter MG, McCorkell BE, Schneider DA, Hinch, GN 2015. Characterising activities of free-ranging Merino ewes before, during and after lambing from GNSS data. Small Ruminant Research 131: 12-16.

Fogarty ES, Swain D, Cronin GM, Moraes LE, Bailey DW, Trotter M 2020. Potential for autonomous detection of lambing using Global Navigation Satellite System technology. Animal Production Science 60: 1217–1226.

Fogarty ES, Swain DL, Cronin GM, Moraes LE, Bailey DW, Trotter MG 2020. Potential for autonomous detection of lambing using global navigation satellite system technology. Animal Production Science 60: 1217-1226.

Fogarty ES, Swain DL, Cronin GM, Moraes LE, Bailey DW, Trotter M 2021. Developing a Simulated Online Model That Integrates GNSS, Accelerometer and Weather Data to Detect Parturition Events in Grazing Sheep: A Machine Learning Approach. Animals 11: 303.

O'Connor CE, Jay NP, Nicol AM, Beatson PR 1985. Ewe maternal behaviour score and lamb survival. Proceedings of the New Zealand Society of Animal Production 45: 159-162.

SAS. 2005: SAS/STAT Software. Release 9.11. SAS Institute Inc., Cary, North Carolina, USA.

Smith D, McNally J, Little B, Ingham A, Schmoelzl S. 2020. Automatic detection of parturition in pregnant ewes using a three-axis accelerometer. Computers and Electronics in Agriculture 173: 105392.

BRIEF COMMUNICATION: Investigating the pattern of lameness in dairy goats across the first two years of life

L Deeming[abc*], N Beausoleil[c], K Stafford[c], J Webster[b] and G Zobel[b]

[a]*Waikato Institute of Technology (Wintec), 51 Akoranga Road, Hamilton, 3200, NZ;* [b]*AgResearch Ltd., Ruakura Research Centre, 10 Bisley Road, Private Bag 3123, Hamilton 3214, NZ;* [c]*Massey University, Private Bag 11 222, Palmerston North, New Zealand.*

**Corresponding author:* Email: laura.deeming@wintec.ac.nz

Abstract

Lameness is a painful condition with significant welfare and economic implications. There are few data about the high-risk periods of lameness in dairy goats, therefore, the aim of this study was to evaluate how lameness occurrence changes across the first two years of life. Eighty female goats of approximately five months of age were enrolled. Gait-scoring assessments were completed at 9, 13, 17, 21 and 25 months of age. Gait scores were assigned using a 5-point gait scoring system. As there were a low number of 3, 4 or 5 gait scores recorded, a binary outcome variable was created to compare non-lame (score 1) to an impaired gait (score ≥ 2). A logistic regression (PROC GLIMMIX) was used to test if there was a difference in the proportion of goats with an impaired gait among assessments. When compared with the nine-month assessment, the odds of a goat having an impaired gait were greater by a factor of 2.15 (95% CI: 1.02 – 4.54, P<0.05) at the 13-month assessment and 3.79 (95% CI: 1.90 – 7.57, P<0.001) at the 25-month assessment; as these assessments were both following kidding events, this suggests that parturition may increase the risk of lameness in dairy goats.

Keywords: gait score; uneven gait; impaired gait; parturition; welfare

Introduction

Lameness is a painful condition (Whay et al. 1997) that results in an altered gait due to the animal attempting to avoid weight-bearing on the affected limb (Leach et al. 2009). In its milder forms, a lame animal may exhibit an uneven gait, whereby there is a decreased symmetry of limb movement (e.g., shortening of stride) (Winckler & Willen 2001; Flower & Weary 2006). In a case of severe lameness, an affected animal may be unwilling or unable to bear any weight on the affected limb (Flower & Weary 2006; Dyer et al. 2007).

Lameness in dairy cows has been researched extensively; it is one of the most significant welfare and economic issues due to its high prevalence (39%: Haskell et al. 2006; 55%: von Keyserlingk et al. 2012), its associated impacts on production (Willshire & Bell 2009) and the serious negative impact it has on individual animals (von Keyserlingk et al. 2009). While the effect of lameness on goats has received less attention, available literature suggests that it does impact production (Christodoulopoulos 2009) and lameness prevalence on farms can be high (19.2%: Anzuino et al. 2010; 67%: Groenevelt et al. 2015).

Lameness is a complex and often multifactorial problem (Chesteron et al. 1989; Solano et al. 2015), with numerous associated management and animal-related risk factors. One of the high-risk periods of lameness in dairy cows occurs after calving (Offer et al. 2000; Tarlton et al. 2002). Metabolic and hormonal changes associated with calving weaken the connective tissue of the hoof suspensory apparatus, leading to an increased risk of lameness due to sole ulcers and white line disease (Tarlton et al. 2002). The first calving may be a critical period as high numbers of dairy heifers become lame early in their first lactation (Webster 2002). Additionally, animals that have previously been lame are more likely to go lame in the future (Hirst et al. 2002; Randall et al. 2015). Similar, robust work is lacking in goats, particularly in a New Zealand context; however, the findings in cows suggest that management around first parturition may be important for lameness prevention. Indeed, Groenevelt et al. (2015) suggested that the difference in the incidence of lameness in lactating goats and young goats was due to a parturition effect. Unfortunately, the young goat stock in this study were assessed for lameness in pens rather than on concrete; this is a significant confounder, as goats do not often show visible lameness until they start walking on solid hard flooring substrate (Groenevelt 2017). Thus, the aim of this study was to investigate how lameness occurrence changes in the first two years of life in a herd of New Zealand dairy goats.

Materials and methods

This study was part of a larger study which was approved by AgResearch Ltd., Ruakura Animal Ethics Committee (#13686). Eighty female goats of approximately five months of age from one commercial farm in the Waikato region of New Zealand were enrolled in the study. The herd was maintained indoors and bedded on wood shavings. Housing and husbandry management was maintained as per the farm's routine protocol. Goats were first mated at approximately 7-8 months of age and first kidded at approximately 12-13 months of age, at which point they entered the milking herd. Goats were dried off at approximately 21 months of age and had their second kidding at 24-25 months.

Gait-scoring assessments were completed at 5, 9, 13,

17, 21 and 25 months of age. At each assessment, goats were recorded (HC-V270, Panasonic Camcorder, Osaka, Japan) walking along a concrete raceway from the milking parlour back towards their pens. The video camera was set up on a tripod to allow a length of approximately 4.5 m of the walkway to be viewed. This allowed at least four full strides to be recorded. From the 13-month assessment onwards, gait scoring was completed following morning milking to reduce any effect of milk fill and udder distention on gait. The videos were watched separately by two trained observers and gait scores assigned using the 5-point gait score described by Deeming et al. (2018) (Table 1). Each scoring session occurred four months apart which minimised the risk of observers being familiar with the goats and the previously assigned gait scores. Additionally, inter-observer and intra-observer reliability was determined following the completion of each assessment to ensure $K_w \geq 0.80$ (almost perfect agreement; Dohoo et al., 2003). Study goats that were assigned a gait score of 2 or above at any of the assessments were investigated and promptly treated by a veterinarian if necessary.

There was a low number of 3, 4 or 5 gait scores recorded, thus a binary outcome variable comparing non-lame (score 1) to an impaired gait (score 2-5) was created. A logistic regression (PROC GLIMMIX) using a binary distribution and a logit link function was used to test if there was a difference in the proportion of goats with an impaired gait among the gait assessments. Goat within assessment was included as a random effect. The assessment at five-months of age was excluded from analysis as accurate gait scores could not be assigned due to goats moving faster than a walk. Therefore, the nine-month assessment was used as the reference category. The results are presented as odds ratios and 95% confidence intervals. In addition to the logistic regression, the number and percentages of goats with a non-lame gait (score = 1), an uneven gait (score = 2) and lame gait (score > 3) at each assessment are presented.

Results and discussion

Few goats were clinically lame (gait score ≥ 3) at each of the assessments: the majority were either not lame (gait score = 1, assessment range: 52 – 77%) or showed an uneven gait (gait score = 2, assessment range: 12 - 43%) (Table 2). The prevalence of clinical lameness was less than 9% of goats at all assessments over two years. Nevertheless, it is suggested that a clinical lameness prevalence of 5% or more in sheep should prompt an investigation of the cause (Winter 2004).

The highest proportions of goats classified as having an impaired gait were observed at the 13-month (37.3%) and 25-month assessment (47.5%). The odds of a goat having an impaired gait were greater by a factor of 2.15 (95% CI: 1.02 – 4.54, $P<0.05$) at the 13-month assessment and 3.79 (95% CI: 1.90 – 7.57, $P<0.001$) at the 25-month assessment compared to the nine-month assessment. These assessments were following kidding, suggesting a potential parturition effect.

There are a number of factors during the period when dairy cows transition from dry to the milking herd that can lead to lameness (Bell 2015; Tarlton et al. 2002; Bergsten et al. 2015). Physiological and hormonal changes involved with the weakening of the suspensory apparatus in the hoof, metabolic stressors through feeding of transition diets, changes to social dynamics and changes in housing and husbandry have been identified as lameness risk factors during the transition period in dairy cows (Tarlton et al. 2002; Bergsten et al. 2015). The higher prevalence of an impaired gait following kidding observed in the current study suggests that the time around parturition could be relevant in dairy goats as well. Our study was limited to lameness observations, and we were not able to collect data regarding changes in housing, diet, and social structure. Nonetheless, given that these management variables are typically altered following parturition, we suggest goats could be helped to adjust to the upcoming changes. For

Table 1 Description of the 5-point gait scoring system previously developed for use in dairy goats (Deeming et al. 2018).

Gait scoring system		Assessment criteria					
Category	5-point	Limp	Moving forward	Weight bearing	Head nod	Identify affected leg(s)	Other descriptors
Normal gait	1	No	Yes	Yes	No	-	Even stride on all four legs, tracking up, walks with a fluid motion.
Uneven gait	2	No	Yes	Yes	No	No	Shorter stride, not tracking up, joints slightly stiff, inward or outward swinging of a hoof at each stride.
Mildly lame	3	Yes	Yes	Yes	No	Possibly	One or more legs may be affected. Observer may not be able to determine affected leg(s). Mild limp.
Moderately lame	4	Yes	Reluctant	Reluctant	Possibly	Yes	One or more legs may be affected. Moderate limp or slight goose stepping.
Severely lame	5	Yes	Unwilling/ unable	Unable	Yes	Yes	One or more legs may be affected. Severe limp or walking on knees, or pronounced high goose stepping.

Table 2 Number and percent (presented in brackets) of goats that were scored as being not lame, having an uneven gait or a lame gait using a 5-point gait scoring system at six assessments.

		Gait score			n[b]
Assessment	Age (months)	1, not lame	2, uneven gait[a]	≥ 3, lame gait[a]	
1[c]	5	-	-	-	-
2	9	60 (76.9)	17 (22.8)	1 (1.3)	78
3	13	42 (62.7)	19 (28.4)	6 (8.9)	67
4	17	57 (86.4)	8 (12.1)	1 (1.5)	66
5	21	47 (74.6)	16 (25.4)	0 (0.0)	63
6	25	32 (52.5)	26 (42.6)	3 (4.9)	61

[a]For analysis, goats with an uneven gait and lame gait were grouped to create a binary variable comparing non-lame to impaired gait; [b]Numbers decline due to goats being removed from the herd for health and production reasons. No animals were removed due to lameness; [c]Assessment 1 was excluded from analysis as accurate gait scores could not be assigned due to goats moving faster than a walk.

instance, goats spend up to 15 h/d lying down prior to kidding, however, this pattern decreases following kidding (Zobel et al. 2015) due to the sudden change in management (e.g., twice-daily milking). Therefore, farmers could move the goats through the milking parlour during the dry period to increase activity levels and to ensure the goats' hooves are exposed to concrete during this time. Unlike in cattle, where concrete is an established risk factor for lameness (Somers et al. 2003; Dippel et al. 2009), access to hard surfaces is suggested to be beneficial for goats (Zobel et al. 2019). Exposure to the milking parlour and the concrete walkways may be particularly important for primiparous goats kept entirely indoors, as they will have had minimal prior contact with concrete. In addition to increasing activity, keeping groups of goats as stable as possible during the dry period and transition period may reduce the stress on the goats (Patt et al. 2012), thus reducing the risk of antagonistic interactions. In cows, reducing aggression has been linked to reduced lameness risk (Mahendran & Bell 2015).

We caution that it was not within the scope of the present study to investigate the exact cause of the observed gait impairment. Further work is required to determine the environmental and animal-related risk factors associated with the transition period. Nevertheless, management interventions around kidding may be considered to minimise lameness risk at this potentially critical time point.

Acknowledgements

Funding for the research and the student stipend was provided by the NZ Ministry of Business, Innovation and Employment (MBIE, Wellington, New Zealand), and the Dairy Goat Cooperative (NZ) Ltd. (DGC, Hamilton, New Zealand). Special thanks are given to the participating DGC farmer.

References

Anzuino K, Bell NJ, Bazeley KJ, Nicol CJ 2010. Assessment of welfare on 24 commercial UK dairy goat farms based on direct observations. Veterinary Record 167: 774-780.

Bell NJ 2015. Evidence-based claw trimming for dairy cattle. Veterinary Record 177: 220-221.

Bergsten C, Telezhenko E, Ventorp M 2015. Influence of soft or hard floors before and after first calving on dairy heifer locomotion, claw and leg health. Animals 5: 662-686.

Chesterton RN, Pfeiffer DU, Morris RS, Tanner CM 1989. Environmental and behavioural factors affecting the prevalence of foot lameness in New Zealand dairy herds – A case-control study. New Zealand Veterinary Journal 37: 135-142.

Christodoulopoulos G 2009. Foot lameness in dairy goats. Research in Veterinary Science 86: 281-284.

Deeming LE, Beausoleil NJ, Stafford KJ, Webster JR, Zobel G 2018. The development of a reliable 5-point gait scoring system for use in dairy goats. Journal of Dairy Science 101: 4491-4497.

Dippel S, Dolezal M, Brenninkmeyer C, Brinkmann J, March S, Knierim U, Winckler C 2009. Risk factors for lameness in cubicle housed Austrian Simmental dairy cows. Preventive Veterinary Medicine 90: 102-112.

Dohoo IR, Martin W, Stryhn H 2003. Veterinary Epidemiologic Research; AVC Incorporated Charlottetown: PEI, Canada.

Dyer RM, Neerchal, NK, Tasch U, Wu Y, Dyer P, Rajkondawar PG 2007. Objective determination of claw pain and its relationship to limb locomotion score in dairy cattle. Journal of Dairy Science 90: 4592-4602.

Flower FC, Weary DM 2006. Effect of hoof pathologies on subjective assessments of dairy cow gait. Journal of Dairy Science 89: 139-146.

Groenevelt M 2017. Foot lameness in goats: like sheep, like cattle or completely different? Livestock 22: 98-101.

Groenevelt M, Anzuino K, Smith S, Lee MR, Grogono-Thomas R 2015. A case report of lameness in two dairy goat herds; a suspected combination of nutritional factors concurrent with treponeme infection. BMC Research Notes 8: 791.

Haskell MJ, Rennie LJ, Bowell VA, Bell MJ, Lawrence AB 2006. Housing system, milk production, and zero-grazing effects on lameness and leg injury in dairy cows. Journal of Dairy Science 89:4259-66.

Hirst WM, Murray RD, Ward WR, French NP 2002. A mixed-effects time-to-event analysis of the relationship between first-lactation lameness and subsequent lameness in dairy cows in the UK. Preventive Veterinary Medicine 54: 191-201.

Leach KA, Dippel S, Huber J, March S, Winckler C, Whay HR 2009. Assessing lameness in cows kept in tie-stalls. Journal of Dairy Science 92: 1567-1574.

Mahendran S, Bell, N 2015. Lameness in cattle 2. Managing claw health through appropriate trimming techniques. In Practice 37: 231-242.

Offer J, McNulty D, Logue D 2000. Observations of lameness, hoof conformation and development of lesions in dairy cattle over four lactations. The Veterinary Record 147: 105-109.

Patt A, Gygax L, Wechsler B, Hillmann E, Palme R, Keil NM 2012. The introduction of individual goats into small established groups has serious negative effects on the introduced goat but not on resident goats. Applied Animal Behaviour Science 138: 47-59.

Randall LV, Green MJ, Chagunda MGG, Mason C, Archer SC, Green LE, Huxley JN 2015. Low body condition predisposes cattle to lameness: An 8-year study of one dairy herd. Journal of Dairy Science 98: 3766-3777.

Solano L, Barkema HW, Pajor EA, Mason S, LeBlanc S, Zaffino Heyerhoff J, Nash C, Haley D, Vasseur E, Pellerin D, Rushen J, de Passille AM, Orsel K 2015. Prevalence of lameness and associated risk factors in Canadian Holstein-Friesian cows housed in freestall barns. Journal of Dairy Science 98: 6978-6991.

Somers JG, Frankena K, Noordhuizen-Stassen EN, Metz JH 2003. Prevalence of claw disorders in Dutch dairy cows exposed to several floor systems. Journal of Dairy Science 86: 2082-2093.

Tarlton J, Holah D, Evans K, Jones S, Pearson G, Webster A 2002. Biomechanical and histopathological changes in the support structures of bovine hooves around the time of first calving. The Veterinary Journal 163: 196-204.

von Keyserlingk MAG, Barrientos A, Ito K, Galo E, Weary D.M 2012. Benchmarking cow comfort on North American freestall dairies: Lameness, leg injuries, lying time, facility design, and management for high-producing Holstein dairy cows. Journal of Dairy Science 95: 7399-7408.

von Keyserlingk MAG, Rushen J, de Passillé AM, Weary DM 2009. Invited review: The welfare of dairy cattle – Key concepts and the role of science. Journal of Dairy Science 92: 4101-4111.

Webster A 2002. Effects of housing practices on the development of foot lesions in dairy heifers in early lactation. Veterinary Record 151: 9-12.

Whay HR, Waterman AE, Webster AJF 1997. Associations between locomotion, claw lesions and nociceptive threshold in dairy heifers during the peri-partum period. Veterinary Journal 154: 155-161.

Willshire J, Bell N, 2009. An economic review of cattle lameness. Cattle Practice 17: 136-141.

Winckler C, Willen S 2001. The reliability and repeatability of a lameness scoring system for use as an indicator of welfare in dairy cattle. Acta Agriculturae Scandinavica, Section A-Animal Science 51: 103-107.

Winter A 2004. Lameness in sheep 1. Diagnosis. In Practice 26: 58-63.

Zobel G, Leslie K, Weary DM, Von Keyserlingk MAG 2015. Ketonemia in dairy goats: Effect of dry period length and effect on lying behavior. Journal of Dairy Science 98: 6128-6138.

Zobel G, Neave HW, Webster J 2019. Understanding natural behavior to improve dairy goat (Capra hircus) management systems. Translational Animal Science 3: 212-224.

Genome-wide association study of UV-protectant ambilateral circumocular pigmentation in Hereford cattle.

S Jivanji[a*], T Kosch[a], M Littlejohn[a] and DJ Garrick[a]

[a]School of Agriculture and Environment, Massey University, Palmerston North, New Zealand

*Corresponding author: Email: swati.jivanji.1@uni.massey.ac.nz

Abstract

Hereford cattle have a dominantly inherited white face that is characteristic of their breed. A small proportion of Hereford cattle present with regions of pigmentation around the eyes, known as ambilateral circumocular pigmentation (ACOP), which is hypothesised to provide UV-protectant properties to the underlying tissue. ACOP has been associated with reduced incidence of two diseases, bovine ocular squamous cell carcinoma, and bovine infectious keratoconjunctivitis, making it a desirable trait for selection. We conducted a genome-wide association study for ACOP in 605 American Hereford cattle using 43,789 genetic markers, and investigated the phenotypic association between face-colour and ACOP. The most highly associated variant from our genome-wide association study mapped to Chr6 g.71059814G>A and explains most of the variation for this trait. This variant falls in close proximity to the well-known coat-colour patterning gene *KIT*, suggesting it may be the causal gene for this trait. Our results suggest that ACOP is not directly influenced by face-colour, and may be influenced by non-additive genetic effects. These results provide an opportunity to select for Hereford cattle with ACOP, enhancing welfare, and reducing the incidence of bovine infectious keratoconjunctivitis and bovine ocular squamous-cell carcinoma.

Keywords: Hereford cattle; ambilateral circumocular pigmentation; ocular squamous cell carcinoma; infectious kerato-conjunctivitis; GWAS; pigmentation traits

Introduction

Hereford cattle are a popular beef breed easily identifiable by their dominantly inherited white face, a striking trait actively selected for by pedigree breeders as an indicator of 'true breeding'. The white-face trait manifests due to a complete lack of pigmentation (melanin), attributable to the lack of functional pigment cells (melanocytes) occupying the skin and hair within the face region. Melanin functions as an antioxidant and radical scavenger to protect the skin and underlying tissues against damage from ultraviolet (UV) radiation, so although the white-face trait is cosmetically attractive to breed enthusiasts, the lack of eyelid pigmentation increases the risk of damage from the sun and at least two diseases: ocular squamous cell carcinoma (OSCC), and bovine infectious keratoconjunctivitis (BIK; Heeney & Valli 1985; D'Mello et al. 2016; Seid 2019). The most common form of malignant tumour affecting cattle is OSCC which occurs on the eye and eyelids. This disease occurs at a high frequency in Hereford cattle (Heeney & Valli 1985). The occurrence of OSCC causes reduced longevity and economic loss as cancer-stricken cattle are usually rejected, or condemned, at the meat-processing plant as unfit for human consumption. Russell et al. (1976), followed 396 Hereford cows over five years and observed that 51% of these cows were identified to have a lesion on their eye at least once in their life, and White & Moore (2009) reported that OSCC accounted for approximately 11.6% of the beef-carcass condemnations between 2003 to 2007.

Some white-faced Hereford, Simmental and Fleckvieh cattle have regions of pigmentation around the eyes known as ambilateral (both sides) circumocular (around the eyes) pigmentation. In Fleckvieh cattle, the frequency of ambilateral circumocular pigmentation (ACOP) is estimated to be around 36% (Mészáros et al. 2015), and in Hereford cattle it is suggested that approximately two thirds of the population have at least some pigmentation around the eyes or on the eyelids (Anderson et al. 1957). ACOP is associated with reduced incidence of OSCC and BIK in white-faced cattle (Ward & Nielson 1976; Davis 2013). ACOP has a reported heritability of between 0.41 and 0.50 (Anderson et al. 1957), indicating that selection for this trait in white-faced cattle would increase the occurrence of ACOP and would likely reduce the incidence of disease. Although the heritability of ACOP and its correlation with OSCC and BIK have been extensively studied, to our knowledge only one study has investigated the genetics of this trait. Pausch et al. (2012) estimated breeding values for ACOP in a Fleckvieh cattle population of 320,186 cows and 3,579 genotyped bulls. That study reported twelve quantitative trait loci (QTL) associated with ACOP across chromosomes 2, 5, 6, 11, 13, 14 and 22, and implicated well-established pigmentation genes *paired box 3* (*PAX3*), *proto-oncogenne receptor tyrosine kinase* (*KIT*), and *micropthalmia-associated transcription factor* (*MITF*), among others. They also reported higher heritability values for ACOP (0.79±0.04) than previously reported by Anderson et al. (1957).

The aim of our study was to investigate the genetic architecture of ACOP in Hereford cattle. We report a single significant signal on chromosome 6, and suggest *KIT* as the likely causal gene at this locus, with a single nucleotide polymorphism (SNP)-based heritability of 0.75. The results reported in this paper may inform future breeding decisions to select for ACOP in Hereford cattle, thereby enhancing welfare and profitability, especially in pasture-based farming systems where cattle are directly exposed to potentially damaging UV radiation.

Materials and methods

Study population

Face colour (white, splotchy or speckled) and the presence of pigmentation around the eye (one eye, two eyes, or none) were visually scored from photographs of 706 Hereford cattle taken in August of 2019. The cattle were sired by 102 bulls in a progeny test programme, with 38 bulls having more than five offspring within the dataset. The cattle were raised on the Olsen Hereford Ranch in Harrisburg, Nebraska, USA. DNA samples from ear tissue biopsies were extracted and genotyped by GeneSeek (Lincoln, NE, USA) using an Illumina Bovine 50k SNP Chip. All SNP positions used and mentioned in this study are based on the ARS-UCD1.2 bovine genome build (Rosen et al. 2020).

Population structure adjustments and GWAS

A genomic relationship matrix (GRM) was generated using 50k SNP Chip genotype data for each bovine autosome in GCTA (v1.93.2beta) to address possible population stratification in the association model due to relatedness. We treated the ocular pigmentation trait as a binary variable and used a 'leave one chromosome out' approach to conduct a mixed-linear-model-based association analysis in GCTA, which incorporated the GRM as previously described by Jivanji et al. (2019). We investigated the potential functional impact of all variants that surpassed the Bonferroni-corrected genome-wide significance threshold ($P=5\times10^{-8}$) using the Ensembl Variant Effect Predictor (VEP; McLaren et al. 2016). In subsequent analyses, we fitted individual face-colour classes (white face, splotchy face, and speckled face), and the tag SNP genotype as separate covariates in the model.

Mode of inheritance and trait heritability

Dominant and recessive modes of inheritance for the ACOP trait were tested on the most significant signal identified in the GWAS (chromosome 6) by using the '--model' function in PLINK (v1.9; Purcell et al. 2007). In the dominant and recessive models, the minor allele (in this example the G allele) was used as the reference allele to define recessive and dominant encoded genotypes. The dominant model tested the effect of GG and AG versus AA, the recessive model tested the effect of GG versus AG and AA, and the additive model tested the regression of the number of G alleles on the presence or absence of ACOP. The ACOP trait was treated as a binary variable, as described above, and the association model was conducted on genetic markers across chromosome 6 (n=2,102). A principal-component-analysis-based approach was used to account for stratification due to relatedness in this model. Eigenvectors were calculated using all 50k SNP Chip markers, and the top ten principal components were fitted as covariates, based on visualisation of principal component clusters and a scree plot of the eigenvalues. The top ten principal components explain more than 90% of the genotype variance in our population. A genome-wide GRM was used to estimate the SNP-based trait heritability in GCTA-GREML (v1.93.2beta), where the trait prevalence was set to 0.34 based on a study by Mészáros et al. (2015).

Results

Genome-wide association analysis

Phenotypes of the Hereford population comprised 101 cattle with pigmentation around just one eye (unilateral circumocular pigmentation; UCOP), 240 cattle with pigmentation around both eyes (i.e., ACOP), and 365 cattle with no pigmentation around their eyes. The genome-wide association study (GWAS) for the absence or presence of any eye pigmentation (UCOP or ACOP; n=706 cattle), and that for the presence or absence of ACOP (n=605 cattle), both revealed the same signal. The results for both GWAS revealed a single signal on chromosome 6 that exceeded our multiple-hypothesis-testing threshold ($P=5\times10^{-8}$), and mapped to the same top variant at Chr6 g.71059814G>A (UCOP or ACOP $P=5.2\times10^{-19}$; just ACOP $P=1.7\times10^{-22}$; Fig. 1). Since the P value was more significant for the second analysis, where cattle with ACOP were compared to cattle with no circumocular pigmentation, this phenotype became the focus of the analyses that are presented below.

Influence of face colour on ACOP

Within our population, we observed Hereford cattle to have one of three different face-colour classes: white face, splotchy face, or speckled face (Table 1), where UCOP or ACOP were not considered in the allocation of animals to these three classes. The occurrence of ACOP did not appear to segregate with any one face-colour class, appearing to manifest regardless of face colour. In our association model, we accounted for face-colour class and found that the top variant (Chr6 g.71059814G>A) remained at the top of the peak with very marginal shift in significance (white face $P=3.2\times10^{-22}$; splotchy face $P=2.7\times10^{-21}$; speckled face $P=1.2\times10^{-22}$).

Analysis of chromosome 6 locus

The top variant at the chromosome 6 locus was a SNP at Chr6 g.71059814G>A (rs134749374; $P=1.7\times10^{-22}$), which maps to intron 1 of the *Neuromedin-U* (*NMU*) gene, and is 805 kb downstream of the *proto-oncogene receptor tyrosine kinase* (*KIT*) gene. We detected 43 additional variants that exceeded the genome-wide wide significance threshold ($P=5\times10^{-8}$), spanning approximately 20 Mb in the vicinity of the top variant. Our investigation of all genes that mapped to this region identified the *KIT* gene as the most likely causal gene. Assessment of the predicted functional impact of the top (n=44) variants using Variant Effect Predictor (VEP) did not present any variants with obvious impacts on gene expression and/or protein function: one variant was predicted to be a 3' untranslated region variant, two variants were upstream gene variants, 11 variants were intron variants, and the remaining 30 variants were intergenic. Likewise, the top associated variant was not found to map to a highly conserved region that could

Figure 1 Manhattan plot based on genome-wide association analysis for the absence or presence of ambilateral circumocular pigmentation in Hereford cattle. There is a single signal that exceeds the significance threshold indicated by the dashed line ($P=5\times10^{-8}$) on chromosome 6, and the top variant maps to Chr6:71,059,814bp with a P value of 1.7×10^{-22}.

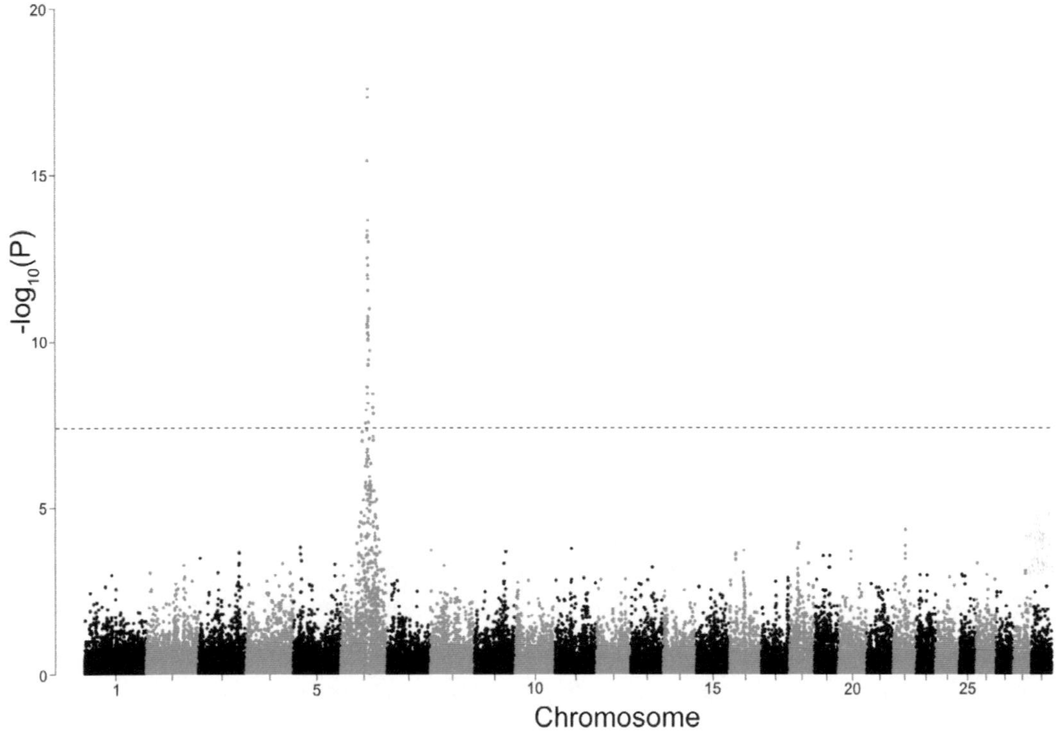

Table 1 Image of each trait scored and number of cattle scored as having no pigment around their eyes versus pigment around both eyes (ambilateral circumocular pigmentation) for each face colour trait: white face, splotchy face, and speckled face. Cattle with both splotchy faces and speckles (n=7) were counted twice.

	White Face 	Splotchy Face 	Speckled Face 
No pigment around the eyes 	332	2	28
Pigment around both eyes 	151	71	28

Table 2 Top 10 variants for the chromosome 6 locus detected in the genome-wide association analysis for the presence or absence of ambilateral circumocular pigmentation in Hereford cattle, with minor allele frequency, predicted variant effects from Ensembl Variant Effect Predictor, and gene the variant maps to.

Variant reference ID	Genomic position on Chr6	P value	Minor allele frequency	Predicted variant effect	Gene symbol
rs134749374	g.71059814G>A	1.7×10^{-22}	0.27	Intron variant	*NMU*
rs135243955	g.69234941C>T	2.6×10^{-18}	0.19	Intergenic variant	-
rs110185900	g.69224236C>T	4.6×10^{-18}	0.19	Intergenic variant	-
rs109877370	g.67582129C>T	3.7×10^{-16}	0.22	Intergenic variant	-
rs110547897	g.69596878C>T	2.2×10^{-14}	0.27	Upstream gene variant	*CHIC2*
rs132937166	g.68028962T>C	4.7×10^{-14}	0.2	Intergenic variant	-
rs133388993	g.68428584G>A	6.5×10^{-14}	0.32	Intergenic variant	-
rs109242003	g.67159718C>T	7.7×10^{-14}	0.13	Intergenic variant	-
rs29017695	g.71028209G>T	1.0×10^{-13}	0.17	Upstream gene variant	*PDCL2*
rs41653781	g.67991798A>C	3.1×10^{-13}	0.27	Intergenic variant	-

support functional impact, as might be expected given the variants represented 50k SNP chip data only. Table 2 lists the top ten associated variants from chromosome 6 and their predicted functional effects. After fitting the top variant as a covariate in our association model, most of the significance was removed from the locus (smallest P value was 0.0005 for Chr6 g.27370287A>G), suggesting a single biallelic QTL as responsible.

Mode of inheritance

The model used to test the contribution of additive, dominant, and/or recessive inheritance on the manifestation of ACOP identified the same top SNP from the previous additive model conducted in GCTA: Chr6 g.71059814G>A. The smallest P value for this variant came from the additive model ($P=1.6\times10^{-37}$), followed by the dominant model ($P=2.0\times10^{-36}$), and then the recessive model ($P=4.8\times10^{-10}$). Although these results support an additive mode of inheritance, the small number of cattle with ACOP that were homozygous for the minor allele of this variant (n=29, minor allele frequency=0.27) may have limited our ability to resolve the mode of inheritance. When we examined the difference between phenotype (ACOP or no ACOP) by genotype class, we observed a larger difference in phenotype for the AA genotype, compared to the AG and GG genotypes. These results may allude to a non-additive mode of inheritance (Fig. 2). The heritability of ACOP estimated by GCTA-GREML was 0.75±0.07.

Figure 2 The number of cattle observed to have ambilateral circumocular pigmentation (grey) or no pigmentation around the eyes (black) by genotype class for tag variant Chr6 g.71059814G>A.

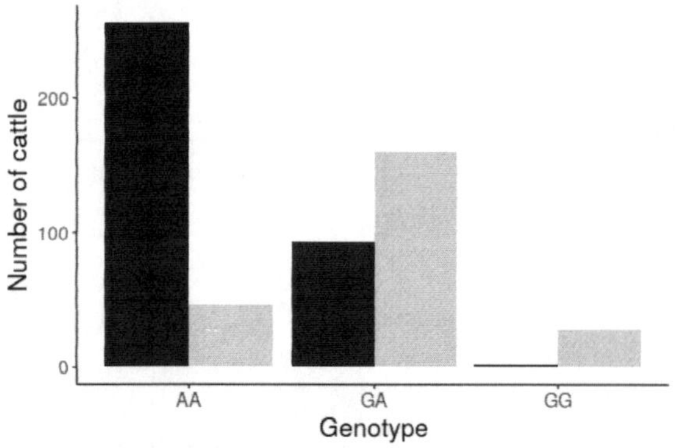

Discussion

To the best of our knowledge, we present the first genome-wide association analysis for ACOP in Hereford cattle. Using phenotype scores for 605 American Hereford cattle, and 43,789 genome-wide genetic markers, we identified a single significant association signal on chromosome 6, mapping to Chr6 g.71059814G>A ($P=1.7\times10^{-22}$). Unexpectedly, the strength of association for this signal was not affected by face colour, suggesting that ACOP is controlled by different alleles to those that influence the colour of the face in Hereford cattle. We suggest that the *KIT* gene is the most likely causal gene for ACOP, and given that the white-face trait has also been attributed to dominant mutations of the same gene (Whitacre 2014), these findings suggest multiple segregating pigmentation QTL in these animals.

There were 44 variants that exceeded the genome-wide significant threshold ($P=5\times10^{-8}$) on chromosome 6, with the most significant variant mapping 805 kb downstream of the *KIT* gene. Although this variant is nearly 1 Mb away from *KIT*, other variants with P values of a similar magnitude mapped within 240 kb of the *KIT* gene. Given the use of a relatively sparse genotyping platform (43k SNP) for mapping, and the status of *KIT* as one of the most-frequently implicated and best-characterised pigmentation genes in mammals, these findings suggest *KIT* as the likely causal gene for the ACOP trait. The KIT protein is a tyrosine kinase mast/stem cell growth factor receptor that is involved in melanocyte stem-cell migration out of the neural crest during embryonic development. KIT facilitates melanocyte migration along the neural crest to the final destination in the skin. Mutations that alter *KIT* expression or activity are hypothesised to impact stem-cell migration, proliferation and/or survival. This impairs the ability of melanocyte precursors to colonise their final destination, causing regions of the skin and hair to lack pigmentation and appear white (Adameyko et al. 2009; D'Mello et al. 2016; Mort et al. 2016). Bioinformatic prediction of the significant GWAS variants for functional impacts did not detect any likely causal mutations. However, when the most highly associated SNP Chr6 g.71059814G>A, was fitted as a fixed effect in the association model, this SNP accounted for the majority of the signal at this locus. Although it is possible that Chr6 g.71059814G>A may have some effect on gene expression or regulation, this is unlikely as the variant does not appear to be highly conserved amongst vertebrates, and represents but one of many candidates that would likely be represented if GWAS was performed with higher-density genotype data (e.g., whole-genome sequence). More likely, Chr6 g.71059814G>A tags the causal effect for the manifestation of ACOP and is in linkage disequilibrium with the causal mutation. The ACOP trait has also been studied in Fleckvieh cattle, where Pausch et al. (2012), reported several significant signals across the genome. The most significant signal was reported on chromosome 6, mapping to Chr6 g.70362480T>C ($P=1.1\times10^{-11}$). That variant was not captured in our genotype data, but maps within 700 kb of our tag SNP and is likely tagging the same causal mutation. The heritability reported in our population (0.75±0.07), falls close to that reported by Pausch et al. (2012) (0.79±0.04), perhaps supporting a common genetic architecture. Future studies could use whole-genome sequencing of cattle with ACOP from both Fleckvieh and Hereford to fine-map the causal mutation for ACOP, and establish if ACOP is caused by the same genetic effect in both breeds.

The dominantly inherited white-face trait in Hereford cattle has also been mapped to chromosome 6, locating upstream of the *KIT* gene. The white-face trait manifests if cattle carry one or more copies of a large tandem duplication approximately 50 kb upstream of the *KIT* gene. The tandem duplication is hypothesised to impair *KIT* expression, preventing functional melanocytes from occupying the skin and hair of the face, resulting in the lack of pigmentation in this region (Whitacre 2014). As ACOP is superimposed on the white-face trait, we investigated if face-colour influenced the presence/absence of ACOP. Fitting each face colour class (white, splotchy and speckled) as covariates in the model did not affect the GWAS signal. These results suggest that face colour cannot account for the signal observed at the chromosome 6 locus, and the manifestation of ACOP is caused by a different allele, perhaps at the same locus. Due to the molecular mechanism proposed to cause the white-face trait, and its dominant mode of inheritance, we investigated the possibility of a non-additive mode of inheritance for ACOP. Comparison of cattle with and without the ACOP trait by genotype class possibly supports a non-additive mode of inheritance, although the small number of cattle with ACOP that were homozygous for the minor allele limited our ability to formally resolve mode of inheritance. An epistatic mode of inheritance would not be entirely surprising as the GWAS signal maps near the *KIT* gene, the causal gene for the white face trait. It is possible that while the white-face mutation impairs *KIT* expression, the ACOP mutation may have a positive effect on *KIT* expression and/or function, allowing functional melanocytes to migrate to, and occupy the eye area, pigmenting the surrounding skin and hair. These results are speculative and would require a larger population size to further investigate the mode of inheritance, and possible epistatic interaction between the white-face mutation and ACOP mutation.

Our results identify a strong association between Chr6 g.71059814G>A and the ACOP trait in American Hereford cattle, and implicate the likely involvement of the *KIT* gene. Our results also suggest genetic complexity with regard to the mode of inheritance, and we suggest that ACOP in Hereford cattle and Fleckvieh cattle may be caused by the same mutation, but more work would be required to support this. Selection for ACOP in white-faced cattle could enhance welfare and profitability, by reducing the incidence of OSCC and BIK, especially in cattle raised in pasture-based grazing systems. Although this work has been done in American Hereford cattle, ACOP has also been observed in New Zealand Hereford cattle, and future studies could aim to investigate if the same genetic variants influencing ACOP segregate in local populations. To our knowledge, New Zealand Hereford breeders do not actively select for ACOP, but may benefit from considering ACOP-associated genetic variants in their selection index. Future work may aim to investigate the causal mutation for ACOP in New Zealand Hereford populations and develop a gene test to select appropriate sires and increase the frequency of ACOP in future generations.

Acknowledgements

The authors would like to acknowledge the American Hereford Association for providing the genotype data and photographs used in this research.

References

Adameyko I, Lallemend F, Aquino JB, Pereira JA, Topilko P, Müller T, Fritz N, Beljajeva A, Mochii M, Liste I et al. 2009. Schwann cell precursors from nerve innervation are a cellular origin of melanocytes in skin. Cell 139: 366-379.

Anderson DE, Chambers D, Lush JAYL 1957. Studies on bovine ocular squamous carcinoma ("cancer eye") III. Inheritance of eyelid pigmentation. Journal of Animal Science 16: 1007-1016.

D'Mello S, Finlay G, Baguley B, Askarian-Amiri M 2016. Signaling pathways in melanogenesis. International Journal of Molecular Science 17: 1144.

Davis KM 2013. Digital analysis of eye pigmentation of Hereford, Hereford x Bos indicus or Hereford × Bos taurus cattle. Doctoral dissertation: Texas A&M University.

Heeney JL, Valli VEO 1985. Bovine ocular squamous cell carcinoma: An epidemiological perspective. Canadian Journal of Comparative Medicine 49: 21-26.

Jivanji S, Worth G, Lopdell TJ, Yeates A, Couldrey C, Reynolds E, Tiplady K, McNaughton L, Johnson TJJ, Davis SR et al. 2019. Genome-wide association analysis reveals QTL and candidate mutations involved in white spotting in cattle. Genetics Selection Evolution 51: 1-18.

McLaren W, Gil L, Hunt SE, Riat HS, Ritchie GRS, Thormann A, Flicek P, Cunningham F 2016. The Ensembl Variant Effect Predictor. Genome Biology 17: 122.

Mészáros G, Petautschnig E, Schwarzenbacher H, Sölkner J 2015. Genomic regions influencing coat color saturation and facial markings in Fleckvieh cattle. Animal Genetics 46: 65-68.

Mort RL, Ross RJH, Hainey KJ, Harrison OJ, Keighren MA, Landini G, Baker RE, Painter KJ, Jackson IJ, Yates CA 2016. Reconciling diverse mammalian pigmentation patterns with a fundamental mathematical model. Nature Communications 7: 1-13.

Pausch H, Wang X, Jung S, Krogmeier D, Edel C, Emmerling R, Götz KU, Fries R 2012. Identification of QTL for UV-protective eye area pigmentation in cattle by progeny phenotyping and genome-wide association analysis. PLoS One 7: e36346.

Purcell S, Neale B, Todd-Brown K, Thomas L, Ferreira MAR, Bender D, Maller J, Sklar P, De Bakker PIW, Daly MJ, Sham PC 2007. PLINK: A tool set for whole-genome association and population-based linkage analyses. American Journal of Human Genetics 81: 559-575.

Rosen BD, Bickhart DM, Schnabel RD, Koren S, Elsik CG, Tseng E, Rowan TN, Low WY, Zimin A, Couldrey C et al. 2020. De novo assembly of the cattle reference genome with single-molecule sequencing. Gigascience 9: giaa021.

Russell WC, Brinks JS, Kainer RA 1976. Incidence and heritability of ocular squamous cell tumors in Hereford cattle. Journal of Animal Science 43: 1156-1162.

Seid A 2019. Review on infectious bovine keratoconjunctivitis and its economic impacts in cattle. Journal of Dairy & Veterinary Sciences 9: 555774.

Ward JK, Nielson MK 1976. Pinkeye (bovine infectious keratoconjunctivitis) in beef cattle. Journal of Animal Science 49: 361-366.

Whitacre L 2014. Structural variation at the KIT locus is responsible for the piebald phenotype in Hereford and Simmental cattle. Doctoral dissertation: University of Missouri, Columbia.

White TL, Moore DA 2009. Food for thought for food animal veterinarians. Journal of American Veterinary Medical Association 235: 937-941.

Live weight and bone growth from birth to 15 months of age in pure-bred and cross-bred Jersey and Friesian heifers.

MJ Gibson[a*], CW Rogers[a,b], RE Hickson [a], KE Dittmer [b] and PJ Back[a,b]

[a]*School of Agriculture and Environment, Massey University, Private Bag 11-222, Palmerston North, New Zealand;* [b]*School of Veterinary Sciences, Private Bag 11-222, Massey University, Palmerston North, New Zealand*

[*]*Corresponding author:* Email: m.gibson@massey.ac.nz

Abstract

Live weight and liveweight gain of dairy heifers has been described in New Zealand and overseas, however, the change in stature and pattern of proximal and distal bone growth has, to date, received little attention. This study describes the proportional changes in stature with live weight, in a cohort of commercial dairy heifers (Holstein-Friesian, Jersey, and Holstein-Friesian-Jersey crossbred) from birth to 15 months of age. Comparisons of height, length, girth, and distal and proximal forelimb length (stature) revealed that at the same live weight, these parameters differed among breeds ($P<0.05$), however, the magnitude of the differences among breeds was consistently less than 1 cm. As heifer live weight increased, the contribution of the distal limb to increases in height decreased, which is consistent with knowledge that physeal closure in cattle occurs earlier in distal limb bones, such as the metacarpus, than in proximal limb bones, such as the humerus. In addition, the understanding of proximal and distal growth in cattle identifies when heifers are most susceptible to growth checks.

Keywords: bone; stature; fracture; dairy heifer

Introduction

In dairy heifers, the improvement in genetics for milk production has caused an increase in demand for calcium during lactation (Grace et al. 2010; LIC & DairyNZ 2019). Spontaneous humeral fracture in first-lactation dairy heifers is a condition unique to the New Zealand dairy industry. Affected heifers present with osteoporosis (reduced circumference of the bone and reduced bone wall thickness) which is hypothesised to be the result of insufficient peak bone mass prior to first lactation (Dittmer et al. 2016). It has been estimated that affected farmers in New Zealand could be losing 2% to 3% of heifer replacements within a herd due to humeral fractures (J. Hunnam, unpublished data).

New Zealand dairy cows are predominantly Holstein-Friesian (33.1%), Holstein-Friesian-Jersey crossbred (48.5%), and Jersey (8.6%) (LIC & DairyNZ 2019). Holstein-Friesians are known to have a larger mature weight than Jersey and Jersey-Friesian crossbreds, requiring a faster growth rate (Handcock et al. 2019b). However, the relative growth rate of Jersey calves is greater than Holstein-Friesians (Handcock et al. 2019b). Jersey calves are born smaller than Friesian calves and remain smaller at maturity (Hickson et al. 2015; LIC & DairyNZ 2019). Animals with heavier mature weight tend to mature later than those with lighter mature weight (Taylor 1965), however, a larger size and faster growth rate requires higher dietary energy intake making them more susceptible to disruptions in growth from feed shortages (Rattray et al. 2007).

Growth in New Zealand dairy heifers is typically non-linear due to seasonal variation in feed quality and quantity (Handcock et al. 2021; Litherland et al. 2002). Times of poor feed quality or quantity such as late summer with low rainfall and mid-winter with low temperatures results in slowed heifer growth rates compared with spring (Handcock et al. 2019b). To understand how the heifer grows at a whole-body level, research has been undertaken to examine changes in stature up until 21 months of age (Handcock et al. 2021). Handcock et al. 2021 proposed a relationship between height, length, and girth but did not report the proportion of height change contributed by leg (metacarpus and radius) *vs* thoracic growth (humerus and scapula). It is known that mature height is achieved before thoracic growth is completed, however, the timing of this growth pattern is not fully described (Guilbert & Gregory 1952). To gain a better understanding of key risk periods for humeral growth, the timing of proportional changes in stature and, therefore, bone growth needs to be investigated. By identifying risk periods of humeral growth, emphasis can be placed on interventions at these times to reduce the likelihood of humeral fractures occurring. Therefore, the aim of this study was to describe proportional changes in stature with live weight in Holstein Friesian, Jersey, and Holstein-Friesian Jersey crossbred calves, from birth to 15 months of age when puberty is likely to have been achieved.

Materials and methods

This experiment was conducted with the approval of Massey University Animals Ethics Committee (MUAEC 19/81).

Animals and management

The cohort consisted of 57 Friesian, Jersey, and Friesian-Jersey crossbred singleton heifer calves born at Massey University's Dairy 1 farm during the 2019 spring calving season. Dam identification was confirmed by DNA parentage and dam parity was recorded for each calf as heifer (primiparous, n=11) or cow (multiparous, n=50). Calves were housed for the first three weeks of life and at pasture thereafter. Calves were reared under commercial management, that involved *ad-libitum* colostrum until three days of age, then 7L/day of milk split over two feeds for approximately six weeks, then calves were reduced to 5L of milk fed once daily until weaning. Calves were weaned at approximately 100 kg live weight according to on-farm practice, which occurred at 71-95 days of age. *Ad-libitum* access to a proprietary calf meal was provided until five months of age (SealsWinslow 20% muesli followed by SealesWinslow 18% pellets; SealesWinslow, Whanganui). Once they exited the calf shed (3 weeks of age), calves were managed in a single group, except around weaning when calves fed milk were grazed separately to weaned calves, and also when the biggest 52 calves were grazed separately to five smaller calves. At 185-209 days (April 2020) of age calves were moved to a grazier in Kimbolton, Manawatu and were grazed under commercial management, which involved drafting into up to three grazing herds, based on breed (Jersey, Crossbred and Holstein-Friesian) to minimise competition among different-sized heifers and allow each group to be managed to meet breed-specific target live weight.

For analysis, pedigree information was used to classify heifers in to one of five breed groups based on the criteria defined by Handcock et al. (2019a). Breed group was classified as F (F ≥ 14/16, n=15), FX (10/16 ≤ F ≤ 13/16, n=9), FJ (F < 10/16 and J < 10/16, n=6), JX (10/16 ≤ J ≤ 13/16, n=11) and J (J ≥ 14/16, n=16).

Sampling and measurements

Calves were measured within 48 hours of birth and then fortnightly until weaning. After weaning, calves were measured monthly until 15 months, except for the ages of 7-8 months, when a nationwide COVID-19 lockdown resulted in two missed measurements. Measurements included live weight, wither height, girth, wither-rump length (wither to tuber ischii), leg length (from ground to point of olecranon), toe length (from ground to the lateral aspect of the metacarpophalangeal joint), knee height (from ground to the lateral aspect of the antebrachiocarpal joint) and the cannon circumference (circumference of the metacarpus bone mid-length). All leg measurements were made on the left forelimb. Metacarpal length was equal to knee height minus toe height, thoracic length was equal to wither height minus leg length and radius height was equal to leg length minus knee height.

Statistical analysis

Statistical analysis was conducted using SAS version 9.4 (SAS Institute Inc., Cary NC, USA).

Second-, third- and fourth-order Legendre polynomials were considered, and third order was selected based on R^2, AIC and problems of convergence using a fourth order polynomial. A third-order Legendre polynomial was fitted to age using random regression to obtain an average curve for live weight for the population and for each heifer. For this, live weight at 0, 30, 61, 91, 122, 152, 183, 213, 243, 274, 304, 335, 365, 395, 426, 456 days of age was estimated for each heifer.

Second-, third- and fourth-order Legendre polynomials were considered, and third order was selected based on R^2, AIC and problems of convergence using a fourth-order polynomial. A third-order Legendre polynomial was fitted to the raw stature data using random regression on live weight to obtain an average curve for each stature parameter for the population and for each cow. From this, stature parameters at 50, 100, 150, 200, 250, 300 and 350 kg live weight were estimated for each heifer.

Predicted stature parameters were analysed using general linear models. Each stature measurement was analysed separately at each live weight (50-350 kg, in 50-kg increments). Models included the fixed effect of breed group. The effect of dam parity (primiparous or multiparous) was tested but was not significant so was removed.

Results

The polynomial for live weight based on age was a good predictor of actual live weight (R^2=0.996) (Fig. 1), and live weight increased with age (Fig. 2). At 15 months of age, heifers were approximately 78% of their mature weight.

Figure 1 Fit plot for weight. Weight vs predicted weights from third-order polynomial.

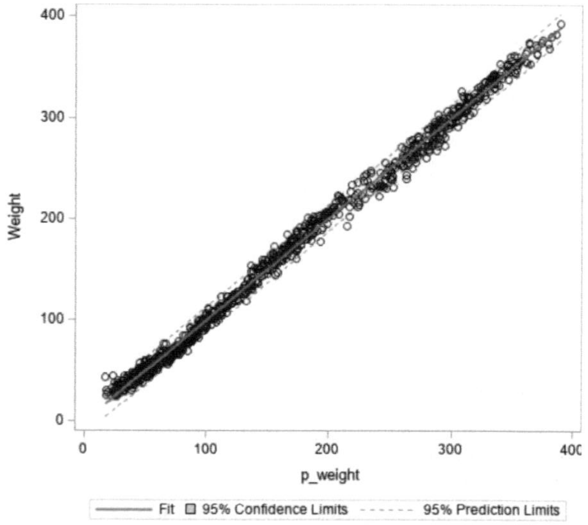

Figure 2 Live weight (kg) vs age (days) in Friesian (F) heifers (black), Friesian-cross (FX) heifers (purple), Friesian-cross-Jersey (FJ) heifers (Green), Jersey-cross (JX) heifers (red) and Jersey (J) (Yellow) heifers from birth to 15 months of age.

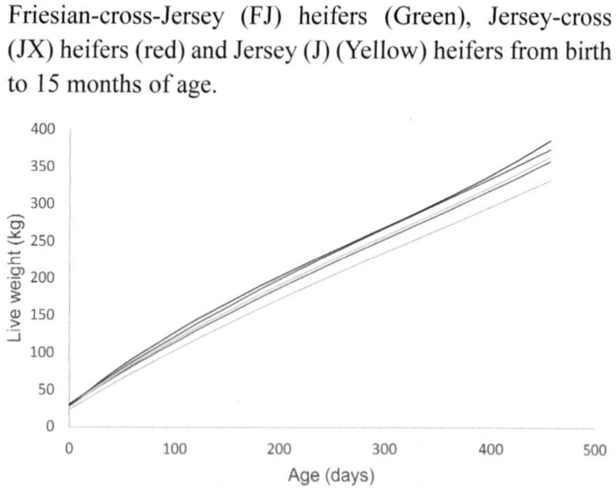

Breed had a significant effect on girth, height, wither-rump length, leg length, radius height, metacarpus length, cannon circumference and toe length, however, the magnitude of the differences among breeds was consistently less than 1 cm, or 5%, so was not considered biologically significant (Fig. 3). The greatest rate of increase in all stature measures occurred between 50 kg and 100 kg live weight.

As live weight increased, the proportion of radius, metacarpal and toe length in the total height decreased whilst the proportion of thoracic length increased (Fig. 4).

Figure 3 Stature (height, girth, thoracic length, and wither rump length) and leg (leg length, radius height, metacarpal length, toe length and cannon circumference) measures in Friesian (F) heifers (black), Friesian-cross (FX) heifers (purple), Friesian-cross-Jersey (FJ) heifers (Green), Jersey-cross (JX) heifers (red) and Jersey (J) (Yellow) heifers from birth to 15 months of age with standard error bars.

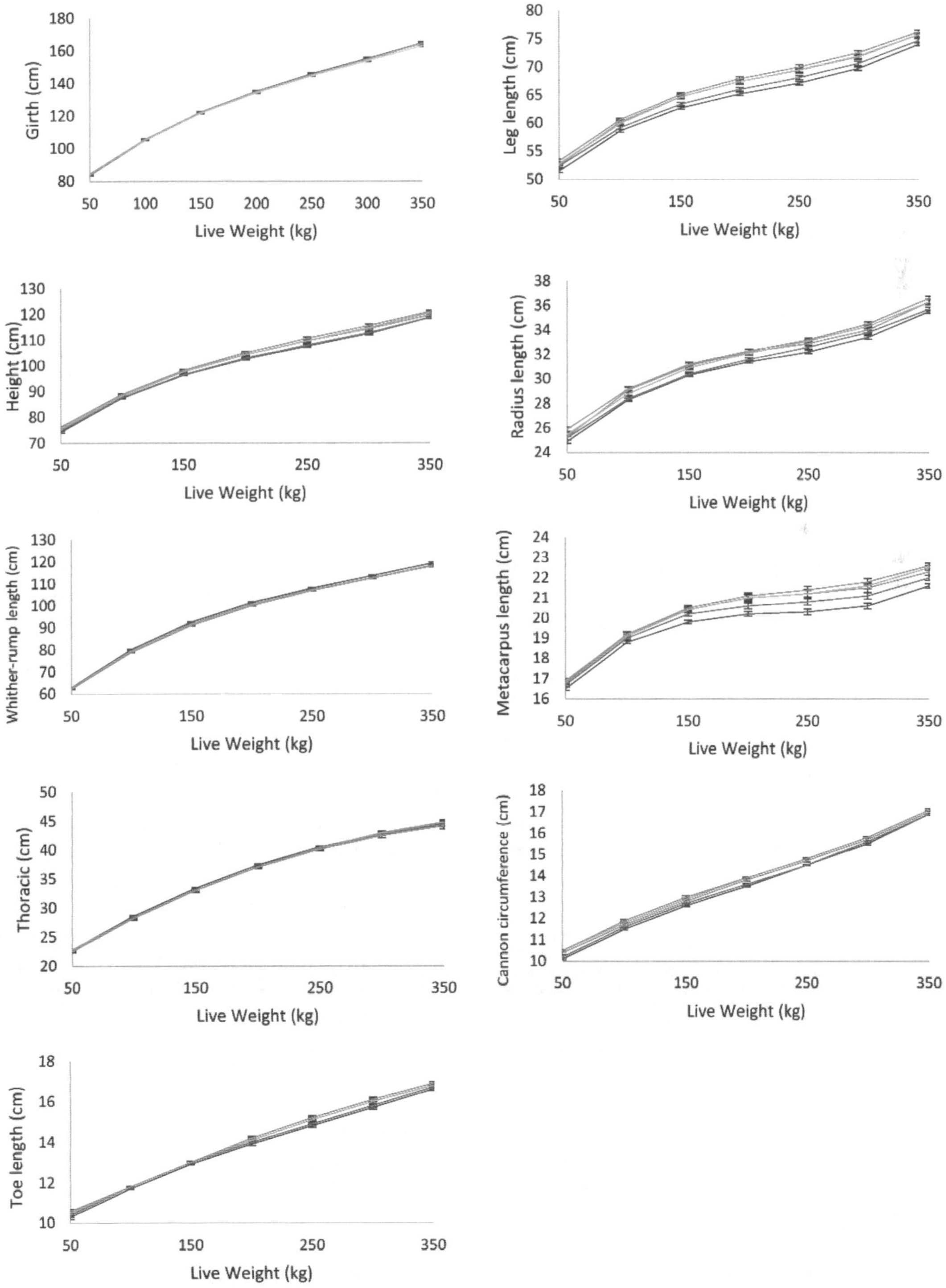

Figure 4 Proportion of thoracic length (Blue), radius height (Orange), metacarpal length (Grey) and toe length (Yellow) in relation to height of dairy calves from birth to 15 months of age with standard error bars.

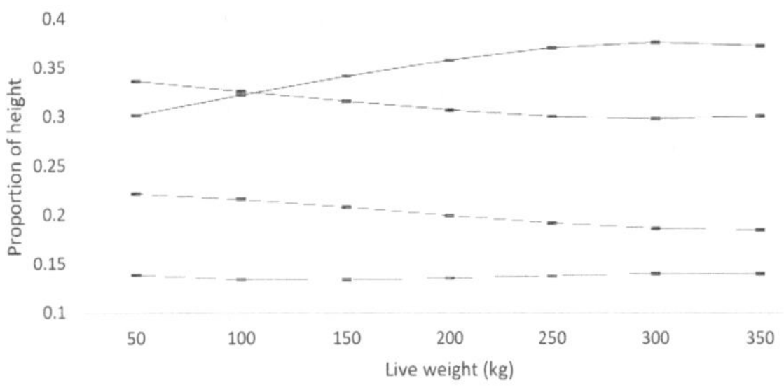

At 50 kg, radius height made up the largest proportion of height, however, at 100 kg the proportion of thoracic length to total height had increased to be equal to that of radius, and thoracic height made up the largest proportion of overall height from 150 kg onwards. There was very little variation among animals as illustrated by the standard error of means. For example, at 50 kg the toe length ranged from 13.3-14.3% of height and at 350 kg the MC3 length ranged from 17.9-19.4% of height.

Discussion

Feed supply on New Zealand farms is seasonal, when pasture availability changes dependent on the time of year (Rattray et al. 2007). Dairy-industry data has indicated that during winter (June-August) a growth check in live weight often occurs, due to lower pasture quality and availability (Handcock et al. 2016). In the current study, liveweight gain in heifers was linear, and live weight at 15 months was between 333 and 387 kg for Jersey and Holstein-Friesians respectively. This corresponds to approximately 78% of mature live weight, well above the industry target of 60% of mature live weight at 15 months (Troccon 1993). Therefore, the heifers in this experiment were well grown by New Zealand standards.

A linear growth pattern is likely to promote continuous appositional and longitudinal bone growth and mineralisation to achieve peak bone mass (Cooper et al. 1995). Increases in animal size cause a greater strain on bone, requiring an increase in bone strength to maintain strain within physiological limits (Rubin & Lanyon 1985). Given the relationship of bone strength with live weight, live weight, rather than age, was used as the predictor for stature measurements (Gibson et al. 2021; Gibson et al. 2019). This allowed comparison of how the body changed as the animal grew.

In the current experiment, increases in leg measures and height slowed at approximately 200 kg, whilst increases in capacity (i.e., girth) remained constant. Early in a heifer's life, increases in live weight are mainly comprised of growth in long bones (linear skeletal growth) resulting in an increase in height (Guilbert & Gregory 1952). However, at puberty, longitudinal bone growth ceases due to increasing oestrogen secretion promoting physeal closure (Harada & Rodan 2003). At seven months of age, 80% of mature height is achieved but only 35-45% of mature weight has been attained (Hammack & Gill 2009). Longitudinal growth after seven months of age slows as 90% of mature height is achieved by one year of age, so contributions to liveweight gain are the result of increases of capacity (axial skeleton) (Hammack & Gill 2009).

In the current study, proportional growth in relation to height in the distal parts limb, such as the metacarpus and toe, decreased, due to cessation of growth in the distal limb. At birth, metacarpal length is approximately 90% of mature length, so longitudinal growth of the bone is limited (Guilbert & Gregory 1952). Bones within the proximal limb, such as the radius and humerus, continue to grow at 15 months of age, especially at the proximal growth plate in the humerus and distal growth plate in the radius (Bartosiewicz 1984). The growth pattern described in the current study is in agreement with previous descriptions of the pattern of growth in cattle where bones in the distal limb mature earlier than bones in the proximal limb (Bartosiewicz 1984; Berg et al. 1978). The difference in bone maturation also reflects differences in the size and direction of the forces placed on areas of the limb. In cattle, the distal sections of the limb, such as the metacarpus, primarily undergo compression forces from stance requiring less bone strength (SSI) compared to the humerus (Berg et al. 1978). After a year of age, increases in compressional forces is moderate so changes in bone size in the metacarpus is limited. However, in the humerus, torsional forces from the surrounding muscles act on bone resulting in a much stronger bone than the metacarpus and the humerus continues to increase in size with increases in axial body growth (Bouza-Rodríguez & Miramontes-Sequeiros 2014).

In the current study, Jersey calves were consistently lighter than Friesian calves of the same age. At the same live weight, there were minimal effects of breed on stature measures, demonstrating a common breed stature shape. However, a lighter mature weight in Jersey calves results in an earlier age at puberty (Hickson et al. 2011). The cessation of longitudinal growth with puberty occurs at a lesser live weight in Jerseys and results in divergence of metacarpus length among breeds as distal portions of the limb mature earlier (Berg et al. 1978). The beginning of this divergence in stature measures was observed in the current study when measures in the radius, MC3 length and height diverged between Friesian and Jersey heifers at approximately 150 kg. Divergence of breed fits with previous observations by Bijen et al. (2020) where breed affected the relationship between bone size and live weight in older animals.

With proximal limb development still occurring at 15 of months age, bones within the proximal limb, such as the humerus, will be susceptible to changes in growth trajectory. Proximal limb development in the current study is consistent with previous research in steers in which growth in the humerus was still occurring at 24 months of age (Gibson et al. 2021). This pattern of growth aligns with observations of mammalian growth most-well described in human studies because of the association of growth trajectory with peak bone mass, fracture risk and osteoporosis (Cooper et al. 1995).

The effect of the first winter on liveweight gain reported by Handcock et al. (2016), showed how times of poor quality and/or lack of pasture can change the trajectory of longitudinal bone growth in the form of an altered wither height. Therefore, if the humerus is continuing to grow during the second winter, it is likely that if adequate feed quality and quantity is not supplied then growth of the humerus may be adversely affected. Heifers with humeral fractures often have growth-arrest lines indicating a period of poor nutrition prior to the fracture occurring, which possibly coincides with their second winter.

Conclusion

Contributions of liveweight gain from height and length slowed at approximately 200-250 kg, whilst capacity measures such as girth continued to increase, reflecting the known growth pattern of cattle in which 90% of height is achieved in the first year. Therefore, growth in the metacarpus is restricted after 12 months, whereas growth in the humerus continues. This raises the possibility that growth restriction due to feed availability after 15 months of age could influence the development of the humerus without affecting the metacarpus. In addition, results from this study suggests that the second winter is a potential risk period for growth in the humerus. Therefore, to reduce the risk of humeral fractures in the replacement herd, farmers should ensure adequate feed supply is provided in the second winter of a heifer's life, up until calving.

The minimal effect of breed on stature at the same live weights reflect the convergence of body morphology towards an ideal type at a given live weight. However, the smaller mature size and lighter weight at puberty with greater Jersey genetics would have resulted in the divergence of stature measures after 15 months of age. It is expected that increases in height will slow earlier in more-Jersey breed groups than for breed groups with more Friesian genetics due to differences in mature size and age.

Acknowledgements

Authors acknowledge Massey University's Dairy One Farm and Keeble farm and technical staff for managing the animals. Many thanks to Trevor and Stephanie Rudman for their managing of animals and assistance in data collection.

References

Bartosiewicz L 1984. Sexual dimorphism of long bone growth in cattle. Acta Veterinaria Hungarica 32: 135-146.

Berg R, Andersen BB, Liboriussen T 1978. Growth of bovine tissues 4. Genetic influences on patterns of bone growth and distribution in young bulls. Journal of Animal Science 27: 71-77.

Bijen L, Gibson MJ, Back PJ, Hickson RE, Van Knegsel A, Rogers CW 2020. Effect of breed and paddock activity on bone mass and strength in three different beef cattle breeds. New Zealand Journal of Animal Science and Production 80: 29-33.

Bouza-Rodríguez JB, Miramontes-Sequeiros LC 2014. Three-dimensional biomechanical analysis of the bovine humerus. Applied Bionics and Biomechanics 11: 13-24.

Cooper C, Cawley M, Bhalla A, Egger P, Ring F, Morton L, Barker D 1995. Childhood growth, physical activity, and peak bone mass in women. Journal of Bone and Mineral Research 10: 940-947.

Dittmer K, Hitchcock B, McDougall S, Hunnam J 2016. Pathophysiology of humeral fractures in a sample of dairy heifers. New Zealand Veterinary Journal 64: 230-237.

Gibson M, Hickson R, Back P, Dittmer K, Schreurs N, Rogers C 2021. The Effect of Sex and Age on Bone Morphology and Strength in the Metacarpus and Humerus in Beef-Cross-Dairy Cattle. Animals 11: 694-704.

Gibson M, Rogers C, Dittmer K, Hickson R, Pettigrew E, Back P 2019. Can bone measures of the bovine metacarpus predict humeral bone structure? New Zealand Journal of Animal Science and Production 79: 8-12.

Grace N, Knowles SO, Sykes A 2010. Managing mineral deficiencies in grazing livestock, New Zealand, New Zealand Society of Animal Production.

Guilbert H, Gregory P 1952. Some features of growth and development of Hereford cattle. Journal of Animal Science 11: 3-16.

Hammack SP, Gill RJ 2009. Texas adapted genetic strategies for beef cattle X: Frame score, frame size, and weight. Texas Farmer Collection.

Handcock R, Lopdell T, McNaughton L. More dairy heifers are achieving liveweight targets. Proceedings of the New Zealand Society of Animal Production, 76: 3-7.

Handcock R, Lopez-Villalobos N, McNaughton L, Back P, Edwards G, Hickson R 2019a. Positive relationships between body weight of dairy heifers and their first-lactation and accumulated three-parity lactation production. Journal of Dairy Science 102: 4577-4589.

Handcock RC, Jenkinson CM, Laven R, McNaughton LR, Lopez-Villalobos N, Back PJ, Hickson RE 2021. Linear versus seasonal growth of dairy heifers decreased age at puberty but did not affect first lactation milk production. New Zealand Journal of Agricultural Research 64: 83-100.

Handcock RC, Lopez-Villalobos N, McNaughton LR, Back PJ, Edwards GR, Hickson RE 2019b. Live weight and growth of Holstein-Friesian, Jersey and crossbred dairy heifers in New Zealand. New Zealand Journal of Agricultural Research 62: 173-183.

Harada S-i, Rodan GA 2003. Control of osteoblast function and regulation of bone mass. Nature 423: 349-355.

Hickson R, Balcomb C, Fraser K, Lopez-Villalobos KP, Morris S 2011. The effect of breed on the onset of puberty in heifers. Proceedings of the Association for the Advancement of Animal Breeding and Genetics: 51-54.

Hickson R, Zhang I, McNaughton L 2015. Brief communication: Birth weight of calves born to dairy cows in New Zealand. Proceedings of the New Zealand Society of Animal Production 75: 257-259.

LIC & DairyNZ 2019. New Zealand Dairy Statistics 2018-2019. Hamilton, New Zealand: Livestock Improvement Corporation Dairy NZ.

Litherland A, Woodward S, Stevens D, McDougal D, Boom C, Knight T, Lambert M 2002. Seasonal variations in pasture quality on New Zealand sheep and beef farms. Proceedings of the New Zealand Society of Animal Production 62: 138-142.

Rattray PV, Brooks I, Nicol AM 2007. Pasture and supplements for grazing animals. New Zealand, New Zealand Society of Animal Production.

Rubin CT, Lanyon LE 1985. Regulation of bone mass by mechanical strain magnitude. Calcified Tissue International 37: 411-417.

Taylor C 1965. A relation between mature weight and time taken to mature in mammals. Animal Science 7: 203-220.

Troccon J 1993. Effects of winter feeding during the rearing period on performance and longevity in dairy cattle. Livestock Production Science 36: 157-176.

BRIEF COMMUNICATION: Changes to cow behaviour when transitioning from twice a day to a 3-in-2 milking schedule

LS Hall[a]*, RH Bryant[a] and JP Edwards[b]

[a]*Department of Agriculture and Life Science, Lincoln University, PO Box 85084, Lincoln, New Zealand;* [b]*DairyNZ Ltd, Private Bag 3221, Hamilton, New Zealand.*

**Corresponding author:* Email: lucy.hall@dairynz.co.nz

Abstract

The objective of this study was to assess changes to cow behaviour when decreasing milking frequency from twice a day (TAD) to three milkings in two days (3-in-2). CowManager SensOor™ ear tags were attached to cows (n=29) for 15 days before, and 30 days after, transitioning to 3-in-2. To investigate general trends, paired t-tests were used to compare means between activity types (time spent active, highly active, not active, eating or ruminating) and milking frequency. After transitioning to 3-in-2 milking, the total time active was greater (+32.3 mins/cow/day; SEM 6.46), whilst the time spent eating (-16.2 mins/cow/day; SEM 7.57) and ruminating (-12.7 mins/cow/day; SEM 3.59) decreased. These differences were inconsistent between daylight and dark hours. These results indicate that the extra time spent in the paddock by the cows due to a 3-in-2 milking schedule was spent active, potentially due to social interaction and feed-searching behaviours, with a decrease in rumination and eating behaviours.

Key words: Milking frequency; behaviour; activity; transition.

Introduction

An increasing number of dairy farmers are using a 3-in-2 milking schedule, where cows are milked three times in two days (DairyNZ 2018). A 3-in-2 milking schedule results in 25% fewer milkings than traditional twice-a-day (TAD) milking, changing the structure of the day for people and cows on farm. For cows, reducing milking frequency results in less time walking, standing in the yard and milking parlour, and allows more time in the paddock every second day. As cows prioritise certain behaviours, such as lying time over social interaction and grazing (Munksgaard et al. 2005), it is unknown whether 3-in-2 milking frequency affects the proportion of time per day spent on certain types of behaviour compared with TAD. This study aimed to investigate the differences in cow behaviour when the milking schedule was changed from TAD to 3-in-2.

Materials and methods

Data were collected from 29 cows between 15/11/2019 and 30/12/2019 at the Lincoln University Research Dairy Farm, Canterbury, New Zealand (approved by the Lincoln University Animal Ethics Committee). The study was part of a wider research programme investigating the effect of milking interval on farming systems. The herd was established from calving and was milked TAD at 6 am and 4 pm until 30/11/2019 through a 12-aside double-up herringbone parlour (DeLaval, Tumba, Sweden). On 1/12/2019 the milking schedule was changed to 3-in-2, with milking times of 5 am, 5 pm (day 1) and 11 am (day 2). Cows were on average 85 days in milk and producing an average of 21.2 kg milk/cow/day over the TAD experimental period and 18.1 kg milk/cow/day over the 3-in-2 period. Cows were exhibiting signs of oestrus over this time and were being artificially inseminated.

The herd consisted of nine heifers and 20 mixed-aged cows. The farm's available mixed-aged cows were randomly selected and then blocked for milk production in the previous milking season, and the whole herd for genetic merit, calving date and breed and randomly selected for the experimental herd. Cows were stocked at 3.5 cows/ha on paddocks containing predominantly perennial ryegrass and white clover. Cows were given a fresh pasture allocation after each milking proportional to the milking interval, therefore timing of a fresh allocation changed with milking frequency but daily allocation did not change. Pasture allowance was 0.38 ha/day with pre-grazing cover target averaging 3000 kgDM/ha before and after the change in milking frequency. When pasture cover was insufficient to meet the target allocation, cows were supplemented with pasture silage in the paddock.

Each cow was fitted with a CowManager SensOor™ ear tag (Cowmanager BV, Harmelen, Netherlands) which partitioned the day into high-active, active and not-active, eating and ruminating time, in minutes per hour (Periera et al. 2018). Active behaviour is defined as when the cow is stood upright and walks or moves her body. High active is associated with oestrus behaviour. Lying behaviour is included in both the rumination and not-active categories.

Data were summarised for individual cows as daily averages of the time spent in each recorded behaviour (in minutes per day) for the TAD (15 days) and 3-in-2 (30 days, which allowed for any adjustment to behaviour) period.

To investigate general trends, paired t-tests (Genstat, 21st Edition, VSN International, Hemel Hempstead, UK) were used to compare means between types of behaviour and milking frequencies. For each cow, behaviour data were also separated into daylight and dark hours, to analyse any effect of milking time changes. Daylight was defined as the hours between 6 am and 9 pm. Paired t-tests were also used to compare average time (minutes per hour) spent on each recorded behaviour during daylight and dark hours for the two milking frequencies.

Results and discussion

On average, the cows spent about a third of their day eating, and a third ruminating (Table 1), which is similar to previous results for grazing cows wearing CowManager Sensors (Mangwe et al. 2020). Small but consistent changes in behaviour ($P<0.05$) occurred when cows changed from a TAD to a 3-in-2 milking frequency, with 25% fewer milkings. Cows milked 3-in-2 spent more time active (+32.3 mins/cow/day; SEM 6.46) and less time eating (-16.2 mins/cow/day; SEM 7.57) and ruminating (-12.7 mins/cow/day; SEM 3.59) compared with when they were milked TAD. The time spent not active did not change significantly after transition to 3-in-2.

Potential reasons for a cow to be active include social interactions, feeding, and walking to the dairy. With fewer milkings per day during a 3-in-2 milking schedule, the amount of time spent walking to the dairy decreases, so it was expected that the overall time active would also decrease, however, this was not supported by the results. One possible explanation for more time active is the change in pasture allocation when switching from TAD to 3-in-2. Cows were given a fresh allocation of pasture after each milking, so for cows milked on 3-in-2 the number of fresh pasture allocations per day decreased and the size of the allocation offered increased. This may account for increased activity as cows increase their feed searching behaviour when given feed once a day compared to twice (DeVries et al. 2005). Further, cows prioritise lying and eating behaviours over social interaction (Munksgaard et al. 2005), therefore, the extra time in the paddock, as a result of fewer milkings, may allow for more social interaction which is reflected by the longer time spent active. Another plausible explanation for greater activity is the adjustment to the change in the daily routine and resulting disturbance when other cows on the farm (still on a TAD schedule) were being moved for milking. Udder distention, as a result of the increased milking interval of 3-in-2 should not significantly alter behaviour (Stelwagen et al. 2013). The greater time spent active suggests that the cows in this study, when milked TAD, were not fatigued at this stage of lactation as a result of milking, although the average distance walked per day was relatively small, at 2.8 km during TAD and 1.9 km during the 3-in-2 schedule.

On average, the cows were spending approximately 120 minutes per day (8% of total time) in behaviour categorised as high active, regardless of milking frequency (Table 1). These data indicate a potential for misclassification between high active and active behaviours, as high active is thought to relate mainly to oestrous behaviour (Pereira et al. 2018). The measurement period coincided with the mating period so the cows were displaying oestrous behaviour over this time, but with 8% of total time in this category, it appears high. The high-active category has not been validated, so it is possible that non-oestrous behaviours are being attributed to this category. Despite this, differences between TAD and 3-in-2 for this category were not significant and, therefore, should not affect conclusions drawn about the other behaviours.

The time spent eating (-16.2 mins/cow/day; 3.5%) and ruminating (-12.7 mins/cow/day; 2.6%) decreased when cows were transitioned to the 3-in-2 milking schedule. These results differ to those of O'Driscoll et al. (2010) who compared the behaviour of cows on either once-a-day (OAD) or TAD milking over a whole lactation and recorded longer grazing times in the OAD group. Those authors noted that cows established their grazing cycles and adjusted grazing time and intake rate according to disruptions such as being milked. The reduction in grazing time in the current study may be indicative of carry-over of established grazing patterns from TAD milking and a small adjustment of grazing time to meet a declining energy demand. Alternatively, a decrease in production and time spent eating and ruminating may also be attributed to a change in pasture quality at this time of year, particularly increasing neutral detergent fibre percentage (Beauchemin 2018). More analysis is needed to investigate this further.

When cows were milked 3-in-2, eating time was less (-1.99 mins/hour/cow) during daylight hours and greater during dark hours (+1.57 mins/hour/cow) when compared with when they were on the TAD milking schedule (Table 2). Ruminating time was not different during daylight but was less during dark hours (-1.64 mins/hour/cow) for cows milked 3-in-2 (Table 2). The diurnal changes and total time spent grazing is supported by other studies, indicating behaviour was still in a normal range (Sheahan et al. 2011). Eating bouts most frequently occur after a milking event and when fresh feed is offered (Beauchemin 2018). When cows were milked TAD a new pasture allocation was provided every afternoon instead of every other afternoon during the 3-in-2 schedule. This could justify a change in

Table 1 Average time (minutes per day) spent on each behaviour (active, not active, eating, ruminating, and high active) for cows at pasture (n=29) milked under two different milking frequencies (twice a day (TAD) and 3 milkings in 2 days (3-in-2)).

Behaviour (total mins/day)	Milking Frequency		Difference between means	SEM	P value
	TAD	3-in-2			
Active	155	187	+ 32.3	6.46	<0.001
High Active	123	121	-1.49	2.82	0.602
Not Active	224	222	-2.15	4.72	0.652
Eating	456	439	-16.2	7.57	0.042
Ruminating	488	475	-12.7	3.59	0.001

Table 2 Average time (minutes per hour) spent on each behaviour (active, not active, eating, ruminating, and high active) for cows at pasture (n=29) milked under two different milking frequencies (twice a day (TAD) and 3 milkings in 2 days (3-in-2)) in daylight and dark hours.

	Daylight					Dark				
Behaviour (mins/hour)	TAD	3-in-2	Diff	SEM	P value	TAD	3-in-2	Diff	SEM	P value
Active	8.49	10.3	+1.83	0.392	<0.001	3.04	3.68	+0.64	0.102	<0.001
Not active	4.99	5.33	+0.34	0.181	0.073	16.5	15.9	-0.64	0.296	0.039
High Active	6.22	6.06	-0.16	0.139	0.247	3.33	3.41	-0.07	0.144	0.620
Eating	27.6	25.6	-1.99	0.458	<0.001	4.60	6.17	+1.57	0.212	<0.001
Ruminating	12.9	12.9	0.00	0.133	0.998	32.7	31.1	-1.64	0.333	<0.001

timing for eating behaviour and perhaps a preference for lying and ruminating whilst environmental temperatures are higher during the day and eating during the cooler temperatures of dark hours.

The time spent active increased during both daylight (+1.83 mins/hour/cow) and dark hours (+0.64 mins/hour/cow) when cows were milked 3-in-2 compared with TAD (Table 2). The mean minimum and maximum temperature over the TAD and 3-in-2 experimental period were not significantly different (The National Climate Database 2019), suggesting that environmental temperature was not a factor that influenced active behaviour change in this experiment. The increase in active behaviour was greater during daylight hours compared to dark hours, indicating the timing of the 3-in-2 milkings influenced this, as the increased paddock time provided by 3-in-2 was during daylight hours.

Overall, the extra time in the paddock provided by the 3-in-2 milking schedule was spent active by cows in this study, suggesting the cows studied were not fatigued at this stage of lactation. Possible explanations for the increase in active time include social interactions and feed-searching behaviours in cows that are milked less frequently and therefore, spend more time in the paddock. Time spent eating and ruminating decreased, potentially due to decreased pasture intake and quality, or a decreased energy demand for milk production. Differences in time spent within each behaviour classification were not consistent between daylight and dark hours. This study was an initial exploration of the differences in behaviour when cows are milked TAD or 3-in-2. A more detailed analysis and broader dataset is required to explore these further.

Acknowledgments

Data for this study were collected as part of the Flexible Milking project, funded by the Sustainable Farming Fund (project 405879) and the dairy farmers of New Zealand through DairyNZ Inc (contract TW2001). We thank N. McMillan and H. Neave and the LURDF farm team for their assistance in data collection and Barbara Kuhn-Sherlock for statistical guidance.

References

Beauchemin KA 2018. Invited review: Current perspectives on eating and rumination activity in dairy cows. Journal of Dairy Science 101:4762-4784.

DairyNZ. 2018. Once a Day Milking. Available: https://www.dairynz.co.nz/milking/once-a-day-milking/ [Accessed 3/03/2021].

DeVries TJ, von Keyserlingk MAG, Beauchemin KA 2005. Frequency of Feed Delivery Affects the Behavior of Lactating Dairy Cows. Journal of Dairy Science 88:3553-3562.

Mangwe M, Bryant R, Gregorini P 2020. Rumen Fermentation and Fatty Acid Composition of Milk of Mid Lactating Dairy Cows Grazing Chicory and Ryegrass. Animals (Basel) 10:169.

Munksgaard L, Jensen MB, Pedersen LJ, Hansen SW, Matthews L 2005. Quantifying behavioural priorities – effects of time constraints on behaviour of dairy cows, Bos taurus. Applied Animal Behaviour Science 92:3-14.

O'Driscoll K, O'Brien B, Gleeson D, Boyle L 2010. Milking frequency and nutritional level affect grazing behaviour of dairy cows: A case study. Applied Animal Behaviour Science 122:77-83.

Pereira GM, Heins BJ, Endres MI 2018. Technical note: Validation of an ear-tag accelerometer sensor to determine rumination, eating, and activity behaviors of grazing dairy cattle. Journal of Dairy Science 101:2492-2495.

Sheahan AJ, Kolver ES, Roche JR 2011. Genetic strain and diet effects on grazing behavior, pasture intake, and milk production. Journal of Dairy Science 94:3583-3591.

Stelwagen K, Phyn CVC, Davis SR, Guinard-Flament J, Pomiès D, Roche JR, Kay JK 2013. Invited review: Reduced milking frequency: Milk production and management implications. Journal of Dairy Science 96:3401-3413.

The National Climate Database. 2019. Combined Daily Data Lincoln. Available: https://cliflo.niwa.co.nz/pls/niwp/wgenf.genform1_proc [Accessed 5/04/2021].

BRIEF COMMUNICATION: Assessing the behaviour of pastoral poultry in an integrated system on a commercial dairy farm

SS Smalling*, B Greig, C Bunt, J Tanner and RH Bryant

Faculty of Agriculture and Life Sciences, PO Box 85084, Lincoln University, Lincoln 7647, Christchurch, New Zealand

**Corresponding author*: Email: Sebrena.Smalling@lincolnuni.ac.nz

Abstract

The aim of this study was to determine the ranging behaviour and dispersal patterns of layer hens on a commercial pastoral dairy farm. One thousand seven hundred and sixty layers housed across three different flocks varying in age and shed type were observed weekly. Their outdoor area was classified into three zones based on the distance from their shed. Hens in each zone were observed for one hour at dawn, midday and dusk. There was no effect of time of day on bird dispersal patterns, but when hens were outdoors, the effect of location was significant ($P<0.001$) with the majority (26.77%) of hens in the apron with the remaining visible hens observed in the outer range (21.99%) and mid-range (16.33%). In integrated pastoral systems with hens, management of hen excreta dispersal can be achieved through regular relocation of housing, feeders and water.

Keywords: free-range; flock; location; zones; integrated farming system

Introduction

As a result of Government legislation for better welfare conditions, the battery-cage egg-production system in New Zealand will soon be phased out, encouraging alternative egg-production systems such as free-range (MPI 2018). However, conversion from cage eggs to free-range may present other challenges such as access to land and unwanted environmental consequences as dispersal of excreta is shifted from an indoor managed area to soils in the outdoors.

For some producers, accessing land may present an opportunity to integrate with other farming enterprises such as ruminant production systems. By diversifying existing farming systems, there may be benefits to both enterprises, for example, hens in pastoral systems contribute to fertilisation, weed control, insect control, and scratching and separation of animal manure (Miao et al. 2005). Hens in pastoral free-range systems are unlikely to have fenced boundaries, as subdivision used for cattle and sheep will not limit hens, so voluntary stocking densities of hens in these systems are likely to enable their natural behaviour.

In New Zealand, the code of layer hen welfare (MPI 2018) stipulates that free-range hens have a maximum outdoor stocking density of 2,500 hens/ha (0.25 hens/m^2) with recommended best practice at no more than 900 hens/ha (0.09 hens/m^2, MPI code of welfare). While these stocking densities seem relatively low, the behaviour of hens would indicate that they are unlikely to randomly use all areas of a range. Campbell et al. (2020) found that 3% of free-range hens never ventured outside the shed while Pettersson et al. (2016) noted that hens from larger flocks limited their use of the range compared to smaller flocks. However, no study to date has investigated the behaviour and dispersal patterns of pastoral hens in an integrated farming system. Behaviour and dispersal patterns of hens has implications on soil nutrient load and on the range and subsequent environmental impacts if nutrients run off the land or leach into waterways. Therefore, the purpose of this research was to investigate the dispersal patterns of layer hens in a pastoral free-range system.

Materials and methods

Research Site

As part of a wider investigation into the impact of hens in ruminant pastoral systems, a case study approach was used for this research to document hen dispersal behaviour under large-scale, natural conditions. A relatively unique relationship between a dairy farmer and free-range egg producer has provided the platform to conduct the analysis. The owner of the poultry enterprise pays a lease ($/hen/year) to the dairy farmer who has permitted the housing of up to 1800 hens on his property. The dairy farm is located in Greenpark, Canterbury and has an area of 125 ha and supports 400 dairy cows in a conventional rotational grazing system. All paddocks on the farm are between 2.5 and 5.0 ha in size and are separated using permanent fencing material consisting of double wire electric fencing designed to contain cattle. The pastures are of variable age, established with perennial ryegrass and white clover pasture. The paddocks are rotationally grazed according to pasture growth and cow numbers.

The free-range enterprise consists of approximately 1,760 brown shaver hens over three flocks varying in size, age and shed management. The poultry farmer replaces each flock approximately annually, purchasing point-of-lay hens at 18 weeks of age and replacing them after 60 to 70 weeks of production. Consequently, the 'batch' buying of hens is a means of managing risk and generating age distribution across all three flocks. These flocks are then housed in sheds which are either fixed or mobile. Each shed (21 m^2 floor area) has perch space for up to 400 hens. Details of flocks and shed management are provided in Table 1.

The sheds were constructed of wood and used corrugated galvanised steel for a pitched roof. Each shed was furnished with perches and nesting boxes, the fixed

Table 1 Description of hen and shed management for the three flocks used in an integrated poultry and commercial dairy farm system between October 2020 and January 2021.

	Flock 1	Flock 2	Flock 3
Shed type	Stationary	Mobile weekly shift	Mobile 3 weekly
No. of sheds	1	2	2
Paddock area (ha)	4.6	2.6	2.6
Peak number of hens	360	600	800
Age of hens 9 October, 2020	25 weeks old	81 weeks old	48 weeks old
Flock replacement during observation period	none	19 November, 2020	20 January, 2021

shed had an inbuilt floor while the four mobile sheds were floorless. All sheds had openings (pop holes of 0.0045 m^2) cut into the walls to allow hens access to outside, as well as doors (1.8 x 1.4 m) which always remained open. The stationary shed had six pop holes, and the mobile sheds had four pop holes each. The four mobile sheds were built on skids and moved up to 20 m within the paddock every seven to 21 days by pulling the shed along the ground with a tractor. Two feeders, which had a holding capacity of 180 kg (0.9 m high and 0.46 m in diameter), were filled by the owner every four days, and were located within 10 m of each shed. Sufficient feed was always available in the feeders to allow *ad libitum* intake. Hens had access to drinking water through livestock reticulated trough water which were located within 100 m of all sheds.

Measurements

Observations reported here were collected between October 2020 and January 2021 (Lincoln University AEC 2020-23) to determine hen's behaviour. Weekly observations were carried out by observing each flock for 60 minutes at dawn, midday and sunset. Using the approach of Chielo et al. (2016), the area around the sheds were classified into distinct zones representing increasing distances from the shed and recording the number of visible hens in those zones. The zones were: the apron (0- 10 m from the shed); the mid-range (10-50 m from the shed); and the outer range (beyond 50 m from the shed). Flagged markers to the north, south, east and west of the sheds were used to identify the zone boundaries. There was always a large proportion of the flock inside the shed during the observations and, in the case of Flock 3, also inside a pump shed which was located within 50 to 100 m of the shed. The hens indoors were not counted. To account for the total number of hens in each flock, a combination of three methods where used: the number of hens that were purchased and the proportion remaining at slaughter, along with a general head count once per month, and by the hen-day egg production. Hens outdoors were counted by visual assessment and a video of the hens on the range was also used to ensure accuracy in the number of hens in each location.

Statistical analysis

A linear mixed-effects model was used to analyse the location of hens around the shed and range using: flock, time of day, location, season and their interaction terms as fixed terms in the model. Subject was fitted as a random effect to account for the repeated measures of observations. Statistical significance was set at $P<0.05$ and all data analyses were performed in STATA 12 (StataCorp, Texas, USA).

Results and discussion

Irrespective of flock-housing system, hens spent most of their time in and around the shed and feeder in the apron area as the proportion of hens in the mid and outer range was less than half the number of hens in each flock. During the morning, 45% of hens were in the mid and outer range, which declined to 40% and 30% of hens at midday and dusk (Fig. 1). Consequently, there were differences in the density of hens in these areas which were 0.325±0.235, 0.009±0.008 and 0.004±0.004 hens/m^2 in the apron, mid, and outer range respectively ($P<0.001$). These results agree with other research in commercial free-range laying systems where most of the hens were located close to the shed (Hegelund et al. 2005; Larsen et al. 2017) and fewer in the mid-range (Chielo at al. 2016). The low percentage of hens in the mid-range in the present study may be due to lack of overhead cover and, therefore, fear of predation from the local swamp harriers.

In this study, flocks varied slightly in their dispersal patterns across the mid-and-outer range depending on nearby enrichment (e.g., trees, feeders, troughs and other farm sheds). For instance, Flock 3 which had pine trees within 100 m of the shed, had a higher proportion of the hens located in the outer range compared with Flock 1 or 2 where most hens were observed in the apron ($P<0.01$, data not shown). Flock 1 hens in the stationary shed had the fewest hens further from the shed which suggests lack of motivation to roam either as a result of lack of cover or being a young flock they were still learning a new environment. These findings support the conclusions of Rault et al. (2013) who observed that variation in structures on the range attracted hens and encouraged wider distribution. Additionally, we also noted differences in behaviour associated with the age group of hens where young hens were observed less frequently in the mid-and-outer range compared with older hens ($P<0.05$). For maturing hens, increasing use of the range probably reflects growing familiarity with their environment (Campbell et al. 2020).

The findings from this study show that the use of

Figure 1 showing the mean percentage of hens in each location (apron, mid-range and outer range) at different time of the day (dawn, midday and dusk).

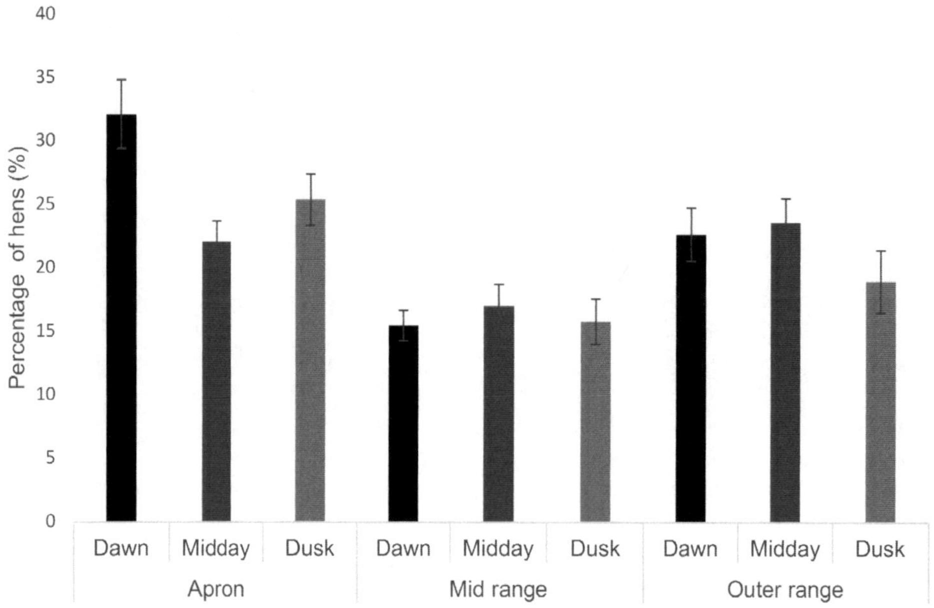

the range varies with different flocks. Hens outside the sheds were not evenly distributed on the range, with most of them located in the apron zone. Singh et al. (2016) recorded a higher-than-normal concentration of N, Ca and S required for pasture production on free-range poultry farms, with higher levels of Ca in flocks that have a fixed shed compared to ones with a mobile shed. Dispersal of hens, and therefore their excreta is not random. Hens staying close to the shed can contribute greatly to nutrient loading around the shed: having mobile sheds can allow for distribution of nutrients across the range. Consequently, the location of hens and stocking density is likely to impact nutrient loading, leaching and runoff and future research will need to evaluate these risks. The absence or presence food pathogens was not carried out in this study; this is an area for future research that will add confidence to consumers.

Acknowledgements

We are grateful to New Zealand Aid Scholarship for financial support to the first author. We would like to thank the farmer from Thulani Free Range Eggs and Sibusio Hadebe for discussion on statistical analysis.

References

Campbell DL, Dyall TR, Downing JA, Cohen-Barnhouse AM, Lee C 2020. Rearing enrichments affected ranging behavior in free-range laying hens. Frontiers in Veterinary Science 7: 446 doi:10.3389/fvets.2020.00446.

Chielo LI, Pike T, Cooper J 2016. Ranging behaviour of commercial free-range laying hens. Animals 6(5): 28-41.

Hegelund L, Sørensen JT, Kjaer JB, Kristensen IS 2005. Use of the range area in organic egg production systems: effect of climatic factors, flock size, age and artificial cover. British Poultry Science. 46(1): 1-8.

Larsen H, Cronin GM, Gebhardt-Henrich SG, Smith CL, Hemsworth PH, Rault JL 2017. Individual ranging behaviour patterns in commercial free-range layers as observed through RFID tracking. Animals 7(3): 21 doi 10.3390/ani703002.

Miao ZH, Glatz PC, Ru YJ, Wyatt SK, Rodda BJ 2005. Integrating free-range hens into a wheat stubble. International Journal of Poultry Science 4(8): 526-530.

MPI. 2018. Animal Welfare (Layer Hen) Code of welfare. https://www.mpi.govt.nz/dmsdocument/36732/direct [accessed 2 April 2021].

Pettersson IC, Freire R, Nicol CJ 2016. Factors affecting ranging behaviour in commercial free-range hens. World's Poultry Science Journal 72(1): 137-150.

Rault JL, Van De Wouw A, Hemsworth P 2013. Fly the coop! Vertical structures influence the distribution and behaviour of laying hens in an outdoor range. Australian Veterinary Journal 91(10): 423-426.

Singh M, Ruhnke I, De Koning C, Drake K, Skerman A 2016. Nutrient loading on free-range layer farms. In Proceedings of the 27th Annual Australian Poultry Science Symposium. Poultry Research Foundation, University of Sydney. pp 29-29.

StataCorp LP. 2011 Stata Data Analysis Statistical Software: Release 12. College Station, TX: StataCorp LP.

Temperament of hand-reared beef-cross-dairy cattle does not influence production traits

NP Martín[a]*, NM Schreurs[a], ST Morris[a], N López-Villalobos[a], J McDade[b] and RE Hickson[a]

[a] *School of Agriculture and Environment, Massey University, Private Bag 11222, Palmerston North 4442, New Zealand;*
[b] *Greenlea Premier Meats Ltd., P.O. Box 87, Hamilton 3240, New Zealand*

**Corresponding author:* Email: N.P.Martin@massey.ac.nz

Abstract

Evaluation of temperament of calves early in life is associated with growth and meat quality in extensively farmed beef cattle. However, the relationship between temperament and performance is unclear for hand-reared beef-cross-dairy calves. This experiment analysed temperament of 1077 Angus- and Hereford-sired cattle born to dairy-breed cows and hand-reared, and evaluated the value of these measurements as predictors of growth and meat quality traits. Cattle in this study were calm, with mean exit velocity (EV) between 0.76 and 1.15 m/s from 200 d to 800 d, crush score (CS) between 1.38 and 1.79 from 200 d to slaughter (1-5 scale), and measurements were highly correlated at all ages ($P<0.05$). Temperament at 200 d and 400 d was not related to growth, but CS was related to growth from 600-800 d (-0.011 kg/d for a 1-point increase in CS at 600 d, $P<0.05$). Temperament at 800 d or before slaughter was not related to meat ultimate pH or colour score. The small variation in temperament among animals was probably the consequence of cattle being acclimatised to the presence of humans due to hand-rearing and frequent handling. Therefore, temperament is unlikely to be an issue in hand-reared beef-cross-dairy cattle and should not influence growth and meat quality traits.

Keywords: Artificial rearing; beef-on-dairy; beef production; crush score; exit speed; exit velocity; flight time

Introduction

Cattle temperament is related to stress responsiveness and can be associated with animal performance and meat quality (Behrends et al. 2009; Cafe et al. 2011; del Campo et al. 2021; Ponnampalam et al. 2017). This is particularly true for *Bos indicus* cattle, entire bulls, animals that have been handled infrequently or that had negative experiences (Behrends et al. 2009; Bonin et al. 2014; Grandin et al. 2015; Haskell et al. 2014). Studies reporting on *Bos taurus* breeds, on the other hand, show variable results in the associations between temperament and performance traits (Boles et al. 2015; Burnham et al. 2005; Cafe et al. 2011; Coombes et al. 2014; Della Rosa et al. 2019; Haskell et al. 2014). The temperament of beef-cross-dairy cattle has never been explored to determine whether this is an issue affecting meat quality, and the incidence of aggressive and nervous animals within these crosses has never been quantified.

The temperament of cattle can be assessed by exit velocity (EV) and crush score (CS), traits that are persistent over time (Cafe et al. 2011). Temperamental cattle (faster EV or greater CS) can have lower growth rates, produce smaller carcasses with less fat cover, and have darker meat that is tougher to eat, all economically detrimental parameters (Behrends et al. 2009; Burrow et al. 1997; Cafe et al. 2011; Cooke et al. 2018; Olson et al. 2019). The slower growth may be a result of lower feed intake and/or more use of energy in avoidance behaviour. The inferior meat quality is partially a result of metabolic mechanisms. Temperamental cattle have a greater stress response to handling and transport, resulting in depletion of muscle glycogen before slaughter, more transport of lactate from the muscle into the bloodstream (Boles et al. 2015; Cafe et al. 2011), and hence, greater carcass pH and the associated negative characteristics. Consequently, evaluation of temperament early in life may be a helpful tool, because animals could be allocated to different management groups for yarding, feeding, production systems or end markets (Behrends et al. 2009; Burdick et al. 2011). Specific management practices for temperamental cattle could include being handled only by experienced staff, and go to slaughter without reaching their ideal fattening conditions, given that they would unlikely have good meat quality, nonetheless.

In most beef systems, one of the first calf-human interactions happens at weaning, when the calf is separated from its dam. This is one of the most stressful times in a beef animal's life, and it is argued that the variation in temperament measured at weaning may re-emerge at the time of harvest, when cattle are exposed to a novel experience and a new environment (Behrends et al. 2009; Cafe et al. 2011). However, animals do get used to being handled (Della Rosa et al. 2019; Parham et al. 2019). Calves born on a dairy farm are usually taken off their dams within 24 hours of birth and reared by humans. These calves are intensively handled at early ages and see the human as a source of feed rather than a threat. Phenotypic temperament may change because of this hand-rearing, and therefore, the relationship between temperament and performance is unclear for hand-reared beef-cross-dairy calves.

The aim of this experiment was to analyse measurements of temperament throughout the lifetime of Angus- and Hereford-sired cattle born to dairy-breed cows, and to evaluate the value of these measurements as predictors of growth and meat quality traits for beef-cross-dairy cattle reared by humans.

Materials and methods

This experiment uses the same animals for which growth and carcass traits were previously reported (Martín et al. 2020, 2021). This experiment was conducted at Limestone Downs, near Port Waikato, New Zealand (37°28'S, 174°45'E) with approval from the Massey University Animal Ethics Committee (15/65 and 18/50).

Animals and management

Angus-sired and Hereford-sired cattle born on Limestone Downs farm, in spring 2016 (n=564) and 2017 (n=513) were included in the study. The calves were born to dairy-breed cows which were predominantly Holstein-Friesian or Holstein-Friesian-cross-Jersey crossbred. For full details, refer to Martín et al. (2020). Angus (n=31) and Hereford (n=34) sires for the mixed-aged cows were selected on the basis of their estimated breeding values for birth weight, gestation length and live weight at 600 days of age (d), but docility was not included in the selection process, with very few sires (n=17 Angus) having a record for this trait. Angus (n=6) and Hereford (n=6) sires for the 15-month-old heifers were selected to be in the lightest 15% of breed for birth weight.

Mean birth date was 3 August 2016 and 5 August 2017 for the calves included in the present experiment, with an overall mean birth weight of 36.4 kg (SD 4.7). Calves were artificially reared on an allowance of 4-6 litres of milk/head/day, fed twice daily for the first three weeks and once daily afterwards, and calf meal was offered during the transition from milk to pasture (Coleman 2020). Calves were weaned at a minimum of 85 kg live weight, resulting in a mean age at weaning of 82 d (SD 11). Once weaned, calves were moved from the dairy platform to the sheep and beef hill-country platform of the same farm. Male calves were castrated before four months of age.

At four months of age, at a mean age of 131 d (SD 17), calves were allocated to six grazing herds based on live weight (light, intermediate and heavy) and sex (female and male) and balanced for sire so that, where possible, all sires were represented in each grazing herd within year. In total, there were 12 grazing herds (two years x two sexes x three liveweight groups), and animals remained in those herds throughout the experiment until slaughter. All cattle were grazed on summer-dry hill-country pasture on the coastal farm under commercial conditions (Martín et al. 2020).

Each grazing herd was slaughtered as a complete group on the same day, when the herd reached the slaughter target weight of 500 kg for heifers and 600 kg for steers (full details in Martín et al. (2021)). Heifers included in this study were slaughtered at a mean age of 823 d (range 693-934 d, 27 months old) and 520 kg (SD 38) live weight on-farm, whilst steers were slaughtered at a mean age of 887 d (range 821-955 d, 29 months old) and 614 kg (SD 42). Animals were processed commercially through Greenlea Premier Meats Ltd, Hamilton plant, according to standard New Zealand industry practice (Animal and Animal Products Directorate 2017), with Halal certification.

Measurements

Live weight. Animals were weighed from entry to the beef platform at a minimum of two-monthly intervals, as described by Martín et al. (2020). The live weights closest to 200 d, 400 d, 600 d and 800 d for each herd were used to estimate growth as average daily gain (ADG, in kg/d), calculated as the difference in live weight divided by the difference in days between the two ages.

Temperament. Temperament was assessed as a visual crush score (CS) and exit velocity (EV). The CS was assessed while animals were loosely restrained for weighing (20 seconds) in the weigh crate (cattle crush model Cattlemaster Titan, made by Te Pari Products Ltd, Oamaru, New Zealand; internal dimensions H 203.2 cm x L 300.6 cm x W 75.0 cm) at 200 d, 400 d, 600 d, 800 d and on the day of slaughter (mean age 856 d), within one hour after yarding. All CS assessment were made by the same person, on individual cattle based on the behavioural scoring system described by Grandin (1993) and adapted by Cafe et al. (2011): [1] Calm – none or slow movements, head mostly still; [2] Slightly restless - shifting, looking around more quickly, moving feet; [3] Restless - moving backwards and forwards, occasional shaking crate, squirming, may try to put head through bale; [4] Nervous - continuous vigorous movement backwards and forwards, snorting, shaking crate; and [5] Very nervous - continuous violent movement, attempting to jump out, rearing, twisting or violently struggling.

Exit velocity was measured as the time required to traverse 1.83 m distance after exiting the crush, once animals had been weighed at 200 d, 400 d, 600 d and 800 d. The time was measured using infrared photogates (TCi Wireless Timing System, Brower Timing Systems, UT, USA), and then converted to EV (m/sec). The first gate was 1.7 m in front of the weigh crate and the second was 1.83 m distance subsequently. This methodology was based on that of Burrow et al. (1988) and adapted by Behrends et al. (2009) and Boles et al. (2015).

Carcass and meat traits. Animals were slaughtered and bodies were dressed to commercial specifications as described by Martín et al. (2021). Carcasses were chilled (4±1°C) overnight and, the following morning, one side of the carcass was cut between the 12th and 13th rib to expose the eye muscle (*M. longissimus thoracis*) for in-chiller assessment of ultimate pH and meat colour score. Ultimate pH was measured by pH spear (Eutech Instruments, Singapore) on the eye muscle and the mean of three measurements was used for analysis. Meat colour was scored against the AUS-MEAT / MSA reference standards (AUS-MEAT Limited 2018). Possible meat colour scores ranged from 1 (light) to 7 (dark).

Statistical Analysis

Data cleaning. Animals born to sires with a minimum of five progeny were included in analysis (excluded n=9 progeny of three sires in 2016 and n=3 progeny of one sire in 2017). Animals that went missing, were recorded to have ill health, were removed from their grazing herd for more than two months, or that died, were excluded from analysis of traits measured after they first left their herd (n=55 from 200 d to slaughter). For animals that had to be chased out of the exit race because they stopped within the 1.83 m, EV was considered 0 for that age (n=8 at 200 d, n=36 at 400 d, n=93 at 600 d, n=43 at 800 d). Animals that turned around and went through the exit race backwards were not considered for measures of EV at that age (n=17 at 200 d, n=4 at 600 d). The final dataset consisted of 1077 animals.

Statistical models. Means and histograms were used to analyse the spread and distribution of the individual temperament measurements (EV and CS) at 200 d, 400 d, 600 d, 800 d and pre-slaughter, with the MEANS and SGPANEL HISTOGRAM procedures (SAS 9.4, SAS Institute Inc., Cary, NC, USA). Pearson correlations coefficients among individual temperament measurements across age, were calculated with the CORR procedure.

General linear models were used to estimate the association of temperament measurements (either EV or CS) at 200 d with ADG between 200-800 d and 200-400 d, temperament at 400 d with ADG between 400-600 d, and temperament at 600 d with ADG between 600-800 d, with the GLM procedure. Similarly, the GLM procedure was used to estimate the association of temperament measures at 800 d and pre-slaughter (mean age 856 d), with ultimate pH and meat colour scores. All models included the fixed effects of breed of sire (Angus and Hereford) and grazing herd (12 herds), and the random effect of sire within breed (73 sires).

Table 1 Number, mean (±SD) and range of temperament measurements as exit velocity (in metres per second) and crush score (scale 1 to 5) of beef-cross-dairy cattle at 200, 400, 600, 800 and 856 (mean age at slaughter) days of age.

Trait	Age	N	Mean	Range
Exit velocity (m/s)	200 d	492	1.15 ± 0.44	0 - 3.05
	400 d	1036	1.06 ± 0.49	0 - 3.10
	600 d	1004	0.76 ± 0.50	0 - 2.65
	800 d	744	0.96 ± 0.51	0 - 2.69
Crush score	200 d	512	1.43 ± 0.62	1 - 3
	400 d	1067	1.79 ± 0.78	1 - 5
	600 d	1036	1.50 ± 0.70	1 - 4
	800 d	763	1.40 ± 0.65	1 - 4
	856 d	1025	1.38 ± 0.63	1 - 4

Results

Cattle in this study grew at 0.58 kg/d (SD 0.07) from 200 d to 800 d, achieving a live weight of 505 kg (SD 47) at 800 d and a carcass weight of 278 kg (SD 31) at a mean age of 856 d. Growth rates varied over the time of the experiment, with an ADG for 200-400 d of 0.61 kg/d, for 400-600 d of 0.77 kg/d and for 600-800 d of 0.39 kg/d. Meat quality traits were within the expected values for beef, with ultimate pH of 5.68 (SD 0.16, range 5.39-6.55) and a meat colour score of 3.0 (SD 0.6, range 1-6).

Temperament of beef-cross-dairy cattle during their lifetime

Cattle in this study had a mean EV between 0.76 and 1.15 m/s at the different timepoints from 200 d to 800 d (maximum of 3.10 m/s, Table 1), and a mean CS between 1.38 and 1.79 at the different timepoints from 200 d to slaughter (maximum of CS 5 on a 1-5 scale). Distribution of the data was right skewed for both traits (data not shown), but more pronounced for CS, with the majority of animals scoring 1 (calm) at any age, and only 22 records (from 4403 records, <0.5%) of 19 animals scoring 4 or 5 (nervous or violent).

Temperament measurements were correlated at all ages (P<0.05, Table 2). Measurements of EV had Pearson correlation coefficients between 0.16 to 0.35 (P<0.001) with the greater correlations at closer ages and decreasing over time. Measurements of CS had Pearson correlation coefficients between 0.28 to 0.60 (P<0.001), also with greater correlations at closer ages but increasing over time. Correlations between trait measurements had coefficients from 0.14 to 0.34 (P<0.001, except for CS200 d-EV800 d with P<0.01), with the greatest correlations between EV and CS at the same age.

Temperament effects on production

Temperament at 200 d and 400 d was not related with the growth of the cattle in the study (P>0.05, Table 3). There was a small association of CS at 600 d with ADG from 600-800 d, such that cattle with higher CS grew less during this period (-0.011 kg/d for a 1-point increase in CS,

Table 2 Pearson correlation coefficients between temperament measurements as exit velocity (in metres per second) and crush score (scale 1 to 5) of beef-cross-dairy cattle at 200, 400, 600, 800 and 856 (mean age at slaughter) days of age. All values are significant to P<0.001, except for the coefficient indicated with * (P<0.01).

		Exit velocity			Crush score				
	Age	400 d	600 d	800 d	200 d	400 d	600 d	800 d	856 d
Exit velocity	200 d	0.35	0.21	0.16	0.30	0.24	0.17	0.21	0.15
	400 d		0.33	0.25	0.21	0.34	0.24	0.19	0.21
	600 d			0.26	0.18	0.16	0.30	0.22	0.15
	800 d				0.14*	0.25	0.31	0.28	0.23
Crush score	200 d					0.41	0.37	0.28	0.30
	400 d						0.48	0.40	0.39
	600 d							0.51	0.43
	800 d								0.60

Table 3 Estimates of regression coefficients (P value in brackets) of temperament measurements (crush score and exit velocity) on growth (average daily gain, ADG) and meat quality traits (ultimate pH and meat colour scores), and P values for breed, sire and grazing herd effects in beef-cross-dairy cattle at 200, 400, 600, 800 and 856 (mean age at slaughter) days of age.

Trait	Age	Regression coefficient (P value)		Effect P value			N	R^2
		Exit velocity	Crush score	Breed[1]	Sire[2]	Grazing herd[3]		
Growth (kg/d)								
ADG 200-800 d	200 d	-0.003 (0.501)		0.059	<0.001	<0.001	468	0.66
	200 d		-0.001 (0.843)	0.025	<0.001	<0.001	488	0.66
ADG 200-400 d	200 d	-0.004 (0.547)		0.816	0.001	<0.001	487	0.61
	200 d		-0.003 (0.575)	0.626	0.001	<0.001	507	0.61
ADG 400-600 d	400 d	-0.006 (0.320)		0.009	<0.001	<0.001	995	0.33
	400 d		-0.003 (0.360)	0.004	<0.001	<0.001	1026	0.32
ADG 600-800 d	600 d	-0.007 (0.249)		0.428	<0.001	<0.001	914	0.52
	600 d		-0.011 (0.008)	0.665	<0.001	<0.001	939	0.53
Meat quality								
Ultimate pH	800 d	-0.001 (0.899)		0.649	0.651	<0.001	653	0.50
	800 d		-0.004 (0.594)	0.691	0.689	<0.001	670	0.50
	856 d		-0.007 (0.352)	0.253	0.198	<0.001	856	0.47
Meat colour	800 d	0.069 (0.078)		0.142	0.576	<0.001	639	0.22
	800 d		0.015 (0.593)	0.101	0.347	<0.001	658	0.23
	856 d		0.028 (0.367)	0.031	0.151	<0.001	914	0.18

R^2: coefficient of determination. [1]Breed effect P value: n=2 breeds, Angus and Hereford. [2]Sire effect P value: n=73. [3]Grazing herd effect P value: n=3 herds per sex, based on live weight (light, intermediate and heavy) at entry to the beef platform.

P<0.05). Temperament at 800 d or prior to slaughter was not related with meat colour scores or ultimate pH (P>0.05, Table 3).

Sire and grazing herd were highly associated with ADG at all ages (P<0.05), and grazing herd was also associated with both pH and meat colour scores (P<0.05).

Discussion

Cattle in this study were calm. This is supported by previous studies in which the temperamental response to handling of *Bos taurus* breeds (such as Angus, Hereford and their crosses) has been low (Burnham et al. 2005; del Campo et al. 2021), particularly when compared with *Bos indicus* breeds or their crosses (such as Brahman, Nellore or other tropical breeds) (Burrow et al. 1997; Cafe et al. 2011). Mean values of EV reported for *Bos Taurus* cattle ranged from 1.10 to 3.69 m/s and for CS from 1.34 to 3.15 (Boles et al. 2015; Cafe et al. 2011; Coombes et al. 2014; Della Rosa et al. 2019; Parham et al. 2019), while the cattle in this study had mean values in the lower bound of the ranges previously reported (EV from 0.76 to 1.15 m/s and CS from 1.38 and 1.79). In addition to the genetic predisposition, animals' responses depend on previous experiences (Grandin et al. 2015), and these cattle were hand-reared, thus they were used to being handled and unlikely to consider humans as a threat. Consequently, the inherited temperament may already not be expressed by 200 d.

Temperament measurements were correlated at all ages, with similar coefficients compared to previous studies (Boles et al. 2015; Cafe et al. 2011; Parham et al. 2019), indicating that these measures are repeatable. Correlations for EV decreased over time, and this has also been shown in earlier experiments (Cafe et al. 2011; Della Rosa et al. 2019). On the other hand, correlations for CS between ages increased as animals got older, while the actual CS decreased over time. Similar effects have been reported in heifers that acclimatised to frequent handling, presenting lower but more consistent CS across time (Parham et al. 2019).

Overall, temperament was not related with growth or meat quality. There was only a small association of CS at 600 d with growth, representing a total potential 2.2 kg for each CS during the 600-800 d period. This is supported by studies with *Bos taurus* breeds, where correlations between temperament and live weight were typically not strong (Boles et al. 2015; Burnham et al. 2005; Cafe et al. 2011; Della Rosa et al. 2019; Haskell et al. 2014). The lack of temperament effects on meat quality has been reported previously for ultimate pH (Burnham et al. 2005; Coombes et al. 2014) and meat colour (Della Rosa et al. 2019), even in animals with higher EV and when comparing both *Bos taurus* and *Bos indicus* breeds (Coombes et al. 2014). It is likely that the animals in this study were too docile for there to be any effect of temperament on performance traits, and the generally good meat quality obtained did not create a range of pH or meat colour to incur a correlation.

In contrast, the effect of contemporary group (grazing herd) was important across all production traits measured. The effect on growth was expected as animals were allocated to grazing herds based on live weight at 131 d. Further, a contemporary group effect has been shown repeatedly in cattle studies, where yarding and handling can affect growth of the animals, and yarding, handling, transportation, and conditions around slaughter can contribute to post-mortem muscle ultimate pH (Dixon et al. 1996; Njisane et al. 2017; Warner et al. 2010; Woiwode et al. 2016). In turn, ultimate pH has significant effects on all the meat quality measurements, including meat colour (Ertbjerg et al. 2017; Hughes et al. 2014; Purchas 1990; Purchas et al. 1993).

In conclusion, beef-cross-dairy cattle in this study were calm and the small variation in EV and CS among animals was likely a result of hand-rearing on the dairy farm and cattle being acclimatised to handling, yarding and the presence of humans. Overall, temperament was not related with growth or meat quality, probably because the animals in this study were too docile for there to be any effect of temperament on performance traits. Therefore, temperament is unlikely to be an issue in hand-reared and frequently handled beef-cross-dairy cattle, so producers can expect temperament to have minimal influence on growth and meat-quality traits.

Acknowledgements

The authors gratefully acknowledge C. Alma Baker Trust NZ Ltd., Limestone Downs farm staff, Massey University technical staff, Greenlea Premier Meats staff and Lucy Coleman for their valuable support and assistance in data collection.

References

Animal and Animal Products Directorate 2017. Red Meat Code of Practice Chapter 5: Slaughter and Dressing. Ministry for Primary Industries, Wellington, New Zealand. https://www.mpi.govt.nz/consultations/code-of-practice-red-meat-slaughter-and-dressing/

AUS-MEAT Limited 2018. Handbook of Australian Beef Processing. 2018 ed. Queensland, Australia, AUS-MEAT Limited. www.ausmeat.com.au

Behrends SM, Miller RK, Rouquette FM, Randel RD, Warrington BG, Forbes TDA, Welsh TH, Lippke H, Behrends JM, Carstens GE, Holloway JW 2009. Relationship of temperament, growth, carcass characteristics and tenderness in beef steers. Meat Science 81: 433-438. DOI: 10.1016/j.meatsci.2008.09.003

Boles JA, Kohlbeck KS, Meyers MC, Perz KA, Davis KC, Thomson JM 2015. The use of blood lactate concentration as an indicator of temperament and its impact on growth rate and tenderness of steaks from Simmental x Angus steers. Meat Science 103: 68-74. DOI: 10.1016/j.meatsci.2015.01.003

Bonin MN, Ferraz JBS, Eler JP, Rezende FM, Cucco DC, Carvalho ME, Silva RCG, Gomes RC, Oliveira ECM 2014. Sire effects on carcass and meat quality traits of young Nellore bulls. Genetics and Molecular Research 13: 3250-3264. DOI: 10.4238/2014.April.29.3

Burdick NC, Agado B, White JC, Matheney KJ, Neuendorff DA, Riley DG, Vann RC, Welsh TH, Randel RD 2011. Technical note: Evolution of exit velocity in suckling Brahman calves. Journal of Animal Science 89: 233-236. DOI: 10.2527/jas.2010-2973

Burnham DL, Purchas RW, Morris ST 2005. Relationships between on-farm and pre-slaughter behaviour and growth and meat quality for bulls and steers. Proceedings of New Zealand Society of Animal Production 65: 261-265.

Burrow HM, Dillon RD 1997. Relationships between temperament and growth in a feedlot and commercial carcass traits of Bos indicus crossbreds. Australian Journal of Experimental Agriculture 37: 407-411. DOI: 10.1071/ea96148

Burrow HM, Seifert GW, Corbet NJ 1988. A new technique for measuring temperament in cattle. Proceedings of Australian Society of Animal Production 17: 154-157.

Cafe LM, Robinson DL, Ferguson DM, McIntyre BL, Geesink GH, Greenwood PL 2011. Cattle temperament: Persistence of assessments and associations with productivity, efficiency, carcass and meat quality traits. Journal of Animal Science 89: 1452-1465. DOI: 10.2527/jas.2010-3304

Coleman LW 2020. The use of high genetic merit Angus and Hereford bulls in a New Zealand dairy herd. Thesis presented in partial fulfilment of the requirements for the degree of Doctor of Philosophy in Animal Science. Massey University, Palmerston North, New Zealand. 197 p. http://hdl.handle.net/10179/16378

Cooke RF, Moriel P, Cappellozza BI, Miranda VFB, Batista LFD, Colombo EA, Ferreira VSM, Miranda MF, Marques RS, Vasconcelos JLM 2018. Effects of temperament on growth, plasma cortisol concentrations and puberty attainment in Nelore beef heifers. Animal: 1-6. DOI: 10.1017/S1751731118002628

Coombes SV, Gardner GE, Pethick DW, McGilchrist P 2014. The impact of beef cattle temperament assessed using flight speed on muscle glycogen, muscle lactate and plasma lactate concentrations at slaughter. Meat Science 98: 815-21. DOI: 10.1016/j.meatsci.2014.06.029

del Campo M, Manteca X, Soares de Lima JM, Brito G, Hernández P, Sañudo C, Montossi F 2021. Effect of different finishing strategies and steer temperament on animal welfare and instrumental meat tenderness. Animals 11: 859. DOI: 10.3390/ani11030859

Della Rosa MM, Pavan E, Maresca S, Spetter M, Ramiro F 2019. Performance, carcass and meat quality traits of grazing cattle with different exit velocity. Animal Production Science 59: 1752-1761. DOI: 10.1071/an18064

Dixon H, Thomson B, Graafhuis A 1996. Producing quality beef: Practical experience, current practice and future direction. Proceedings of New Zealand Grassland Association 57: 183-188.

Ertbjerg P, Puolanne E 2017. Muscle structure, sarcomere length and influences on meat quality: A review. Meat Science 132: 139-152. DOI: 10.1016/j.meatsci.2017.04.261

Grandin T 1993. Behavioral agitation during handling of cattle is persistent over time. Applied Animal Behaviour Science 36: 1-9. DOI: 10.1016/0168-1591(93)90094-6

Grandin T, Shivley C 2015. How farm animals react and perceive stressful situations such as handling, restraint, and transport. Animals 5: 1233-1251. DOI: 10.3390/ani5040409

Haskell MJ, Simm G, Turner SP 2014. Genetic selection for temperament traits in dairy and beef cattle. Frontiers in Genetics 5: 368-386. DOI: 10.3389/fgene.2014.00368

Hughes JM, Oiseth SK, Purslow PP, Warner RD 2014. A structural approach to understanding the interactions between colour, water-holding capacity and tenderness. Meat Science 98: 520-532. DOI: 10.1016/j.meatsci.2014.05.022

Martín N, Schreurs N, Morris S, López-Villalobos N, McDade J, Hickson R 2020. Sire effects on post-weaning growth of beef-cross-dairy cattle: a case study in New Zealand. Animals 10: 2313-2324. DOI: 10.3390/ani10122313

Martín N, Schreurs N, Morris S, López-Villalobos N, McDade J, Hickson R 2021. Sire effects on carcass of beef-cross-dairy cattle: a case study in New Zealand. Animals 11: 636-648. DOI: 10.3390/ani11030636

Njisane YZ, Muchenje V 2017. Farm to abattoir conditions, animal factors and their subsequent effects on cattle behavioural responses and beef quality - A review. Asian-Australasian Journal of Animal Sciences 30: 755-764. DOI: 10.5713/ajas.16.0037

Olson CA, Carstens GE, Herring AD, Hale DS, Kayser WC, Miller RK 2019. Effects of temperament at feedlot arrival and breed type on growth efficiency, feeding behavior, and carcass value in finishing heifers. Journal of Animal Science 97: 1828-1839. DOI: 10.1093/jas/skz029

Parham JT, Tanner AE, Barkley K, Pullen L, Wahlberg ML, Swecker WS, Lewis RM 2019. Temperamental cattle acclimate more substantially to repeated handling. Applied Animal Behaviour Science 212: 36-43. DOI: 10.1016/j.applanim.2019.01.001

Ponnampalam E, Hopkins D, Bruce H, Li D, Baldi G, Bekhit A 2017. Causes and contributing factors to "dark cutting" meat: current trends and future directions: a review. Comprehensive Reviews in Food Science and Food Safety 16: 400-430. DOI: 10.1111/1541-4337.12258

Purchas RW 1990. An assessment of the role of pH differences in determining the relative tenderness of meat from bulls and steers. Meat Science 27: 129-140. DOI: 10.1016/0309-1740(90)90061-A

Purchas RW, Aungsupakorn R 1993. Further investigations into the relationship between ultimate pH and tenderness for beef samples from bulls and steers. Meat Science 34: 163-178. DOI: 10.1016/0309-1740(93)90025-D

Warner RD, Greenwood PL, Pethick DW, Ferguson DM 2010. Genetic and environmental effects on meat quality. Meat Science 86: 171-183. DOI: 10.1016/j.meatsci.2010.04.042

Woiwode R, Grandin T, Kirch B, Paterson J 2016. Effects of initial handling practices on behavior and average daily gain of fed steers. International Journal of Livestock Production 7: 12-18. DOI: 10.5897/IJLP2015.0277

Grazing behaviour of young cattle around faecal pats

CJ Boom[a], GW Sheath[b], VT Burggraaf[c]

[a]*AgFirst Northland, 549 Kara Rd, Whangarei;* [b]*8 Ellerslie Ave, Hamilton;* [c]*AgResearch Ruakura, Private Bag 3123, Hamilton*

Corresponding author: Email: gavin.sheath@gmail.com

Abstract

Young Holstein Friesian bulls grazed four mixed-contour paddocks for a duration of three days on ten occasions between December 2002 and March 2004. For the first seven grazing events, five faecal pats on both easy- (<25°) and steep-contoured land in each paddock were selected for monitoring herbage removal during the subsequent four to five grazing events. After each day of grazing, pasture height was measured at 5-cm intervals for 50 cm from the centre of each pat. The decrease in pasture height was converted to estimates of herbage dry matter removal. For the first three grazing events following pat deposition, avoidance of grazing herbage around pats was greatest on the first day of the three-day grazing duration. The mean distance of grazing from the centre of the pats was 21, 14, 10 and 7 cm for the first to fourth grazing events following deposition. Avoidance was greatest around pats that were deposited in late spring/early summer with little difference occurring between pats located on easy or steep land. Increased grazing pressure appeared to override the natural behaviour of young cattle to avoid grazing close to faecal pats. Grazing management of young cattle should ensure residuals of at least 1500 kg DM/ha in order to minimise the grazing of herbage surrounding faecal-pat areas which are highly contaminated with parasite larvae.

Keywords: cattle; grazing behaviour; faecal pats; cattle parasites

Introduction

In New Zealand, young cattle are intensively grazed within a number of different pasture-based farming systems. These include dairy-heifer rearing, dairy-beef finishing and traditional beef cow-calf systems. These systems were described by Smeaton (2007). It is recognised in practice, and reported in the literature, that young cattle avoid grazing around fresh and aged faecal pats (Weeda 1967; Nolan & Connolly 1980; Gruner & Sauve 1982; Burggraaf et al. 2009). This avoidance has implications for future pasture quality and for the infection of young cattle with gastrointestinal nematodes. This latter challenge provides the context of this paper.

Gastrointestinal nematode infections can significantly limit growth in young cattle (Bisset 1994; Charleston 1997). Previous studies have shown that infective third-stage parasite larvae (L3) populations are aggregated around historic faecal pats and few larvae migrate more than 30 cm from the centre of the faecal pat in which they developed, even after significant time and rainfall (Rose 1961; Durie 1961; Williams & Bilkovich 1973; Boom & Sheath 2008). Reducing the ingestion of infective nematode larvae from this highly contaminated zone should, therefore, be considered in the design of grazing management strategies.

The ability of cattle to select herbage is strongly affected by herbage availability and grazing pressure (Litherland & Lambert 2007). Studies by Hansen et al. (1981) concluded that increased grazing pressure reduced the ability of the cattle to select against herbage with high parasitic larvae concentrations. Hutchings et al. (2001) demonstrated the same with sheep, and further developed the concept of how ruminants have a trade-off between nutrition and parasitism in relation to grazing around historic faecal pats.

The study reported in this paper is part of a series (Boom & Sheath 2008; Burggraaf et al. 2009) that aimed to improve the understanding of the relationship between cattle faecal pats, animal grazing behaviour and the removal of herbage and nematode larvae. It provides data that describe the grazing behaviour of young cattle around faecal pats of contrasting age during different seasons of the year and on different land contours.

Materials and methods

Trial location and context

The study was conducted during 2002-2004 at AgResearch, Whatawhata Research Centre, New Zealand. This site is located at latitude 38° South, with a mean annual rainfall of 1655 mm. Pastures were predominantly perennial ryegrass (*Lolium perenne*) and white clover (*Trifolium repens*) on easy contoured land (<25°), and a mix of temperate grasses and legumes on sloped land.

Paddocks, grazing conditions and stock

Four mixed-contour paddocks (0.4-0.8 ha) were repeatedly grazed by young cattle between December 2002 and March 2004. Grazing management mimicked New Zealand beef finishing systems, with paddocks grazed when herbage mass reached approximately 2400 kg DM/ha during winter-spring and 2800 kg DM/ha during summer-autumn. Paddocks were grazed for three days and stocked to achieve a herbage residual of approximately 1200 kg DM/ha during winter-spring and 1600 kg DM/ha during summer-autumn. Average paddock herbage mass was assessed prior to grazing (pre-graze), every 24 hours during grazing and after grazing ceased (post-graze) by calibrated visual estimation (Haydock & Shaw 1975). Dates and estimated paddock pre-graze and post-graze herbage mass for each grazing period are outlined in Table 1.

One group of Holstein Friesian bulls was used for all grazing events. These were approximately four months of age at the start of the study (December 2002) and 18 months of age at the completion of the study (March 2004). Mean live weights of these bulls were 110 kg and 420 kg at the start and end of the study period.

Faecal pat selection and monitoring

At the completion of each grazing of each paddock, five freshly deposited faecal pats on easy land (<25° slope) and five on steep land (>25° slope) were identified in each paddock. Chosen pats needed to fit within a 20 cm radius from the centre of the pat and have no other obvious faecal contamination (either new or historic) within 1 m of the edge of the pat. The dry weight of each pat was assessed at the time of identification by visual score and calibrated to ten pats of various sizes which were scored, dried and weighed. The average dry weight of faecal pats increased from 59 to 135 g as the cattle aged. If new faecal material was deposited within a 1 m radius of the historic pat during subsequent grazing, then the pat was considered contaminated and no longer used in the study.

Grazing of herbage adjacent to the faecal pat was measured through changes in herbage height. Herbage height was measured at 5-cm horizontal spacings in four equiangular directions from the centre of the pat, up to a distance of 50 cm. These measures were made just prior to each grazing (pre-graze), then at 24 hours (day 1), 48 hours (day 2) and 72 hours (day 3, post-graze), at which point cattle were removed from the paddock. This approach provided the measurement of herbage height adjacent to faecal pats at three different stages during the grazing period.

These herbage height measurements were made during the four grazing events that followed the deposition of faecal pats in December, August, October and November. Where faecal pats were deposited in February, April and June, herbage height was measured for five subsequent grazing events. To estimate the apparent herbage removed, the decrease in pasture height was converted to pasture mass using the relationships derived by Webby and Pengelly (1986).

Statistical analyses

The patterns of herbage removal at the end of each day during the three-day grazing periods were analysed using a mixed-model smoother Flexi (Upsdell 1994). These patterns combined data for both month of deposition and land contours.

The mean distance of grazing from the pat centre was calculated for each faecal deposition time and the four grazing events that followed. The occurrence of grazing was defined as a decrease in mean pasture height of >0.5 cm from one day to the next. These distance estimates were analysed with Genstat (2003) using covariates of herbage height at the start and end of grazing and faecal-pat dry weight. Aggregated error terms are presented in the tables as raw data were not available for more detailed analysis.

Results

Grazing conditions

The mean herbage mass of the paddocks during each grazing event is given in Table 1. Herbage mass at the start, during and end of the grazing period tended to be greater during the summer months compared with the rest of the year. Available pasture decreased from day one to day three for all grazing events. Herbage removal over the three-day grazing period ranged from 507 to 1219 kg DM/ha and averaged 817 kg DM/ha. The number of days between grazing events varied according to pasture growth rates, being greater during the slower-growing autumn and winter period compared to spring and summer. When averaged across all of the grazing events, the number of days between pat deposition and first, second, third, fourth and fifth subsequent grazing events were 50, 99, 147, 191 and 228 days respectively.

Effects of grazing pressure on herbage removal

The pattern analysis of herbage removal showed that for the first, second and third grazing following pat deposition, the majority of initial herbage removal occurred furthest away from the faecal pat (Fig.1a-c). As paddock herbage mass decreased and grazing pressure increased during the grazing period, the distance between the focal

Table 1 Average herbage mass (kg DM/ha) of paddocks assessed at the start of the grazing event (pre-graze), after 24 and 48 hours, and after grazing finished (post-graze); and the number of days since the previous grazing event.

Grazing event	Pre-graze	24 hours	48 hours	Post-graze	Days since previous grazing
Dec '02	2831	2550	2315	2156	
Feb '03	2756	2488	2271	1990	50
Apr '03	2352	1916	1673	1481	63
Jun '03	1753	1508	1329	1187	68
Aug '03	1619	1402	1258	1112	62
Oct '03	2075	1721	1436	1157	49
Nov '03	2249	1891	1541	1344	33
Dec '03	2420	1980	1746	1536	35
Jan '04	2951	2497	2107	1732	42
Mar '04	2211	1856	1563	1354	43

Figure 1 Calculated disappearance of herbage (kg DM/ha) in relation to distance from the centre of cattle faecal pats (cm) after cattle grazing for 24, 48 and 72 hours over the 1st to 5th grazing events following faecal pat deposition across all seasons and contour.

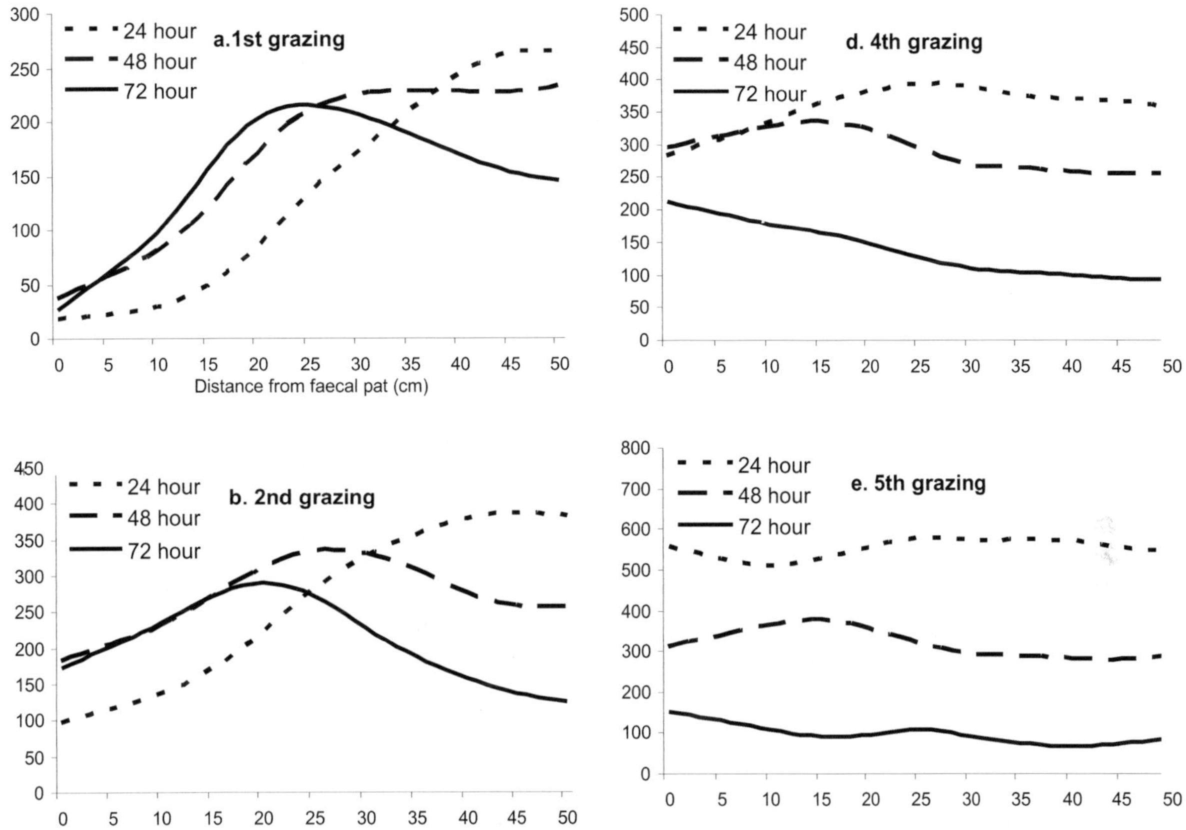

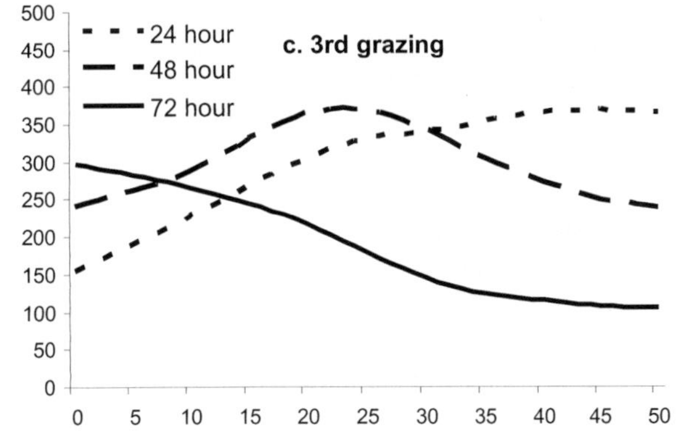

point of herbage removal and the faecal pat decreased. This pattern was not apparent for the fourth and fifth grazing events where herbage removal was uniform across all distances from the faecal pats (Fig.1d-e).

Effect of deposition season and age of pat on herbage removal

Estimates of the mean distance of grazing from the faecal pats during each three-day event is provided in Table 2. When data were combined across all deposition times, the apparent avoidance of faecal pats decreased with each subsequent grazing ($P < 0.05$). The mean distances of grazing from the faecal pat were 21, 14, 10 and 7 cm for the first to fourth grazing events, respectively. However, there were some seasonal differences. It was not until the third grazing of the faecal pats deposited in October and November that distance of grazing significantly reduced. In contrast, the distance of grazing declined sharply during the second and subsequent grazing of pats deposited in April, June and August. Average paddock herbage mass for the second, third and fourth grazing events was 1420 kg DM/ha after two days of grazing and 1260 kg DM/ha at the end of grazing (see Table 1).

Effect of land class and apparent avoidance of herbage

The location of faecal pats on the two land contours had limited impact on the grazing behaviour of the cattle (Table 3). Mean distance of grazing from the pat centre was slightly greater on steep land (>25 degrees) during the first day of grazing, but as grazing pressure increased during the second and third day of the grazing period the difference between pats on easy and steep land disappeared.

Discussion

The patterns of herbage removal that are provided in Figures 1a-e clearly demonstrate that young cattle avoided grazing close to recently deposited faecal pats. This avoidance was most noticeable during the early stages of grazing, particularly during the first and second grazing

Table 2 Mean distance (cm) of grazing from the centre of the faecal pats for each deposition time and each of the subsequent four grazing events with the mean standard error of the difference (mean SED)

	Dec '02	Feb '03	Apr '03	Jun '03	Aug '03	Oct '03	Nov '03
1st grazing	28	22	22	15	17	17	22
2nd grazing	17	16	8	8	10	20	22
3rd grazing	12	9	2	7	11	16	12
4th grazing	9	5	4	9	10	5	10
Mean SED				2.34			

Table 3 Mean distance (cm) of grazing from the centre of faecal pats deposited on either easy or steep land after 1, 2 and 3 days of grazing. Results are the mean of all pat deposition and grazing times with mean standard error of the difference (mean SED).

	Pats on easy land	Pats on steep land
Day 1	17	22
Day 2	12	13
Day 3	7	8
Mean SED	0.99	

events following deposition. As grazing continued and grazing pressure increased during the three days, removal of herbage within the 0-20 cm zone eventually occurred. Grazing pressure appeared to override the natural avoidance of faecal pats when paddock pasture mass declined below 1400-1500 kg DM/ha.

The general patterns of herbage removal also indicated that the impact of faecal pats on grazing behaviour had disappeared by the time of the fourth and fifth grazing following deposition. However, there were some seasonal differences with regard to faecal pat age. The distance of grazing from faecal pats that were deposited in April, June and August was significantly closer to the pats during the second grazing event compared to the first. This was despite the paddocks having a similar post-grazing herbage mass of 1100-1200 kg DM/ha for these grazing events. This result may reflect the more rapid disappearance of faecal pats that occurs under moist conditions (Boom & Sheath 2008; Burggraaf et al. 2009).

In this study, grazing conditions were typical of New Zealand beef-finishing systems and varied greatly through the year. This variation makes the interpretation of seasonal differences somewhat confounded by changes in pasture mass, grazing pressure and duration between grazing events. For those pats that were deposited during October to February, avoidance of grazing around them continued during the second and third subsequent grazing events. While this result may reflect slower disappearance rates of faecal pats during drier periods of the year, it should be noted that herbage mass was higher and grazing pressure was lower during these grazing events. Post-grazing herbage mass ranged between 1500-1700 kg DM/ha. This greater avoidance may also have been exacerbated by the accumulation of lower quality grass seed head and dead material which cattle actively select against (Litherland & Lambert 2007).

There was a small difference in grazing behaviour around faecal pats that were deposited on easy- compared to steep-contoured land. Grazing distance was less on the easy-contoured land after the first day, but this difference disappeared during the subsequent two days. The initial preference of livestock to graze easy-contoured land in mixed topography can explain this result (Sheath 1983).

Given that this study on grazing behaviour was conducted within the context of managing gastrointestinal infection of young cattle, the implications of the results deserve consideration. Larvae of gastrointestinal nematodes are highly concentrated around historical cattle faecal pats (Durie 1961; Rose 1961; Williams & Bilkovich 1973) and these levels can remain high for up to five months following deposition, particularly during dry conditions (Boom & Sheath 2008). Therefore, grazing management practices that do not force young cattle to graze this highly contaminated zone should be considered. While young cattle naturally avoid grazing close to faecal dung pats for at least the two grazing events following deposition, grazing pressure can override this behaviour. When paddock herbage mass falls below approximately 1500 kg DM/ha, young cattle will graze the 0-20 cm zone surrounding pat. Grazing management of young cattle should ensure residuals of at least 1500 kg DM/ha in order to minimise the grazing of highly contaminated herbage that surrounds faecal pats.

Acknowledgements

Appreciation is extended to Anna Brooky for management of data collection and collation. Appreciation is also extended to Catherine Cameron and Martin Upsdell for statistical analysis of data and to Dave Clark and Dave Leathwick for critique of manuscript drafts.

References

Bisset SA 1994. Helminth parasites of economic importance in New Zealand. New Zealand Journal of Zoology 21: 9-21.

Boom CJ, Sheath GW 2008. Migration of gastro-intestinal nematode larvae from cattle faecal pats onto grazable herbage. Veterinary Parasitology 157: 260-266.

Burggraaf VT, Boom CJ, Sheath GW, Brooky AR 2009. The effect of rotational grazing by either calves, cows, lambs or ewes on the removal of herbage contaminated with gastro-intestinal parasite larvae around cattle faecal pats. New Zealand Journal of Agricultural Research 52: 289-297.

Charleston WAG 1997. Internal parasites of cattle in New Zealand: Gastrointestinal nematodes and lungworm. In: Barrell, GK (Ed.) Sustainable control of internal parasites in ruminants - Animal Industries Workshop, June 1997, pp. 23-40.

Durie PH 1961. Parasitic gastro-enteritis of cattle: The distribution and survival of infective strongyle larvae on pasture. Australian Journal of Agricultural Research 12: 1200-1211.

Gruner L, Sauve C 1982. The distribution of Trichostrongyle infective larvae on pasture and grazing behaviour in calves. Veterinary Parasitology 11: 203-213.

Hansen JW, Nansen P, Foldager J 1981.The importance of stocking rate to the uptake of gastrointestinal nematodes by grazing calves. Veterinary Medicine and Animal Science 9: 471-495.

Haydock K.P, Shaw HH 1975. The comparative yield method for estimating dry matter yield of pasture. Australian Journal of Experimental Agriculture and Animal Husbandry 15: 663-670.

Hutchings M R, Gordon I J, Kyriazakis I, Jackson F 2001. Sheep avoidance of faeces contaminated patches leads to a trade-off between intake rate of forage and parasitism in subsequent grazing decisions. Animal Behaviour 62: 955-964.

Litherland AJ, Lambert MG. 2007. Chapter 6. Factors affecting the quality of pastures and supplements produced on farms. In: PV Rattray IM Brookes AM Nicol (ed). Pastures and Supplements for Grazing Animals. New Zealand Society of Animal Production, Occasional Publication No. 14. pp. 81-96.

Nolan T, Connolly J 1980. Research on mixed grazing by cattle and sheep in Ireland. Proceedings of workshop on mixed grazing, Agricultural Institute, Galway, Ireland pp. 1-20.

Rose JH 1961. Some observations on the free-living stages of *Ostertagia ostertagi*, a stomach worm of cattle. Parasitology 51: 295-307.

Sheath GW 1983. Pasture utilisation in hill country 1. Influence of grazing duration and land contour. New Zealand Journal of Experimental Agriculture 11: 309-319.

Smeaton DC 2007. Profitable beef production. Meat and Wool New Zealand Beef Council Publication (3rd edition). pp. 215

Upsdell MP 1994. Bayesian smoothers as an extension of non-linear regression. The New Zealand Statistician 29: 66-81.

Webby RW, Pengelly WJ 1986. The use of pasture height as a predictor of feed level in North Island hill country. Proceedings of New Zealand Grassland Association 47: 249-253.

Weeda WC 1967. The effect of cattle dung patches on pasture growth, botanical composition and pasture utilisation. New Zealand Journal of Agricultural Research 10: 150-159.

Williams JC, Bilkovich FR 1973. Distribution of Ostertagia ostertagi infective larvae on pasture herbage. American Journal of Veterinary Research 34: 1337-1344.

Preliminary investigations into genetic variation in distance travelled by young sheep

Johnson PL*[1], N Cullen[2], SM Hickey[2], K Knowler[1], B Bryson[3], M Hall[3] and P Pletnykov[4]

[1]*Invermay Agricultural Centre, AgResearch Ltd., Private Bag 50034, Mosgiel, New Zealand;* [2]*Ruakura Research Centre, AgResearch Ltd., Private Bag 3115, Hamilton, New Zealand;* [3]*Woodlands Research Station, AgResearch Ltd., Woodlands RD1, New Zealand;* [4]*Lincoln Research Centre, AgResearch Ltd., Christchurch, New Zealand*

**Corresponding author:* Email: tricia.johnson@agresearch.co.nz

Abstract

In pastoral grazing systems, animals need to walk to obtain nutrients (pasture) and water, however, it is also thought that the distance an animal travels is further influenced by factors including the grazing behaviour of the animal, its health status and its general motivation for moving. The purpose of this research was to investigate the repeatability and genetic variation in the trait of distance travelled for New Zealand sheep. Data were collected using global positioning satellite (GPS) collars fitted to 422 animals from three different groups of animals (Group). The animals were allowed *ad libitum* grazing. Progeny from 44 sires were represented in the data set. The GPS collars were fitted for up to 17 days. Given between-day variability in distance travelled depending on paddock attributes (size and feed availability) all data were scaled to a constant mean. A univariate repeated measures model was fitted to the data using ASREML, with Group fitted within the model, and day within Group fitted as the repeated measure. The heritability estimate for daily distance travelled was 0.36±0.09, with a repeatability estimate of 0.51±0.03. The results indicate that distance travelled is a repeatable trait at the individual animal level and is under moderate genetic control.

Keywords: sheep; GPS; distance travelled

Introduction

In pastoral grazing systems, animals need to walk to obtain nutrients (pasture) and water (Coop & Drew 1961). The distance that they travel daily will be influenced by their energy requirements, the type and availability of feed within the area that they are grazing, and their physiological status (Nicol & Brookes 2007). However, other factors are likely to also contribute to the distance that an animal walks daily including social structure (hierarchy), the grazing behaviour of the animal, its health status and general motivation for movement. Walking is an energy expense and equations that predict energy requirements for animals can include specific equations for distance travelled (Nicol & Brookes 2007).

The majority of data sets that consider distance travelled by sheep have been relatively small data sets in the context of the use of landscape features, e.g., Bunyaga et al. (2020) or variability as it relates to climatic variation, e.g., Thomas et al. (2008), and have not considered distance travelled as a trait, and its potential repeatability or genetic origin. There is evidence, however, as reviewed by van Oers and Sinn (2013), that genetic variation exists as to how animals behave and interact with their environment. The purpose of this study was to provide first estimates for the heritability of distance travelled for New Zealand sheep in a pastoral grazing system.

Materials and methods

Trial location and animal type were chosen to ensure that the animals could all be viewed twice daily given that these were newly designed collars, to eliminate risks associated with the animals becoming entangled in fences or other objects. As such, the best resource for this first-stage research was a flat-land property with easy visual inspection of all fences within the paddocks. All data collection was approved by the AgResearch Animal Ethics Committee.

Collars and GPS Data

The collars used were custom-made (manufacturer DataCarter) for use in sheep, with GPS componentry and an 18Ah D cell lithium thionyl battery within 3D printed housing made of thick plastic, with a webbing collar and a plastic clasp used to secure the collar around the neck. The weight of the housing and battery (approximately 290 g) was sufficient for the housing to be held on the underside of the neck of the animal with minimal shifting. The GPS componentry was based on a passive antenna and a UBlox NeoM8 chip. The chips were programmed to record a position when the animal moved more than five meters, or if stationary, a position was recorded at ten-minute intervals. The position was recorded as latitude and longitude coordinates, which were loaded into ArcGIS 9 (ArcMap Version 9.3, USA) for further processing and analysis. The batteries were estimated to last for 18 days, with a draw of 1Ah per day. Data was stored on an onboard SD card, and were downloaded after the final set of data was collected within a contemporary group.

Animals

Data were collected from three contemporary groups of animals within research flocks run at AgResearch Woodlands, Invercargill, New Zealand between October 2018 and March 2020. For all animals, sire and dam information was available, confirmed through DNA parentage testing.

The first data set was based on 2017-born ewes from the AgResearch Progeny Test Research maternal flock (Brito et al. 2017), with data collected on the animals when they were approximately 14 months of age in October 2018. GPS collars were fitted on 300 animals which were randomly allocated to one of three groups (Group 1-3) as only 100 collars were available at any one time. The GPS collars were fitted on a group of animals for one week and then rotated to the next group of ewes. After the collars had been fitted on the third group, collar deployment was repeated starting with the first group. Ultimately each animal had up to two periods of data available (Period 1 and 2) but issues with the collars, GPS componentry and batteries meant only 111 animals had data for both periods. A further 154 animals had data for a single period.

The second data set was based on 2018-born ewe lambs from the AgResearch methane-selection lines (Pinares-Patiño et al. 2013), with data collected on the animals when they were approximately nine months of age in June 2019. The GPS collars were fitted on 100 animals for 17 days, resulting in 74 animals with a full set of data available.

The third data set was based on 2019-born ram lambs from the AgResearch methane-selection lines (Pinares-Patiño et al. 2013), with data collected on the animals when they were approximately five months of age in March February 2020. The GPS collars were fitted on 100 animals for 17 days, resulting in 83 animals with a full set of data available.

Management

All animals were run at AgResearch Woodlands, which is in inland Southland (-45.969738 S, 168.745323 E) and is 34 metres above sea level. The property is nearly exclusively flat and is subdivided into paddocks ranging from less than 0.5 hectares to 3.5 hectares in size. It is in an area of moderate-to-high rainfall year-round (Smith 2012). For these studies, the animals were provided access to several paddocks at one time to allow *ad-libitum* intake for the duration of data collection. The paddocks consisted of ryegrass/white clover swards of high quality (Smith 2012).

Analysis

Data sets were edited to remove records that occurred during the time of turning on and fitting of the collars, together with the movement of animals between the yards and the paddocks and *vice versa*. Data during the grazing period were also processed to remove data points that did not map within the paddock boundaries, however, less than 1% of data points needed to be removed for that reason.

The cleaned GPS data sets were analysed using ArcGIS to determine the distance travelled (sum of individual records) within a 24-hour period starting at midnight the day the collars were fitted and ending at midnight the night before the collars were removed. Although all animals were given *ad libitum* access to feed, there was variability in paddock sizes and differences in offered pasture quality and quantity among groups of animals, even for the same animals during days and weeks. To account for this, the data were expressed as a proportion of their contemporary group (CG) mean multiplied by the global mean for that Group of animals as described by Brito et al. (2017). Data for the 2017-born animals were collected across two time periods, approximately three weeks apart, and for this group of animals CG corresponded to group and week; for the two other groups of animals, all data were collected across a consecutive time period. Variance and covariance components were estimated using restricted maximum likelihood (REML) procedures fitting an animal model in ASReml (Gilmour et al. 2015). The file was structured such that each animal had multiple data points corresponding to the different days for which the collars were fitted. A single repeatability and heritability estimate was generated by fitting day as a repeated measure. Group was fitted as a fixed effect, given the data had been scaled by contemporary group, no additional variables were fitted in the models. Pedigree data for use in the analysis were obtained from Sheep Improvement Limited (Newman 2009).

Results and discussion

As indicated in the materials and methods, not all GPS collars successfully generated data, however, where data sets were generated they were considered to be sound based on producing expected locations and complete data sets for at least 12 days. The average distance travelled (DT) across all data sources was 3.4±0.89 km/day. Phenotypically, there was a large range in coefficients of variation (CV) among days for individual animals, ranging from 4-8% (consistent) to 28-60% (inconsistent) in the different data sets. The largest amount of variation was observed in the first data set where data were collected in two different periods three weeks apart. In all cases, data were not available for all animals on which the collars were fitted. There were several reasons for this loss of data, with a combination of battery connection issues and GPS chip malfunctions reducing the size of the final data set, highlighting that development work is still required to optimise the design of the GPS collars.

For the first data set, there were examples of animals that exhibited consistency in distance travelled between time periods, but some animals were not consistent. Examples of such animals are in Fig. 1, including an example of an animal that consistently travelled low or high distances between the two weeks of measurement respectively, and an animal that was not consistent between the two weeks. A visual representation of examples of animals exhibiting extremes (high and low) for distance travelled per day is in Fig. 2. The animal that was extreme for not travelling long distances averaged 1958 m per day (range 1427 to 2205 m per day), whilst the animal that was extreme for travelling long distances averaged 4534 m per day (range 3366 to 5332 m per day).

Data from the first data set is given in Fig. 3 which compares the average distance travelled for each group for the two measurement periods. Overall, for each of the three groups, there was a good phenotypic correlation between the two data sets for each animal (R^2 range 0.54-0.67). For the data sets collected in subsequent years, there was similar variability between extreme animals, as shown in Fig. 4.

Figure 1 Examples of the distribution of distances travelled per day from the first data set for an animal consistently travelling a low distance per day (left) across the two measurements weeks (W1 or W2), a high distance per day across the two measurement weeks (middle) and one individual that changed the distance travelled per day significantly between the two measurement weeks (right). Each plot represents 12-17 days worth of data.

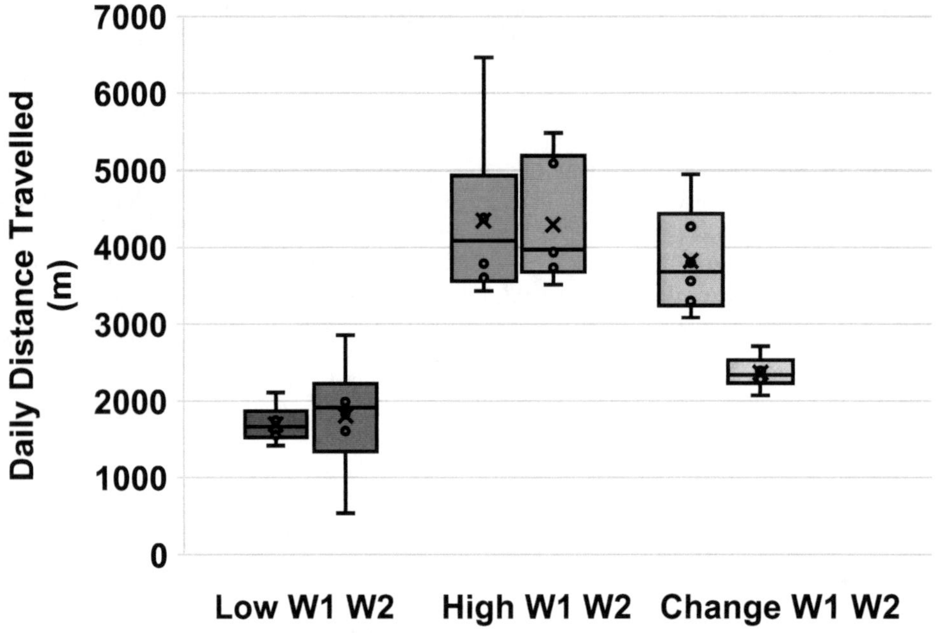

Figure 2 Example of GPS traces for two sheep exhibiting extremes in distance travelled. The animal with the dark-coloured trace travelled on average 1958 m per day (range 1427 to 2205 m per day) whilst in these paddocks, whilst the animal with the light-coloured trace travelled on average 4534 m per day (range 3366 to 5332 m per day).

Figure 3 Raw phenotypic correlations between distances travelled between the first and second periods of measurement (seven-day measurement periods separated by three weeks) for three groups of animals. a) 1st group of animals, R^2=0.67; b) 2nd group of animals, R^2=0.61; c) 3rd group of animals, R^2=0.54.

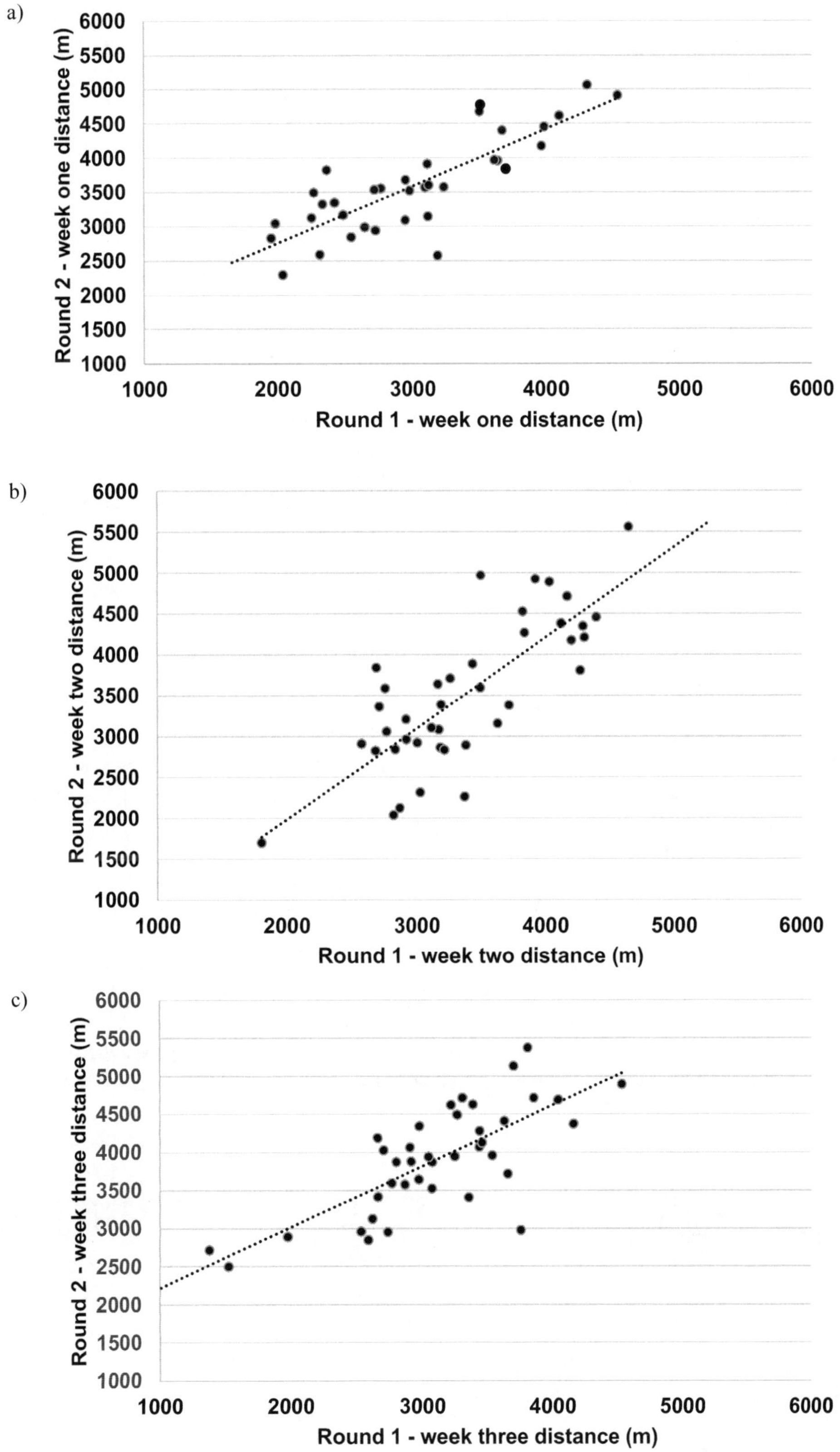

Whilst some technical issues limited the size of the data set (compared to that originally targeted), the amount of data was still sufficient to estimate the heritability and repeatability when all data were considered as a repeated measure in the same analysis. The heritability estimate for daily distance travelled was 0.36±0.09, with a repeatability estimate of 0.51±0.03. The design of studies was targeting the estimation of the repeatability of distance travelled, which means that the heritability estimate benefited from multiple measures on each individual within the data set. The results demonstrate that the trait of distance travelled is under genetic control and is a repeatable (across days within a period of weeks) attribute of an individual.

Whilst several papers describe the distances covered by sheep in different environments (Thomas et al. 2008; Bunyaga et al. 2020), this is the largest data set to date. This has allowed the determination of the repeatability of the trait of distance travelled, and the ability to generate a preliminary heritability estimate. No comparable data sets have been found for any production animal species. A review by Fogarty et al. (2018) reported that most GPS data sets in production animals are based on low numbers of animals without pedigree records. Increased reliability of GPS chips and batteries would have increased the size of the data set. Despite this, the data were adequate to provide accurate repeatability and heritability estimates. There are, however, factors that appear to affect the repeatability of the distance travelled by individual animals, with some animals exhibiting significantly different walking distance behaviours between the two time periods. It is known that disease conditions such as parasites or other sub-clinical issues will impact on distances that animals walk, and this is used as a method for identifying sub-clinically sick animals using technologies such as pedometers.

Given that distance travelled has an energy cost associated with it, the large (over two-fold variation) in distances covered by different animals highlights the need to incorporate such trait data when considering traits such as residual feed intake, where the most common models only account for live weight and liveweight gain. The trait of residual feed intake is the error term in a model and, therefore, traits like distance travelled could account for some of this error.

Conclusion

This data set is the most comprehensive data set to date to consider the distance travelled by individual sheep in a pastoral grazing system. The results demonstrate that distance travelled is a repeatable trait at the individual animal level, but that it can be influenced by factors thought to include parasite burden or other sub-clinical illnesses. The results also demonstrate that it is a moderately heritable trait. Genetic variation in walking distance will influence energy partitioning and potential intake. The potential for GPS data to reveal further information about grazing behaviour and how animals interact with their environment will be the next stage of analysis on the existing data set.

Acknowledgements

Funding for this series of studies was sourced from AgResearch Strategic Science Initiative Funding.

References

Brito LF, McEwan JC, Miller SP, Bain WE, Lee M, Dodds KG, Newman S-A, Pickering NK, Schenkel FS, Clarke, SC 2017. Genetic parameters for various growth, carcass and meat quality traits in a New Zealand sheep population. Small Ruminant Research 154: 81-91.

Figure 4 Examples of the distribution of distances travelled per day over ten to seventeen days from two different sets for animals consistently travelling a high distance per day (left within data set), or a low distance per day (right within data set). Data set two animals were ewe lambs from the AgResearch methane selection lines, with data collected on the animals when they were approximately nine months of age. Data set three animals were ram lambs from the AgResearch methane selection lines, with data collected on the animals when they were approximately five months of age.

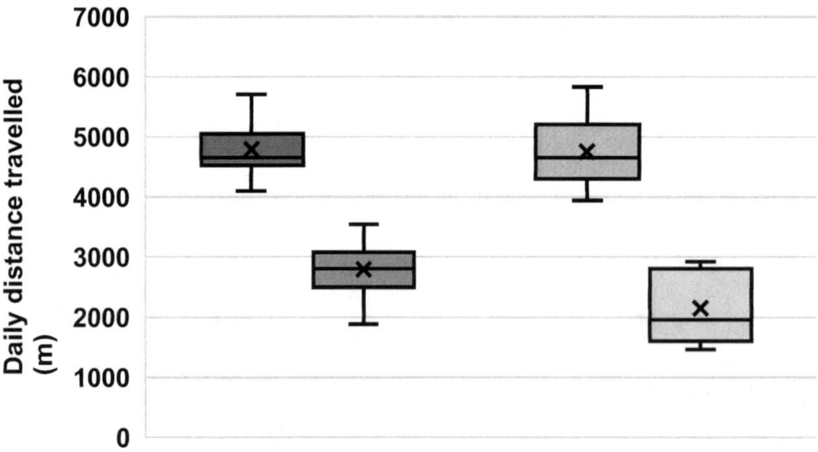

Bunyaga AS, Corner-Thomas RA, Burkitt LL, Draganova I, Kenyon PR 2020. BRIEF COMMUNICATION: The behaviour of sheep around a natural waterway. New Zealand Journal of Animal Science and Production 80: 124-127.

Coop IE, Drew KR 1961. Maintenance and lactation requirements of grazing sheep. Proceedings of the New Zealand Society of Animal Production 21: 53-62.

Fogarty ES, Swain DL, Cronin G, Trotter M 2018. Autonomous on-animal sensors in sheep research: A systematic review. Computers and Electronics in Agriculture 150: 245-256.

Gilmour AR, Gogel BJ, Cullis BR, Thompson R 2015. ASReml User Guide Release 4.1. VSN International Ltd, Hemel Hempstead, Hertfordshire, UK.

Newman S-AN 2009. Sheep Improvement Limited (SIL) – the first 10 years. Introductory remarks. Proceedings of the New Zealand Society of Animal Production 69: 148.

Nicol AM, Brookes IM 2007. The metabolisable energy requirements of grazing livestock. In: Rattray PV, Nicol, AM, Brookes, IM eds Pasture and supplements for grazing dairy cattle New Zealand Society of Animal Production, Hamilton New Zealand Occasional Publication 14. pp 151-172.

Pinares-Patiño CS, Hickey SM, Young EA, Dodds KG, MacLean S, Molano G, Sandoval E, Kjestrup H, Harland R, Hunt C, Pickering NK, McEwan JC 2013. Heritability estimates of methane emissions from sheep. Animal 7: 316-321.

Smith LC 2012. Long term pasture growth patterns for Southland New Zealand: 1978 to 2012. Proceedings of the New Zealand Grassland Association 74: 147-152.

Thomas DT, Wilmot MG, Alchin M, Masters DG 2008. Preliminary indications that Merino sheep graze different areas on cooler days in the Southern Rangelands of Western Australia. Animal Production Science 48: 889-892.

van Oers K, Sinn DL 2013. Quantitative and molecular genetics of animal personality. In: Carere C, Maestripieri D, eds. Animal Personalities: Behavior, Physiology, and Evolution. Chicago, IL: The University of Chicago Press pp 149-200.

Defecation behaviour reduces horse's utilisation of pasture

BR Adams[a], EK Gee[a], PJ Back[ab] and CW Rogers[ab]

[a]*School of Veterinary Science, Massey University, New Zealand;* [b]*School of Agriculture and Environment, Massey University, New Zealand*

**Corresponding author:* Email: b.adams@massey.ac.nz

Abstract

Horses kept at pasture create focal areas with high-density of faeces (latrines), which under commercial conditions can account for 30% of total pasture area. This New Zealand study observed the utilisation of pasture, and establishment of lawns and roughs within one paddock, intensively grazed (semi-set-stocked) alternately by three herds of mixed-age (mature), predominately Standardbred, horses. The paddock was grazed by horses 480 of 730 days, with mean stocking density of four horses per ha (2148 kg live weight/ha). Median grazing duration was 44 days (IQR 37-47 days). The median overall pasture cover for the study period was 1114 kg DM/ha (range 496-2191 kg DM/ha), lawns was 683 kg DM/ha (range 389-1822 kg DM/ha), and roughs was 2058 (range 753-3971 kg DM/ha). There was a trend for greater post-grazing estimated pasture growth rates in roughs relative to the lawns (P=0.054). Latrine areas constituted a median 31% (IQR 24-33%) of available pasture area, with up to two faecal-piles/m^2. The results of the current study suggest that intensive semi-set-stocked management promotes the establishment of easily identified latrine areas of moderate faecal-pile concentration. These data can contribute to modelling the nitrogen-holding capacity of the soil, nitrogen leaching and the efficient utilisation of pasture of grazing horses.

Keywords: equine; faeces; nitrogen; pasture

Introduction

Commercial farming enterprises must comply with a nutrient management plan (NMP) as defined by the respective regional councils. Traditionally equine livestock and equine enterprises have not been included within the NMP requirements due to their relative limited number and contribution to nutrient loss compared to other large-scale farming enterprises (Maseyk et al. 2018; Rogers et al. 2020a). In 2019, the Waikato Regional Council's Healthy Rivers Plan Change 1 (PC1) investigated the inclusion of horses and equine farms within the NMP and the requirement to calculate baseline N losses using Overseer™. Preliminary work identified that the assumptions used in Overseer™ to model horse nitrogen utilisation, and published literature describing nutrient loss from equine properties in New Zealand, had a number of limitations (Chin et al. 2019a; Chin et al. 2019b).

In contrast to many other countries, the temperate climate of New Zealand allows horses to be maintained at pasture year-round. This difference in management system often prevents the direct translation of overseas data generally derived from European intensive systems. The use of pasture within the New Zealand commercial breeding and racing industry has been described (Bengtsson et al. 2018; Rogers et al. 2018; Rogers et al. 2020b). However, there remains a lack of data on the protein digestion and characteristics of nutrient loss for horses at pasture (Chin et al. 2019a).

The characteristics of the commercial breeding sector of the New Zealand equine industry have been summarised in a number of reviews (Rogers et al. 2016; Rogers et al. 2017; Rogers et al. 2020b). Briefly, the data has described moderate stocking densities (~2 horses/ha) during the off season (non-breeding season) with a doubling of stock numbers during the breeding season. During the breeding season stocking rates varied from 1.2 mares/ha to 5.9 mares/ha with mares managed in consistent cohorts with a semi-set-stocked management system (Bengtsson et al. 2018). Total pasture DM on offer has been reported to range from 911 kg DM/ha for non-pregnant mares through to 3823 kg DM/ha for pregnant mares. These reports may not accurately reflect pasture on offer, as pasture utilisation by horses is influenced by selective grazing and latrine behaviour (Odberg & Francis-Smith 1976). These behaviours result in the establishment of preferentially grazed areas (nutritive areas/lawns) with avoidance of areas that are used for excretion of faeces and urine (roughs or eliminative/latrine areas) possibly limiting utilisation by 10 to 30% of effective pasture area (Rogers et al. 2017; Rogers et al. 2018). Early studies estimated that the pasture utilisation on Thoroughbred stud farms could range from 48-80% (Goold et al. 1988; Odberg & Francis-Smith 1976). More-recent on-farm calculations derived from feed budgets has refined this estimate to 70% (Rogers et al. 2017). Similar data for horses other than the commercial breeding industry in New Zealand are scarce (Fernandes et al. 2014; Verhaar et al. 2014).

Further compounding this issue of limited pasture-centric data is the greater numerical representation of the sport and recreation sector within the New Zealand horse population, estimated at ~55% of the 100,000 horses in New Zealand (Rogers et al. 2017). In contrast to the commercial racing and breeding industries, horses within the sport and recreation sector are described as having a thriftier phenotype and management often focuses on relatively high stocking density and limited access to pasture, or provision of limited pasture mass (Fernandes et al. 2014; Goh et al. 2020; Parvage et al. 2013). It is important to obtain accurate data on pasture utilisation and the pattern of defecation of horses managed under these restrictive conditions, to use when quantifying nutrient loss for the sport and recreation sector of the New Zealand equine industry. The current study reports on an observational trial examining pasture utilisation and the pattern of defecation with groups of thrifty phenotype horses managed under conditions representative of those used by the recreational equine sector.

Materials and methods

An observational study was conducted from August 2017 until August 2019 at the Massey University Veterinary Large Animal Teaching Unit (LATU), which is a farm comprised of 57.9 ha (effective 42.1 ha) of flat-to-rolling river terrace, grazing sheep, cattle and horses for the purposes of teaching. The soil type of the farm is predominately Tokomaru silt loam.

Horses

Twenty-four horses (19 Standardbred, three Thoroughbred, one Kaimanawa, and one Stationbred) of which 14 were mares and 10 geldings that were managed as three groups of n=7, 8, & 9 horses, subsequently referred to as G1, G2 and G3 respectively. The horses were mean 13 ± 4 years of age at the start of the trial and were kept at pasture 24 hours/day all year.

Body condition was monitored by the same technician monthly using a modified Henneke (1983) nine-point scale. Live weight (LW) was recorded approximately fortnightly using a Tru-Test weighing system (MP 800 load bars, XRP2 chip panel reader, XR5000 weigh scale indicator). The use and care of the horses was monitored by the Massey University Animal Ethics Committee.

Pasture and supplementation

Observational data were collected from a predominately flat 1.96 ha river terrace paddock with soil type Rangitikei loamy sand specifically family Shalimah*f* (S-Map version 3.1.224 (Manaaki Whenua Landcare Research, 2020)). At the start of the trial there had been no pasture improvement in five years, and the sward consisted of several grass species (predominately perennial ryegrass (*Lolium perenne*), but also cocksfoot (*Dactylis glomerata*), fescue (*Festuca arundinacea*)), white clover (*Trifolium repens*), and weeds (typically dock (*Rumex obtusifolius*) and buttercup (*Ranunculus repens*)). A broadleaf herbicide (Conquest, NuFarm) was applied during the first summer of observation (February 2018). Fertiliser (30% potash super, Ravensdown) was applied at a rate of 450 kg/ha in April 2018 and January 2019.

Farm pasture growth (kg DM/ha) was estimated using the pasture cut from three exclusion cages (one cage per group of horses as they rotated through paddocks, including the study paddock). Each cage was placed in areas representative of average pasture cover, sufficient to measure a 30 x 60 cm quadrat from which pasture cuts were collected monthly. Pasture samples were cut to ground level, to mimic the grazing of lawns, and assessed for botanical composition using previously described methods (Bengtsson et al. 2018), and the remaining unsorted pasture was also dried to obtain pasture growth rates. Pasture mass was measured using a rising plate meter (RPM) every week by the same technician using an L-shaped transect. Pasture mass was estimated for total pasture, lawns and roughs using the Jenquip equation for Manawatu (Jenquip 2017):

Average RPM reading x 158 + 200 =
estimated pasture cover (kg DM/ha)

The diet of the horses was supplemented with hay from May to October each year: an estimate of quantity offered was recorded as part-bale/horse/day. Core samples were taken from the middle of several bales of hay each year for DM calculation. Random bales were routinely weighed to generate an estimated average bale weight.

Faeces

When the paddock was stocked with horses, a representative diagonal transect of 160 m was used to record weekly faecal distribution every 10 m. The number of faecal piles contained within a circle of 1-m radius were counted, centred on each 10-m point (method derived from Hirata et al. 2011). The faecal density measurement was cumulative as no faeces were removed, therefore, old and new faecal piles were recorded. The count data was then converted to per m^2 density. Latrine formations were mapped monthly while the paddock was stocked with horses, based on a grid referenced from painted fence-posts positioned at 40-m intervals (adapted from Edwards & Hollis 1982), then overlaid within Google My Maps© to generate an area measurement.

Statistical analysis

Data were analysed for normality using Shapiro-Wilk's test. Simple descriptive statistics are presented as mean ± standard deviation for parametric and median and inter-quartile range for non-parametric data. Maximum and minimum values are reported when emphasis is on the spread of the data. Differences in mean seasonal botanical composition were tested using Tukey's test and differences between pasture growth rates (roughs and lawns) used the paired t-test. Analysis was conducted using R version 4.0.3.

Results

During the observation period, the paddock was used by all mobs for three rotations each; total time spent grazing in the paddock was 209 days, 159 days and 112 days for G1, G2 and G3 respectfully.

Horses

The mean ± SD horse LW was 537 ± 56 kg and the median BCS was 6/9 (IQR 5 – 7). The mean total LW of each mob of horses over the two-year period was 3818 ± 94 kg, 4511 ± 102 kg and 4444 ± 179 kg (G1, G2 and G3 respectively). Estimates of the total DM on offer (including hay when provided) ranged from 1.8 kg DM/horse/day to 26.4 kg DM/horse/day (~ 0.3% to 5.3% LW).

Figure 1 Grazing management in one paddock during a 730-day observation period on a teaching farm. Pasture (kg DM/ha) is represented as total (square), roughs (diamond) and lawns (circle), and stocking rate (kg LW/ha) represented with cattle (dark grey bars), horses (light grey bars). A guide to season is provided along the top of the chart (bars become darker from autumn through to summer).

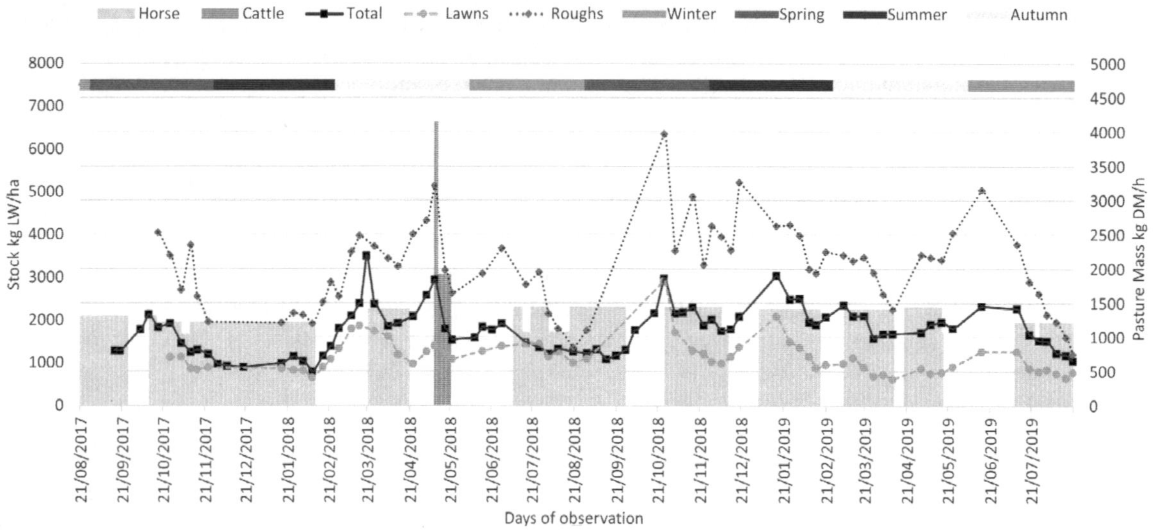

Figure 2 Comparison of the post-grazing pasture growth rates in a paddock grazed by horses (kg DM/ha/day estimated by rising plate meter) showing total (light grey bar), roughs (dark grey bar) and lawns (grey bar). Overall farm pasture growth rate (grey line) measured by quadrat cuts.

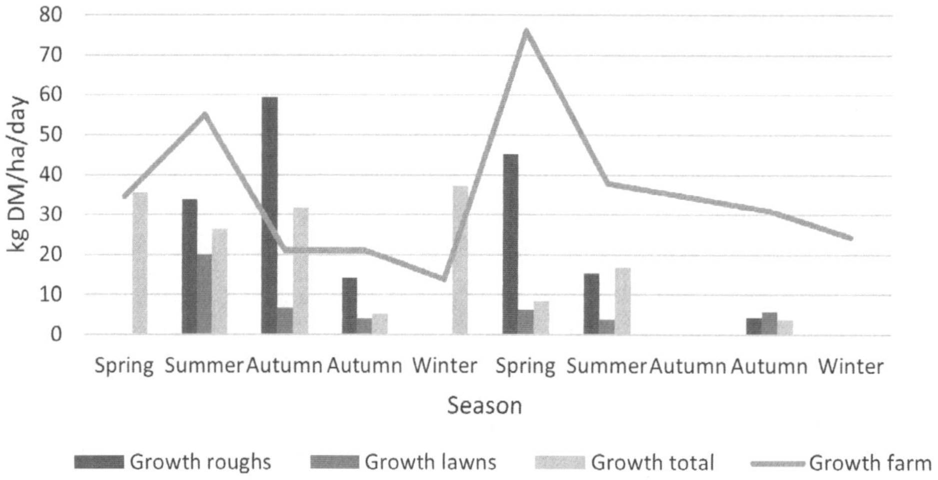

Pasture management

The pattern of horse and ruminant (cattle) grazing during the observation period is presented in Fig. 1. The horses rotated through the observation paddock nine times over the study period with a median grazing period of 44 (range 29-123; IQR 37-47) days, median resting period of 25 (IQR 17-42) days, and the paddock was grazed for 66% of the observation period (480/730 days). Mean stocking density was 4.1 ± 0.4 horses/ha (~ 2202 ± 215 kg LW/ha) in year one and 3.8 ± 0.4 horses/ha (~ 2041 ± 215 kg LW/ha) in year two; 4 ± 0.5 horses/ha (~ 2148 ± 269 kg LW/ha) over both years.

There was one period of cross-grazing by cattle (Fig. 1). In May 2018 (day 263) at the completion of a period of semi-set-stocking and a 19-day non-grazing period, twenty-six mixed age cows grazed for three days (~ 6633 kg LW/ha). The mob size was then reduced to 12 cows which grazed the pasture for a further 10 days (~ 3061 kg LW/ha). During this period, total and roughs pasture mass were reduced by 48% (1843.2– 962.4 kg DM/ha) and 49% (3217.8 – 1643.5 kg DM/ha) respectively, with a 22% reduction in pasture mass of lawns (874.1 – 681.2 kg DM/ha) during this period.

Pasture measurement

Median pasture cover during the observational period for the total area was 1114 (IQR 811 - 1307) kg DM/ha, for the lawns 683 (IQR 544 – 803) kg DM/ha and roughs

Table 1 Seasonal botanical composition of pasture harvested from an exclusion cage in a paddock grazed by horses. Data presented as mean ± standard deviation of dry matter percentage of sample.

	Spring	Summer	Autumn	Winter
Grass % DM	71 ± 19%	60 ± 21%	54 ± 11%	64 ± 17%
Clover %DM	11 ± 12%	9 ± 11%	0 ± 0%	6 ± 7%
Weeds % DM	15 ± 12%^{ab}	7 ± 5%^{a}	17 ± 6%^{ab}	21 ± 9%^{b}
Dead matter %DM	3 ± 2%^{a}	24 ± 11%^{ab}	29 ± 15%^{b}	9 ± 13%^{ab}

^{ab} Different superscripts within a row are significantly different (P<0.05)

2058 (IQR 1596 - 2352) kg DM/ha (Fig. 1). Median total pre-grazing mass was 1760 (IQR 1472 - 1859) kg DM/ha. Estimated median seasonal pasture growth peaked at 59 kg DM/ha/day (IQR 36-70 kg DM/ha/day) in summer and the lowest at 12 kg DM/ha/day (IQR 7-16 kg DM/ha/day) during winter. Growth during spring was 52 kg DM/ha/day (IQR 43-59 kg DM/ha/day) and autumn was 17 kg DM/ha/day (IQR 13-26 kg DM/ha/day).

Pasture mass in the preferred areas of grazing (lawns) remained moderate throughout the observation period and was consistently at, or below, 1000 kg DM/ha with limited pasture mass in the lawns gained before the next period of grazing. At the completion of each rotation, lawns were below 800 kg DM/ha (median 554 kg DM/ha), with the lowest recorded as 390 kg DM/ha. Pasture growth rates after grazing differed between lawns and roughs despite occasions when they had similar post-grazing pasture mass (Fig. 2). There was a trend for greater post-grazing estimated pasture growth rates in roughs relative to the lawns (P=0.054).

The seasonal pasture composition during the observation period is presented in Table 1.

Faecal measurement

Latrine areas were first mapped on Day 58 and consisted of 11% of the paddock. During the two-year observation period the latrine areas consisted of a median 31% (IQR 24-33%) of the pasture area.

Faecal density within the latrine areas along the transect approached a maximum of two faecal piles/m^2 at some sites which were in the centre of the latrine area (Fig. 3). The lower densities recorded were in the transitional areas between latrine and lawn. At least one faecal pile was observed at median 29% (IQR 24-35%) of the measurement points across the 160-m faecal transect.

Discussion

At present there are limited data published on the utilisation of pasture and the pattern of faecal defecation for non-commercial horses managed at pasture under New Zealand's temperate climate. The recording of such information is necessary to provide valid input parameters for regional councils when modelling nutrient turnover on equestrian properties within Overseer™. In contrast to commercial horse-breeding operations, where optimisation of growth rate is a key production goal, many owners of non-commercial or recreational horses are focused on maintaining or reducing weight gain, and thus, manage horses on relatively limited effective pasture area (Fernandes et al. 2014). The pasture management data of the "thrifty phenotype" teaching horses in the current study are believed to provide a suitable model to reflect the management of the recreation horse sector.

The study paddock was grazed by horses for 480 of 730 days, with grazing periods of 29 to 123 days, which is described as semi-set-stocked. The stocking density of four horses/ha (2148 kg LW/ha) was higher than that generally reported on commercial farms and similar to that of competition sport horses which are predominately kept on smaller lifestyle properties with stocking density 4.5

Figure 3 Profile plot of the spatial distribution of median and maximal faecal density (m^2) measured every 10 m across a 160-m transect within a paddock predominately grazed by horses during a two-year observation period.

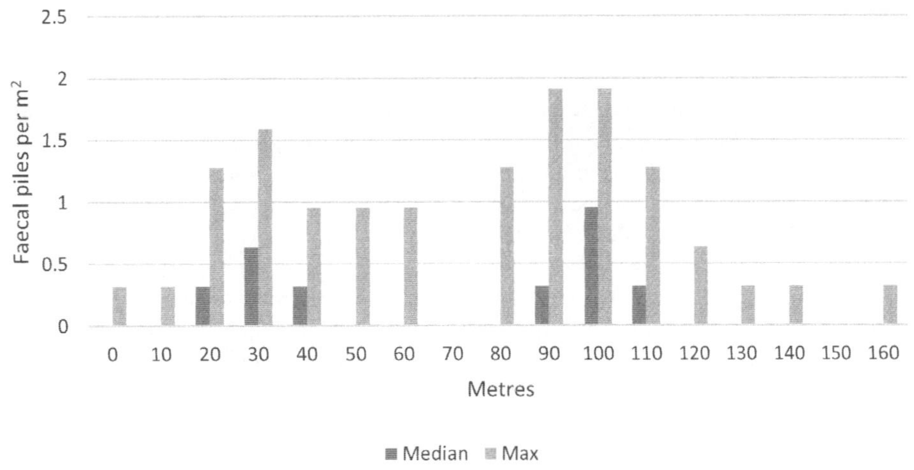

horses/ha (Verhaar et al. 2014; Rogers et al. 2007). The management of the sport horses was different with most at pasture for 12 hours/day, rather than the 24 hours of the teaching horses. A study from Sweden also reported higher stocking densities of competition sport horses with densities of up to 8.7 livestock units (LSU)/ha (1 horse = 0.8 LSU therefore ~ 10- 11 horses/ha) (Parvage et al. 2011). A currently recommended stocking density in Sweden is three horses/ha (2.4 LSU), based on limiting the nutrients (mainly phosphorous) added to pasture through manure to the soil (currently set at 22 kg ha^{-1} yr^{-1}), rather than pasture sustainability (Parvage et al. 2013).

The differences in the management systems between competition and recreation horses makes direct comparison of nutrient loading across systems difficult. Similarly, comparison with other production species is difficult, however, the estimated stocking density reported for competition/recreation horses in New Zealand appears to be higher than that for dairy cows and sheep. DairyNZ figures show a national average stocking density of 2.9 dairy cows per ha, with mature crossbred cows having an average weight of 467 kg resulting in total LW of 1,354 kg/ha (LIC & DairyNZ 2019). National stocking rates of sheep is reported as 6.1 SU/ha with SU equating to a 60-kg mixed-age ewe resulting in total LW of 366 kg/ha (Beef & Lamb 2021).

The pre- and post-grazing pasture mass was lower than those reported for Thoroughbred breeding farms (Bengtsson et al. 2018; Rogers et al. 2007) and for sport horses (based on sward height) (Verhaar et al. 2014). Pasture mass in the current study was similar to parameters reported for commercial Standardbred farms and miniature horses, possibly reflecting the relatively "thrifty phenotype" of the standardbred (Rogers et al. 2020b), and the need for miniature-horse owners to focus on preventing weight gain (Goh et al. 2020). Fernandes et al. (2014) survey results reinforced a focus on preventing weight gain and restricting pasture access in pony-club horses. The sustained grazing pressure combined with the low pre-grazing mass of both the totals and the lawns in this study may explain the limited capability of the pasture to recover after grazing and the lower percentage of grasses and the higher relative weed count in the botanical composition. The increased dead matter percentage in summer and autumn seen in the exclusion cage in this paddock may reflect composition in the roughs, as the protected pasture is able to mature, which would not occur in the closely grazed lawns.

The proportion of pasture classified as roughs/latrines was similar for both the mapping and for the transect measurement and, thus, reinforces that the effective pasture on offer due to horses' preference for grazing lawns is 70% of the pasture mass on offer. The 30% that is not utilised by horses provides pasture mass that could be partially utilised by ruminants, and this was demonstrated by the preferential grazing by cattle of the roughs rather than the lawns. Cross grazing with ruminants is common practice on commercial equine properties (Bengtsson et al. 2018) and may be commonly utilised within non-commercial horse-management systems. A survey of horse properties listed in a New Zealand agricultural database found that 90.9% (698/768) had other livestock; 80.2% (560/698) and 63.0% (440/698) of those contained cattle and sheep respectively, but there was no specific information gathered about their use for cross-grazing (Rosanowski et al. 2012). However, Fernandes et al. (2014) reported that 64% of pony-club owners cross-grazed horse paddocks with ruminants, and 46% (92/198) of miniature-horse owners that had pasture-restriction strategies also utilised mixed-grazing (Goh et al. 2020), but the authors did not identify the species used or the frequency.

Total pasture mass in the latrine areas tended to peak early in the grazing rotation then decrease towards the end of rotation. This would indicate that increasing pressure on the pasture supply resulted in the horses grazing the boundaries of the latrines with a subsequent increase in density of defecations in the central areas. The greatest density of faeces in the centre of the latrine area and decreasing density at the boundaries of the latrine supports this hypothesis and demonstrates how the sites of defecation are relatively consistent, with some or limited spread. The persistence of the areas of lawns and roughs reflects unequal distribution of faeces and thus nutrients within the pasture (Fig. 3). Greater nutrient concentration in the roughs, in combination with greater post-grazing residual, may explain the greater post-grazing growth rates observed in the roughs. Data from Sweden, which has very different climate and management systems to New Zealand, did indicate that excretion areas contained approximately twice the soil concentrations of phosphorus and nitrogen than did the grazing areas (Parvage et al. 2013). At present the lack of data on soil-nutrient distribution due to the presence of lawns and roughs in New Zealand is a limitation with modelling of nutrient turnover within equine pastures. New research is needed to understand what is happening with the soils under these pastures and the potential for nutrient leaching. The collection of such data would contribute to accurate estimate of nutrient management using Overseer™ on equine properties.

Conclusion

Recreation horses are often managed at a relatively high stocking density, intensively and with low pasture mass on offer, to restrict feed supply. Under these conditions, roughs or latrine areas of pasture persisted and accounted for 30% of the effective pasture area. Grazing preference remained for the lawns, the data indicating horses would consume all daily growth in the lawns before starting to graze the outer perimeter of the roughs. Under these management conditions, pasture grazing mass was often reduced to low levels, particularly in lawns and this has implications for pasture recovery, growth and quality. Faeces accumulated in consistent and well-defined latrine areas have the potential to increase the nutrient load and may affect the nitrogen-leaching potential of the soil.

Acknowledgements

The authors would like to thank Mike Reilly, Jalissa Gage, Andrew Smith, and Daniel Sneddon (past and present staff at LATU), for their assistance in gathering data.

References

Beef and Lamb NZ 2021. Benchmarking Tool. Beef and Lamb NZ, viewed 15 April 2021. https://beeflambnz.com/data-tools/benchmarking-tool.

Bengtsson J, Rogers CW, Back PJ, Emanuelson U, Roca J, Gee EK, Bolwell CF 2018. Characteristics of the grazing and farm management of broodmares on commercial Thoroughbred stud farms during spring. New Zealand Journal of Animal Science and Production 78: 88-91.

Chin YY, Back PJ, Gee EK, Rogers CW 2019a. Deterministic modelling of nitrogen utilisation by horses managed under pasture-based, intensive and semi-intensive systems with different levels of pasture intake. New Zealand Journal of Animal Science and Production 79:1-7.

Chin YY, Back PJ, Gee EK, Rogers CW 2019b. Livestock and pasture management on commercial Thoroughbred breeding farms: implications for estimating nitrogen loss. New Zealand Journal of Animal Science and Production 79: 65-70

Edwards PJ, Hollis, S 1982. The distribution of excreta on a New Forest grassland used by cattle, ponies and deer. Journal of Applied Ecology 19: 953-964.

Fernandes KA, Rogers CW, Gee EK, Bolwell CF, Thomas DG 2014. A cross-sectional survey of rider and horse demographics, and the feeding, health and management of Pony Club horses in New Zealand. Proceedings of the New Zealand Society of Animal Production, 74: 11-16.

Goh SY, Gee EK, Thomas DG, Back PJ, Rogers CW 2020. Miniature horses in New Zealand: demographics and feeding management. New Zealand Journal of Animal Science and Production 80: 1-7.

Goold GJ, Baars JA, Rollo MD 1988. Management of thoroughbred stud pastures in the Waikato. Proceedings of the New Zealand Grassland Association 49: 33-36

Hirata M, Ogura S, Furuse M 2011. Fine-scale spatial distribution of herbage mass, herbage consumption and fecal deposition by cattle in a pasture under intensive rotational grazing. Ecological Research 26: 289-299.

Jenquip, 2017. Handbook Manual Platemeter. Retrieved 14 August 2017. https://www.jenquip.nz/wp-content/uploads/2019/05/Handbook-Manual-Generic.pdf

Livestock Improvement Corporation Limited & DairyNZ Limited 2019. New Zealand Dairy Statistics 2018-2019. DNZ30-008. Viewed 15 April 2021. 55 p. https://www.dairynz.co.nz/media/5792471/nz_dairy_statistics_2018-19_web_v2.pdf

Manaaki Whenua Landcare Research 2020. Soil report. Landcare Research, viewed 27 February 2021. https://smap.landcareresearch.co.nz/_

Maseyk F, Brown M, Taylor P 2018. Regional council uses of OVERSEER. A review of the use of the OVERSEER nutrient budget model for nutrient management in regional planning. The Catalyst Group Contract Report No. 2018/140 prepared for the Parliamentary Commissioner for the Environment, viewed 13 August 2021. https://www.pce.parliament.nz/media/196490/regional-council-uses-of-overseer-catalyst-group-report.pdf

Odberg FO, Francis-Smith K 1976. A study on eliminative and grazing behaviour - the use of the field by captive horses. Equine Veterinary Journal 8: 147-149.

Parvage MM, Kirchmann H, Kynkäänniemi P, Ulén B 2011. Impact of horse grazing and feeding on phosphorus concentrations in soil and drainage water. Soil Use and Management 27: 367-375.

Parvage MM, Ulén B, Kirchmann H 2013. A survey of soil phosphorus (P) and nitrogen (N) in Swedish horse paddocks. Agriculture, Ecosystems and Environment 178: 1-9.

Rogers CW, Gee EK, Firth EC 2007. A cross-sectional survey of thoroughbred stud farm management in the North Island of New Zealand. New Zealand Veterinary Journal 55: 302-307

Rogers CW, Gee EK, Bolwell CF 2016. Horse Production. In Stafford, K (Ed.). Livestock Production in New Zealand. Massey University Press. pp 250-279.

Rogers C, Gee E, Back P 2017. Pastures and Supplements in New Zealand commercial equine production systems. In: Rattray PV, Brookes IM, Nicol AM. (Eds.). Pasture and supplements for grazing animals. New Zealand Society of Animal Production. pp 303-315.

Rogers CW, Gee EK, Back PJ, van Zon S, Hirst RL, Pomroy WE 2018. Pasture use and management in commercial equine production systems. New Zealand Journal of Animal Science and Production 78: 40-44

Rogers C, Back P, Gee E, Ying Chin Y, Linton S, Wark A 2020a. Predicting nutrient loss – what to do with equine properties? In: Nutrient Management in Farmed Landscapes. Christensen CL, Horne DJ, Singh R). Occasional Report No. 33. Farmed Landscapes Research Centre, Massey University, Palmerston North, New Zealand. Viewed 24 August 2021. 7 p. http://flrc.massey.ac.nz/publications.html.

Rogers CW, Herbst A, Langhans K, Stone R, Hirst R, Gee EK, Back, PJ 2020b. The management of Standardbred breeding stock on commercial properties. New Zealand Journal of Animal Science and Production 80: 84-89.

Rosanowski SM, Cogger N, Rogers CW, Benschop J, Stevenson MA, 2012. A description of the demographic characteristics of the New Zealand non-commercial horse population with data collected using a generalised random-tessellation stratified sampling design. Preventative Veterinary Medicine 107: 242-252.

Verhaar N, Rogers CW, Gee EK, Bolwell CF, Rosanowski SM 2014. The feeding practices and estimated workload in a cohort of New Zealand competition horses. Journal of Equine Veterinary Science 34: 1257-1262.

The effect of feeding fodder beet or kale during winter on growth and behaviour of rising-one-year-old dairy heifers

P Harris[a], DE Dalley[b] and RH Bryant[c]

[a]168 Beck Road Balfour, 9779, New Zealand; [b]DairyNZ Ltd, PO Box 85066, Lincoln University, Lincoln 7647, New Zealand; [c]Faculty of Agriculture and Life Sciences, PO Box 85084, Lincoln University, Lincoln 7647, Christchurch, New Zealand

**Corresponding author:* Email: dawn.dalley@dairynz.co.nz

Abstract

While fodder beet (FB) has been used by the dairy industry for winter grazing over the last 10 years, there is no published information on its effect on heifer performance. Farmers have recently expressed concern regarding the suitability of fodder beet for wintering growing dairy heifers due to the low crude protein (CP) content of the FB bulb. To compare liveweight gain and grazing behaviour, 191, 9- to 12-month-old heifers at the Southern Dairy Hub were offered either FB (n=93; HFR-FB) or kale (n=98; HFR-Kale), both with pasture baleage as a feed supplement, in winter 2019. Allocations were formulated to achieve similar energy intake, but HFR-Kale were offered a higher proportion of their diet as baleage. There were no differences in apparent DM intake, but HFR-FB consumed a diet with overall lower CP (11.4%) than did HFR-Kale (13.4%). Heifers in both treatments had a similar average daily liveweight gain (0.45 ± 0.083 kg/day) but neither group achieved the 0.6 kg/day average daily gain target for heifer growth. Differences were observed in eating and ruminating behaviour between the two groups, but more research is required to validate the measurement devices used in this study for animals grazing crop. The results indicate the challenges of achieving industry-recommended growth rates over winter in heifers grazing forage crops and highlight the importance of feed testing for nutritive value to ensure nutrient requirements are being met.

Keywords: Live weight; *Beta vulgaris*; *Brassica oleracea*; allocation; intake; feed quality

Introduction

For dairy heifers, meeting their mature liveweight targets to achieve good lifetime productivity and reproductive success requires achieving consistent monthly growth rates. Industry targets suggest reaching 30% of mature weight at 6 months, 60% at 15 months (mating), and 90% at 22 months (pre-calving) (Troccon 1993). To meet these targets, both metabolisable energy (ME) and protein requirements need to be met. The importance of the winter period in achieving growth-rate targets is that in the 19 months between weaning and first lactation, over 30% of the heifer's growth (from 8-11 months and 20-23 months of age) occurs during the cool months of winter. For southern regions of the South Island, winter represents the industry's most-challenging feeding period with low pasture growth (Dalley & Geddes 2012), and corresponds with the lowest level of achievement of liveweight-gain targets (Handcock et al. 2017).

Historically, forage crops such as kale (*Brassica oleracea*) and swedes (*Brassica napus*) have been used for wintering dairy cattle in the South Island of New Zealand (White et al. 1999; Nichol et al. 2003). In recent years, there has been increased interest in feeding fodder beet (FB; *Beta vulgaris*) as an alternative to winter brassicas. This has been driven from a perceived lower cost/kg dry matter (DM) to grow FB compared with kale, its ability to produce high yields, a smaller area requirement (Gibbs et al. 2011), and environmental benefits from a lower nitrogen content in the bulb (Waghorn et al. 2018). Furthermore, as FB yields increase, the quality does not decline, unlike kale, where the stem begins to lignify towards the end of winter (Matthew et al. 2011).

In regions where winter forage crops are the predominant feed source for rising-one-year-old dairy heifers, the extent to which forage type affects animal growth rate is unclear. Many feeding decisions are driven predominately by cost, practicality (ease of feeding and management, fit with the system), and environmental considerations to reduce nutrient losses, rather than the ability of the diet to meet the nutrient requirements, especially protein and energy, with most winter diets being allocated on a DM basis.

The objective of this observational study was to compare the growth of rising-one-year-old dairy heifers grazing either kale or FB from May until August and determine if crop type affected grazing behaviour and rumination. It was hypothesised that the lower crude protein (CP) content of the FB bulb would result in insufficient dietary nitrogen to meet the nutritional requirements for growth.

Materials and methods

The experiment was conducted at the Southern Dairy Hub (SDH), Wallacetown, Southland, New Zealand (46.3298° S, 168.2903° E) between the 30th of April 2019 and the 20th of August 2019, with approval from the AgResearch Animal Ethics Committee (AE Application: 14814).

This study was an un-replicated longitudinal observational study involving 191 Friesian-Jersey crossbred heifers (9-12 months of age). These animals were allocated to winter diets that matched the diet of their dam when *in utero* (93 to FB: HFR-FB, 98 to kale: HFR-Kale) as they were the heifer replacements in a multi-year farm-systems study.

Transition from pasture to FB occurred between the 1st and 13th May. The initial allocation was 0.5 kg DM/head/day for three days with pasture and pasture baleage making up the remainder of the diet. The allocation was then increased by 0.5 kg DM/head every two days until the full target allocation of 5.4 kg DM/head/day was achieved. Pasture allocation was reduced as FB intake increased. The transition from pasture to kale started later on the 22nd May due to limitations in the availability of kale resulting from lower than expected yields. Their allocation started at 2 kg DM/heifer/day and increased by 0.5 kg DM/head per day until the target allocation of 4.1 kg DM/head/day was achieved after five days. Pasture baleage was offered to both groups *ad libitum*, in round-bale feeders, while transitioning onto crop. Diet DM allocations were determined using DairyNZ's winter-crop allocation calculator (DairyNZ 2015) based on the predicted quality of each feed and expected utilisation. The HFR-Kale animals remained on the crop until the 6th August, when all the kale had been consumed, and then grazed pasture while HFR-FB heifers had sufficient crop to graze FB until 19th August. During the experiment HFR-FB, were offered 2.6 kg DM/day baleage and HFR-Kale, 3.7 kg DM/day baleage. The combined allocation of crop and baleage was designed to supply 72 MJ ME/heifer/day. A fresh break of crop and the daily baleage allocation were offered once-daily at approximately 0700 h. Intakes for crop and supplement were pooled for each month and average intake (or allocation) was determined. Reticulated water was provided using portable water troughs located close to the feed face. A back fence was used in both crop paddocks to restrict movement and was moved closer to the feed face twice per week.

Twenty animals were randomly selected from each group to be fitted with a CowManager electronic tag (CowManager SensOor, Agis Automatisering BV, Harmelen, the Netherlands; validated in cattle grazing pasture by Pereira et al. 2018). The tags monitored daily grazing, ruminating, activity and resting (idle) using a proprietary model with the data expressed as percentage of behaviour per hour and per day. These devices automatically downloaded data to a server through readers installed near the paddocks between the 3rd and 12th July. Two tags from HFR-FB failed to collect data, leaving 18 and 20 animals contributing data for HFR-FB and HFR-Kale respectively.

Crop DM yield was determined for kale on the 20th May, 19th June and 12th July and for FB on the 21st May, 18th June and 12th July. For kale, two 1m^2 quadrats were randomly selected from the area to be grazed over the next two weeks. All plants from within the quadrat were cut to ground level and weighed. For FB, yield was assessed from two 2-m double rows of crop in the next area of the paddock to be grazed. All FB plants were pulled from the ground, soil removed, and the fresh weight of leaf and bulb recorded separately (Dalley et al. 2020). Two representative stem/bulb/FB leaf subsamples were retained from each plot at each harvest and used to determine the DM content. Dry weight was determined after drying at 100°C to constant weight. Samples of crop and supplement were sent to Hill Laboratories (Hamilton, NZ) for DM, nutritive quality and mineral analysis by a near infra-red spectrophotometer (NIRS, NIRSystems, Foss, Maryland, USA). Post-grazing crop residual harvests for FB and kale were completed on the 21st June and 22nd July using the same process described for the pre-crop assessment except that samples were not sent for quality analysis. Crop yield and residual measurements were used to estimate apparent intake of DM, energy and CP using feed-quality results from Hill Laboratories.

Unfasted heifer live weight prior to the daily allocation of crop was measured on the 11th April, 30th April (pre-winter), 6th June (early-winter), 12th July (mid-winter), 20th August (post-winter) and 15th October (pre-mating) using Tru-Test weigh scales (Datamars SA, Lamone, Switzerland). Stature measurements of wither height, body length and girth circumference were taken pre-winter, post-winter and pre-mating using a standard tape measure and measuring band.

Statistical analysis

All analyses were performed using Genstat (Version 19.1.0.21390, VSN International Ltd., Hemel Hempstead, UK). Means for crop and animal measurements were compared using a general linear model. Descriptive statistics (means and standard deviations) for diet composition and apparent intake of crop and supplement are presented. Liveweight gain was compared using repeated-measures analysis using crop as the fixed term, animal as the random term across the six measurement dates. Animal behaviour measures for individual heifer means were combined within measurement dates and date used as the random term.

Results

Crop yield and feed quality

The average crop yield for kale (12.6±1.84 t DM/ha; mean±SD) was significantly ($P<0.001$) less than that for FB (22.4±3.83 t DM/ha). The FB crop averaged 19±5.2% leaf.

Fodder beet bulb had higher DM than FB leaf ($P<0.05$), with kale intermediate (Table 1). The baleage DM content was higher than that of both crops ($P<0.05$). The fibre content was greatest in the baleage, which also had the lowest organic matter digestibility (OMD). The dietary components differed in ME ($P<0.001$) with the FB bulb having the highest and the baleage the lowest MJME/kg DM. There was a trend for the CP content of the FB bulb to be less than that of all other feeds ($P<0.10$). There were no statistically significant differences in mineral composition between feeds, though kale tended to contain more calcium ($P<0.10$).

Feed allocation and nutrient intake

Estimated daily DM allocation was similar between the treatment groups, however, HFR-FB received a diet

Table 1 Average composition of fodder beet (FB) bulb and leaf, kale, and pasture baleage eaten by heifer replacements during winter 2019. Means with the same letter (a-c) within a row are not significantly different at the 5% level. SEM = standard error of the mean.

	Baleage	FB bulb	FB leaf	Kale	SEM	P value
Dry matter %	52.2[a]	16.6[b]	8.9[c]	12.0[c]	1.85	<0.001
Organic matter digestibility (%)	67.8[c]	95.7[a]	86.8[b]	85.9[b]	3.28	0.002
Crude protein (%)	14.0[b]	9.1[b]	20.1[a]	13.5[b]	2.48	0.099
Neutral detergent fibre (%)	50.0[a]	10.3[c]	24.1[b]	25.2[b]	1.91	<0.001
Metabolisable energy (MJ ME/kg DM)	9.8[c]	14.7[a]	11.2[b]	12.4[b]	0.60	<0.001
Calcium (%)	0.70[b]	0.12[c]	0.40[bc]	1.33[a]	0.266	0.077
Magnesium (%)	0.22	0.16	0.34	0.21	0.875	0.536
Phosphorus (%)	0.31	0.18	0.19	0.29	0.049	0.182

Table 2 Summary (mean and standard deviation, SD) of average allocation and estimated intake of fodder beet (FB), kale or baleage dry matter, energy and nitrogen offered to 2018-born replacement dairy heifers during winter 2019 and average eating, ruminating and activity (minutes/day).

Diet allocation and intake	HFR-Kale	SD	HFR-FB	SD
Total feed allocation (kg DM/heifer/day)	8.3	0.78	7.8	0.72
Baleage allocation (kg DM/heifer/day)	3.7	0.49	2.5	0.52
Crop allocation (kg DM/heifer/day)	4.5	0.40	5.3	0.39
Baleage intake (kg DM/heifer/day)	3.0	0.39	2.0	0.39
Crop intake (kg DM/heifer/day)	3.4	0.30	4.5	0.33
ME intake (MJ/heifer/day)	71.4	8.40	76.5	6.73
Nitrogen intake (g/heifer/day)	135	7.0	118	10.5
Diet crude protein content (% DM)	13.3	0.67	11.4	0.11
Grazing behaviour	**HFR-Kale**	**HRF-FB**	**SED**	**P-value**
Eating (mins/day)	398	304	19.2	<0.001
Ruminating (mins/day)	315	397	14.5	<0.001
Idle (mins/day)	336	308	15.9	0.120
Active (mins/day)	332	376	13.1	0.030

comprising 68% crop compared with only 54% for HFR-Kale. The HFR-FB animals ate 1.1 kg DM/heifer/day more crop than did HFR-Kale over the winter period (Table 2). Estimated ME intake did not differ between the treatment groups however, HFR-FB had a diet with 14% less CP (Table 2) and, thus, HFR-FB heifers had a lower N intake (13% lower than HFR-Kale).

Live weight and stature

An interaction between crop and time for live weight showed that HFR-Kale were heavier than HFR-FB at the start of the study, both groups had similar weights at the mid-point and HFR-Kale were heavier at the final weighing in October (Fig. 1). Daily liveweight gain between pre- and post-winter measurements did not differ between treatments, averaging 0.46±0.083 kg for HFR-Kale and 0.45±0.093 kg/day for HFR-FB. There were no differences in girth (151.0±0.64 cm) or length (108.7±0.24 cm) between treatments, but HFR-Kale were taller (112.5±0.37 cm), (P<0.001) than were HFR-FB (110.9 cm).

Animal behaviour

HFR-Kale spent, on average, 134 extra minutes/day eating (P<0.001) and 83 fewer minutes/day ruminating (P<0.05) compared with HFR-FB (Table 2). The idle time did not differ between treatment groups, but HFR-FB were more active (P=0.030).

The grazing intensity (i.e., minutes eating per hour) increased for both treatment groups from 07:00 hours, corresponding with the offering of a new break of crop and the daily baleage allocation (Fig. 2a). Grazing intensity gradually decreased throughout the day as crop availability declined. Between 07:00 and 16:00, HFR-Kale spent 10-15 more minutes per hour eating. A decrease in eating time for both treatments occurred at 17:00 hours, corresponding with the onset of darkness.

Average rumination time during daylight hours was 7.9 and 4.4 min/h for HFR-FB and HFR-Kale respectively (Fig. 2b). Rumination time for both treatments was less than 10 min/h between 07:00 and 12:00 hours (Fig.2b). Rumination time increased for both treatments after 17:00 hours to a peak of 24-29 min/h at 04:00 and 05:00 h. HFR-FB were more active between 07:00 to 11:00 hours following the allocation of fresh feed (Fig. 2c).

Discussion

The aim of the study was to monitor the performance of rising-one-year-old dairy heifers when managed in crop-wintering systems typical of those offered in the southern South Island and identify any nutritional or performance risks. While average daily liveweight gain between 30th April and 20th August 2019 for both groups was similar (ADG; 0.45-0.46 kg/day), neither group achieved the 0.6 kg/day ADG target for heifer growth between 6 and 15 months of age (Roche et al. 2015). It was estimated that sufficient DM (7.8 and 8.3 kg DM/heifer/day for HFR-FB and HFR-Kale, respectively; Zhang et al. 2016) and energy

Figure 1 Average live weight of 2018-born dairy heifers before, during and after winter when grazed on either kale (▬▬) or fodder beet (FB ▬▬) during winter 2019.

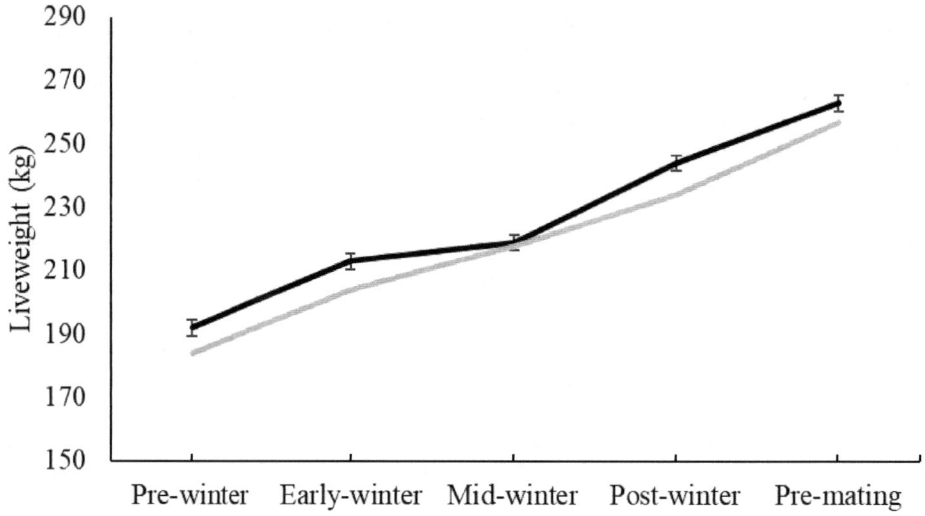

Figure 2 Mean a) eating time (min/hour), b) ruminating time (min/hour), c) idle time (min/hour) and d) active time (min/hour) for each treatment compared over a day for dairy heifers wintered on fodder beet (FB ▬▬) or kale (▬▬) for a nine-day period from the 3rd to the 12th July. Error bars are 1 standard error of the difference (SED).

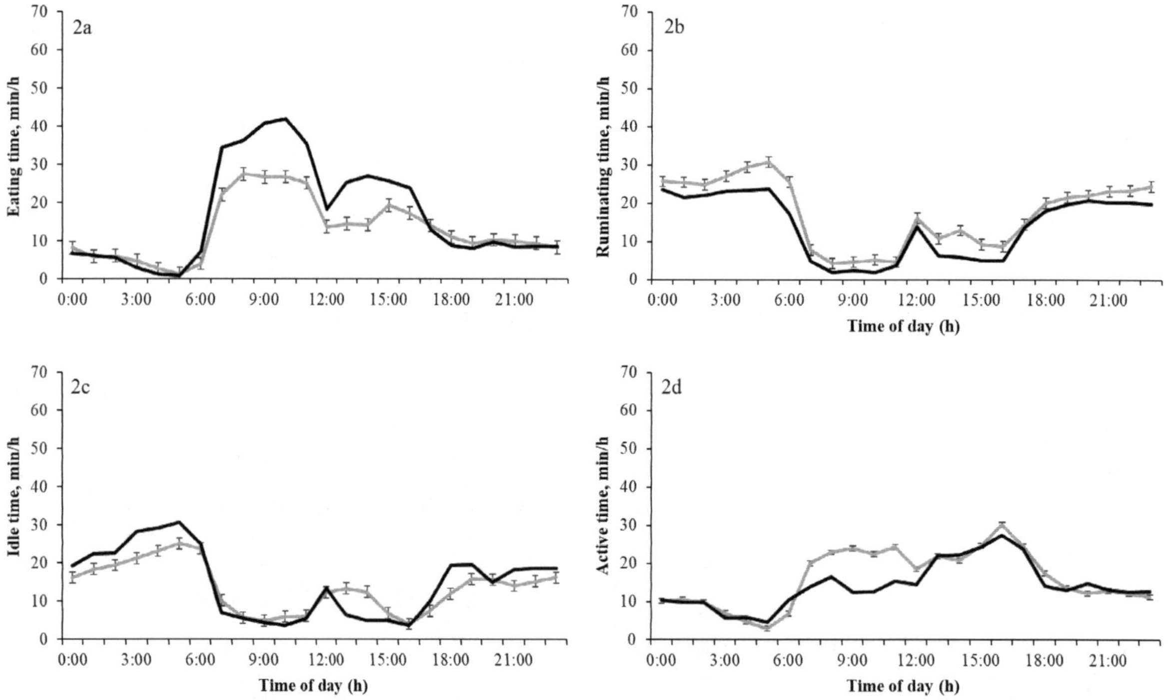

(76 and 71 MJ/head/day) was offered to achieve this gain. Heifers grazing kale were offered more total DM to account for an expected lower kale utilisation (Edwards et al. 2014; Dalley unpublished data).

Heifers in HFR-FB and HFR-Kale consumed 3.0% and 2.9% of their live weight daily, respectively. In contrast, Saldias and Gibbs (2016) reported that beef steers offered *ad libitum* FB consumed 2% of LW but gained 1.0 kg/head/d. The discrepancies in intake and liveweight gain between the current study and that of Saldias and Gibbs (2016) could relate to the age, breed and gender differences of the experimental animals, along with inherent difficulties in determining yield and utilisation of FB crops (Gibbs et al. 2011) which are required to estimate intake.

For both treatments, the CP content of the diet was lower than the minimum requirement for optimising growth of 15-17% DM (Alderman & Cottrill, 1993) with the FB diet only averaging 11.4% CP (Table 2). The low

CP content of the kale diet was surprising given kale is known for its ability to take up large amounts of excess N (Fletcher & Chakwizira, 2012) and the CP content of 71 kale samples from crops grown at SDH from 2017 to 2021 averaged 16.5% with 66% containing more than 15% CP (unpublished results). The kale crops in the current study were lower in DM yield than expected and this may have contributed to the lower CP content. Given neither group achieved the target daily LWG of 0.6 kg/day and both diets were low in CP, we hypothesize that protein intake restricted growth. This result highlights the importance of testing all winter feeds for nutritive value and formulating winter diets to ensure nutrient requirements are being met, rather than just allocating feed on a DM basis.

The heifers used in the current study were part of a larger farm-systems experiment investigating potential cumulative effects of feeding FB to mixed-aged cows in late lactation, the dry (winter) period and early lactation. At one day of age, HFR-FB were, on average, 9% lighter (unpublished data) than HFR-kale and, despite co-rearing of animals from two days of age, they remained lighter and smaller in stature at the start of winter (8-9 months of age; Fig. 1). Hammond et al. (2021) also reported lower body weights and growth rates and higher mortality rates of twin lambs born to ewes fed FB in mid-to-late gestation, indicating dietary impacts on offspring while *in utero*. The numerical differences in LW between the HFR-FB and HFR-Kale in the current experiment persisted until at least pre-mating. To statistically determine the impact of crop type on heifer intake and liveweight gain over winter, a replicated trial would be required where a single cohort of animals were allocated to the treatments prior to winter, ensuring the treatment groups were balanced for age, breed and live weight.

Troccon (1993) recommended that heifers should be 30%, 60% and 90% of mature live weight at 6, 15, and 24 months of age, respectively, requiring a linear trajectory of growth. However, both the availability of feed and the quality of available feed will vary, particularly when heifers are reared in grazing systems. In our study, both groups failed to achieve the 60% of mature liveweight target at 15 months of age, with the HRF-Kale achieving 54% and HFR-FB 52%. McNaughton and Lopdell (2012) reported that 9- to 12-month-old heifers with very low growth rates during their first autumn/winter, were not able to regain the target trajectory. However, an industry review of heifer growth targets by Roche et al. (2015) reported that a 'phased nutrition regime' where periods of feed restriction are followed by periods of re-alimentation may be better suited to heifer rearing in grazed dairy systems. Further research to determine the best option could take two approaches. One would be to determine whether rising-one-year-old heifers who do not reach their 15-month liveweight target following wintering on crop can achieve compensatory growth through improved nutrition to meet their 24-month targets. The second approach would be to investigate options to improve the birth weight and growth performance of calves in their first year, to avoid the need for compensatory growth, given the known challenges of providing sufficient nutrition to this class of stock during winter and/or summer, depending on regional pasture growth supply and quality.

The differences in the crop:baleage ratio for the two groups was intentional and resulted from lower-than-expected kale yields. To balance the winter-feed budget and ensure the required daily DM and energy allocations for HFR-Kale for the winter period were achieved, additional baleage was offered. Even with the revised allocation, HFR-Kale finished their crop on the 6th of August and returned to pasture grazing while HFR-FB stayed on crop until the 19th of August.

In the current observational study, there were differences in grazing, ruminating and activity behaviour between the two groups. Generally, time spent eating/h for both treatments increased rapidly from 07:00 hours, immediately after the daily allocation of fresh feed. Data indicate that HFR-FB ruminated more from 12:00 to 15:00 compared with the HFR-Kale and HFR-FB were more active in the period immediately after the allocation of fresh feed (Fig. 2 a-d). Across the day, HFR-Kale ate for 6 h 36 min and HFR-FB ate for 5 h 4 min, considerably lower than the >7 h/day grazing observed by Saldias and Gibbs (2016). There might be several reasons for differences between the groups in their eating behaviour, but a key driver for variation between visually observed behaviour and the SensOor eartags relates to calibration of the sensors. Periera et al. (2018) validated the tags under pasture grazing, however, animals on crops, especially bulb crops, move their mouths and heads differently when compared with animals grazing pasture, potentially affecting how the algorithms assign motion to grazing and ruminating. Further research is required to validate the SensOor tags in animals grazing crops to verify the behavioural differences observed in this study.

Conclusion

In this observational study, rising one-year-old dairy heifers were unable to achieve the recommended 0.6 kg/day liveweight gain targets between May and August when wintered on diets based on either kale or FB. As a result, they were only 52-54% of mature live weight at 15 months of age rather than the recommended 60%. For both diets the recommended dietary CP content of 15-17% for optimum growth was not achieved, potentially contributing to the lower liveweight gain. The results highlight the importance of testing feeds for nutritive value and adjusting winter diets to ensure nutrient requirements are being met. While differences in grazing and ruminating behaviour between HFR-FB and HFR-Kale were observed more research is required to validate the SensOor tags in animals grazing crop.

Acknowledgements

This project was funded by the dairy farmers of New

Zealand through DairyNZ Inc. (Hamilton, New Zealand), contract RDN1805. We thank C Crack and N Hammond for managing data collection and the farm team at the Southern Dairy Hub for animal management.

References

Alderman G, Cottrill BR 1993. Energy and protein requirements of ruminants: an advisory manual. Agricultural and Food Research Council (Great Britain). Technical Committee on Responses to Nutrients. Wallingford, Oxon, UK. p159.

Dalley DE, Geddes T 2012. Pasture growth and quality on Southland and Otago dairy farms. Proceedings of the New Zealand Grassland Association 74: 237-241.

Dalley DE, Edwards JP, Woods RR 2020. Impact of winter fodder beet or kale allocation on body condition score gain and early lactation performance of dairy cows. Journal of New Zealand Grasslands 82: 73-81.

DairyNZ. 2015. *Winter crop allocation calculator*. Retrieved 28 April 2020 from: https://www.dairynz.co.nz/media/2182574/winter-crop-allocation-calc-may-2015.xls

Edwards GR, de Ruiter JM, Dalley DE, Pinxterhuis JB, Cameron KC, Bryant RH, Di H, Malcolm BJ, Chapman DF 2014. Dry matter intake and body condition score change of dairy cows grazing fodder beet, kale and kale-oat forage systems in winter. Proceedings of the New Zealand Grassland Association 76: 81-88.

Fletcher AL, Chakwizira E 2012. Developing a critical N dilution curve for forage brassicas. Grass and Forage Science 61(1): 13-23.

Gibbs J 2011. Wintering dairy cows on fodder beet. Proceedings of the South Island Dairy Event. pp230-238.

Hammond KJ, Sandoval E, McKenzie CM, Lees S, Pacheco D, McCoard SA 2021. The effect of a fodder beet versus rye-grass grazing regime during mid-to-late gestation twin-bearing ewes on dam and progeny performance and lamb survival. New Zealand Journal of Agricultural Research 1-18. https://doi-org.ezproxy.lincoln.ac.nz/10.1080/00288233.2021.1879175.

Handcock RC, Lopez-Villalobos N, McNaughton LR, Edwards GR, Hickson RE 2017. Growth curves of New Zealand Holstein-Friesians, Jersey and Holstein-Friesian-Jersey crossbred heifers. Proceedings of the New Zealand Society of Animal Production 77: 64-68.

Matthew C, Nelson NJ, Ferguson D, Xie Y 2011. Fodder beet revisited. Proceedings of the Agronomy Society of New Zealand 41: 39-48.

McNaughton LR, Lopdell TJ 2012. Brief communication: Are dairy heifers achieving liveweight targets? Proceedings of the New Zealand Society of Animal Production 72: 120-122.

Pereira GM, Heins BJ, Endres MI 2018. Technical note: Validation of an ear-tag accelerometer sensor to determine rumination, eating, and activity behaviours of grazing dairy cattle. Journal of Dairy Science 101 (3): 2492-2495.

Roche JR, Dennis NA, Macdonald KA, Phyn CVC, Amer PR, White RR, Drackley JK 2015. Growth targets and rearing strategies for replacement heifers in pasture-based systems: a review. Animal Production Science 55 (7): 902-915.

Saldias B, Gibbs SJ 2016. Brief Communication: Ad libitum fodder-beet and pasture beef-finishing systems – intake, utilisation, grazing behaviour and liveweight gains. Proceedings of the New Zealand Society of Animal Production 76: 87-89.

Troccon JL 1993. Effects of winter feeding during the rearing period on performance and longevity in dairy cattle. Livestock Production Science 36: 157-176.

Waghorn GC, Law N, Bryant M, Pacheco D, Dalley DE 2018. Digestion and nitrogen excretion by Holstein-Friesian cows in late lactation offered ryegrass-based pasture supplemented with fodder beet. Animal Production Science 59(7): 1261-1270.

White JGH, Matthew C, Kemp PD 1999. Supplementary feeding systems. In J White & J Hodgson (Eds), New Zealand Pasture and Crop Science (pp175-198). Auckland: Oxford University Press.

Zhang B, Wang C, Liu H, Liu J, Hongyun L 2016. Effects of dietary protein level on growth performance and nitrogen excretion of dairy heifers. Asian-Australasian Journal of Animal Sciences 30 (3): 386-391.

An evaluation of the FeedSmart app for feed planning in a controlled nutrition study

JL Peers-Adams[a]* and P Smith[a]

[a]*AgResearch Invermay Agricultural Centre, Private Bag 50034 Mosgiel 9053, New Zealand*

**Corresponding author:* Email: jacqueline.peers-adams@agresearch.co.nz

Abstract

FeedSmart is an app designed for smartphones, tablets, and computers to help farmers make instant and accurate decisions around feed management. The accuracy of the FeedSmart app was assessed in an intensive controlled nutrition trial, over two consecutive years. The objective of the trial was to compare a maintenance (M) diet to a restricted diet, 0.6 of maintenance (R), over a 55-day period in sheep. In the first year, 48 animals were split between M and R groups. In the second year, 274 ewes were divided into light (L) and heavy (H) groups, then into M and R. The FeedSmart app was used to calculate the area required for both M and R groups. Based on previous data, restricted animals were expected to lose 7.1% of their body weight over the 55-day period. The changes in year 1 M and R groups and year 2 HM and HR groups were not significantly different from the expected changes. The weight changes in the year 2 LM and LR groups were significant, but were within the errors associated with the calculations underlying the FeedSmart app. In these trials, the FeedSmart app provided an accurate tool for feed management.

Keywords: FeedSmart; feed Planning; sheep; nutrition

Introduction

Feed planning is used to improve animals' health and productivity and pasture growth and quality to increase farm profitability (Densley et al. 2004). Having precise management of a farm's feed supply ensures good nutrition for livestock, while maintaining sufficient pasture quality and quantity.

Feed planning and prediction of feed demand increases farm profitability (Beukes et al. 2015) driving and influencing the amount of supplement or cropping required for optimal stock performance (Clark & Woodward 2007; Johns et al. 2016). However, it has also been demonstrated that matching animal performance with pasture performance and supply can also be more important for increasing farm profitability than improving animal performance *per se* (Thompson et al. 2016). Either scenario requires accurate feed planning.

During significant periods of an animal's lifespan, it is important to maintain an animal's health and increase productivity through feed. The requirements for feed supply can vary depending on gender and key activities (Beef and Lamb NZ 2008). In ewes, underfeeding animals during late stages of pregnancy and during lactation will result in poor milk production (Coop et al. 1972), higher ewe and lamb mortality (Everett-Hincks et al. 2004), poor udder development, and lower weaning weights (Litherland et al. 1999), while overfeeding may result in increased number of lambs with dystocia, both causing an economic effect on farmers (Stevens 1999; Wu et al. 2004)

The FeedSmart application (app) is designed to assist in feed management at all stages of the year. FeedSmart is a web app designed for smartphones, tablets, and computers to estimate energy and pasture intake of different classes of livestock in sheep and cattle throughout the year, including during pregnancy and lactation. Using equations from Nicol and Brookes (2007), the FeedSmart app considers different variables (weight, stock class and growth rate) to calculate an animal's nutritional energy requirement. These variables are the used to estimate either the number of days in a paddock, tally of animals, residual pasture cover, or size of paddock break, taking into consideration starting pasture cover, supplementary feed, feed utilisation and land type (flat, rolling, steep or high country). This paper utilised a set of experiments designed to examine the impacts of under-feeding in early gestation on foetal development and subsequent offspring performance. These experiments used the FeedSmart app to determine pasture allocation for maintenance diets. The objective of the current study was to use the data generated in these trials to assess the accuracy of the FeedSmart app. While an above-maintenance diet may have been useful for app validation, the main aim of these trials was related to foetal programming, so an above-maintenance diet was not required. Therefore, this paper reports the outcomes of using the FeedSmart app when feeding at maintenance and 60% of maintenance over two consecutive years.

Material and methods

All experiments were approved by the Invermay Animal Ethics Committee (14000, 14374, 14647 and 14887) and were conducted in accordance with New Zealand Animal Welfare regulations. All experiments were conducted at the Invermay Agricultural Centre (45°51′ S, 170°23′ E) across two years (2017–2018). The animals (ewes and rams) used in the experiments were of composite breeds, comprised of Coopworth, Romney and Texel. All ewes were given a Campyvax4® booster, Ultravac®7in1 booster and Flexidine prior to mating, as per manufacturer recommendations.

The 2017-trial comprised 48 mixed-age ewes randomly allocated to either a maintenance group (M) or a restricted group fed at 60% of the maintenance diet (R). As there was a substantial weight range amongst the ewes in 2018, to ensure more accurate feed calculation, the ewes were initially divided into a light mob (n=130) and a heavy mob (n=144). Each mob was then randomly split into a maintenance and a restricted-nutrition group. Therefore, the trial comprised a light-maintenance group (LM), a light-restricted group (LR), a heavy-maintenance group (HM) and a heavy-restricted group (HR) (Table 1). Further, the 2018 ewes comprised a mixture of two-tooth ewes and mixed-age ewes, the proportion of which ranged from 15% in the HR group to 33% in the LR. Further details are presented in Table 1.

The experiments described here were part of a larger program of research examining the effects of altered nutrition during gestation on the fertility of the female offspring (Smith et al. 2019). The requirements of this program were for either maintenance or restricted (0.6 of maintenance) nutrition for the first 55 days of gestation. Irrespective of year, fertile rams harnessed with crayons were introduced to the ewes and visually assessed for oestrus daily. Once a ewe showed signs of oestrus, as indicated by crayon marking on rump, she was accepted into the trial and randomly assigned to a maintenance or restricted nutritional group. Harnessed fertile rams were then introduced into each group. Ewes showing signs of oestrus were assumed to be not pregnant from the initial mating and withdrawn from the trial.

Ewes were given a new pasture allocation every day for the first 55 days of gestation (ryegrass clover mix) derived on calculations from the Beef and Lamb FeedSmart app based solely on starting weights. To assess the impact of the diets, live weights of ewes were regularly monitored for the 55 days. Further, body condition scores (BCS) were recorded on day 0 and day 55 to help assess the overall impact of the restricted diet.

Calculation of daily feed area

Diets (allocated area) for the maintenance nutrition groups were calculated using the following Feedsmart app settings, pasture growth rate of 0, the default pasture/ crop-quality setting of 10.8 MJME, and flat land. Allocation for the restricted nutritional groups was calculated by first establishing maintenance allocation for those ewes and then multiplying the allocated area by 0.6.

Starting pasture and residual cover (kg DM/ha) were measured daily using a plate meter (tru-test EC-10). The plate meter provides a relative reading which is based on pasture height and density.

The daily area allocated was calculated using the average weight of the mob, tally of animals in each group, average mating date, starting cover of each area, and the residual left the following day. Residual cover was estimated based on farm-management advice and confirmed with daily residual cover readings.

Results from a tightly controlled indoor experiment, in which animals lost 7.1% of live weight when fed at 0.6 of maintenance (Smith 2017), were used as a baseline to assess the accuracy of the FeedSmart app in calculating the restricted diet.

Statistics

Statistical analyses were performed using the Genstat software package. (Genstat 18th edition, VSN International). Comparisons between expected and actual weight changes were analysed using a paired t test. For the maintenance group the hypothesis tested was initial weight (Day 0) – final weight (day 55) =0. For the restricted groups, allowing for the expected 7.1% loss in body weight, the hypothesis tested was 0.93 x initial weight (day 0) – final weight (day 55) = 0.

Table 1 Average starting weights (day 0), finishing weights (day 55), percentage of body weight change over the trial period and the percentage of two-tooth ewes in each group.

Group	Year of Trial	Number of animals	Day 0 weight (kg)	Day 55 weight (kg)	% of body-weight change	% of two-tooths
Maintenance	2017	24	70.7 ± 1.1	70.4 ± 1.2	-0.33 ± 0.07	
Restricted	2017	24	71.5 ± 0.9	67.0 ± 1.1	-6.3 ± 0.17	
Light maintenance	2018	67	63.4 ± 0.3	65.2 ± 0.4	2.90 ± 0.47	22.4
Mixed-age		52	63.8 ± 0.4	66.2 ± 0.4	3.4 ± 0.4	
Two-tooth		15	62.1 ± 0.9	61.7 ± 0.7	-0.5 ± 1.0	
Light restricted	2018	63	63.2 ± 0.3	60.1 ± 0.5	-4.9 ± 0.07	33.3
Mixed-age		42	63.7 ± 0.5	62.0 ± 0.5	-2.8 ± 0.4	
Two-tooth		21	62.0 ± 0.6	56.4 ± 0.7	-9.0 ± 0.8	
Heavy maintenance	2018	71	70.2 ± 0.4	70.6 ± 0.5	0.50 ± 0.06	18.3
Mixed-age		58	70.8 ± 0.4	71.7 ± 0.5	1.3 ± 0.5	
Two-tooth		13	67.9 ±0.5	65.7 ± 1.1	-3.2 ± 1.2	
Heavy restricted	2018	73	70.6 ± 0.4	65.2 ± 0.4	-7.7 ± 0.04	15.1
Mixed-age		62	70.9 ± 0.4	65.6 ± 0.4	-7.4 ± 0.4	
Two-tooth		11	69.4 ± 1.0	62.9 ± 0.9	-9.3 ± 0.9	

Maintenance groups received a maintenance diet based on metabolizable energy requirements for the first 55 days of gestation while restricted groups received of diet equivalent to 0.6 of maintenance. 2018 groups were subdivided into light and heavy based on starting weights. Data are means and standard error of the means.

Figure 1 Average weights of the 2018 ewes on either a maintenance or restricted diet over the first 55 days of gestation for the light group (a) and heavy group (b).

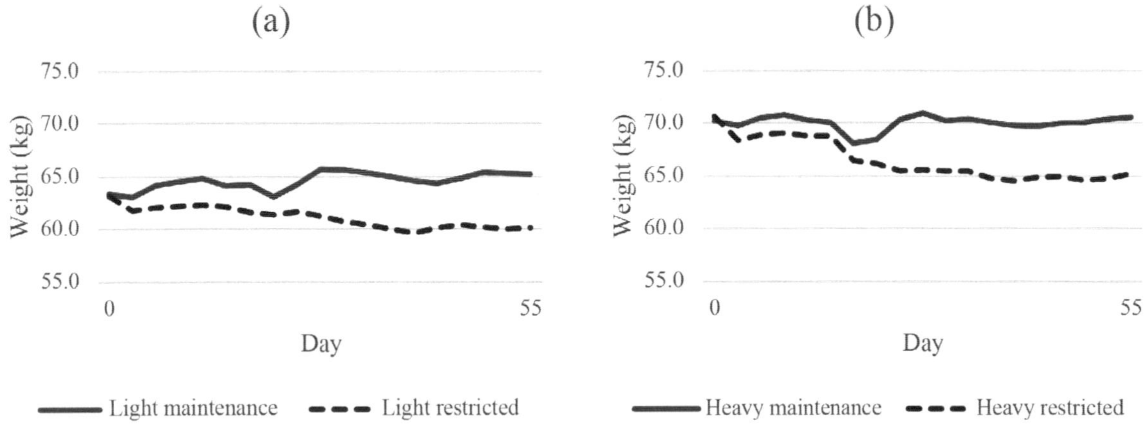

Results

In both 2017 and 2018, all groups were grazed on either adjacent paddocks or, in some cases, in the same paddock. There were no differences in either initial pasture cover, residual pasture cover, or change from initial to residual between years or between nutritional groups. The energy requirements calculated by the app for a 2017 M ewe was 11.37 MJME/day, 2018 LM ewe 10.47 MJME/day and 2018 HM ewe 11.31 MJME/day. The values for the corresponding R ewes were, for a 2017 R ewe 6.8/MJME/day, 2018 LR ewe 6.27 MJME/day and 2018 HR ewe 6.81 MJME/day. Starting pasture cover averaged 3853 (4005± 59 for M, 3696±30 for R groups respectively) while residual cover averaged 1511 (1608±65 for M groups and 1396±31 for R groups respectively). The 2018 heavy ewes, each maintenance ewe was, on average was allocated 3.2 m^2, while the corresponding restricted ewes were allocated 1.9 m^2. For the light ewes in 2018 the allocations were 3.4 m^2 per ewe for the maintenance group and 1.9 m^2 per ewe for the restricted group.

In both years, M animals showed a relatively consistent trend in live weight over the 55-day period with minimal change from their starting weight. R animals followed a pattern consist with the initial indoor-trial ewes to establish the baseline sub-maintenance weights. These animals showed an initial drop in weight, flattening out to a more consistent weight after approximately 25 days (fig. 1)

In the 2017 trial, ewes on the maintenance diet lost 0.3% (0.3 kg) of their starting body weight. Those animals on the restricted diet, expected to lose 7.1% of their body weight, lost 6.3% of their starting body weight (4.5 kg) (Table 1). Neither of these changes in weight were significantly different from the expected change (Table 2).

In the 2018 trial, both the LM and LR groups completed the trial significantly heavier than expected (Table 2). The LM animals gained 0.5% of their starting body weight, an increase of 1.8 kg. The LR group lost only 4.9% of their body weight, a loss of 3.1 kg when the expected loss was 4.4 kg (Table 1). Most of the two-tooth ewes (36/60) were within the light groups and their performance over the course of the trial differed from that of the mixed-age ewes. Whereas overall the LM group gained 2.9% of their initial body weight, the two-tooth ewes lost 0.5% of their body weight. In the LR group, while the overall loss was 4.9% of body weight, the two-tooth ewes lost 9%.

The HM group gained 0.5% of their body weight, a gain of 0.4 kg; within this group the two-tooth ewes lost 3.2% of body weight. The HR group lost 7.7% of their body

Table 2 Analysis of recorded weight changes vs expected weight changes of ewes fed maintenance or restricted diets during pregnancy.

Year/Group	Diet	Mean Difference expected change (kg)	Significantly different from 0 change	95% CI
2017	Maintenance	-0.2	no	(0.62, -1.09)
	Restricted	0.49	no	(1.72, -0.73)
2018 Light	Maintenance	1.83	yes	(2.42, 1.24)
	Restricted	1.37	yes	(2.04, 0.70)
2018 Heavy	Maintenance	-0.37	no	(1.05, -0.31)
	Restricted	-0.52	no	(-0.03, -1.00)

Maintenance animals expected change = 0 kg. Restricted animals expected change was equivalent to 7.3% of body weight lost. Results of paired t-tests testing the hypotheses that for M groups day 55 weight- day 0 weight= 0 and for R groups day 55 weight - 0.93 x day 0 weight = 0. P<0.01 represent significance.

weight, a loss of 5.5 kg: within this group the two-tooth ewes lost 9.3% of body weight (Table 1). These overall losses were not significantly different from the expected changes (Table 2).

Irrespective of year and weight group, those animals on the restricted diet lost on average 0.5 of a BCS, while the animals on a maintenance diet lost an average of 0.1 BCS.

Discussion

FeedSmart app is a tool freely available for farmers to help with feed and nutrition management in sheep and cattle. To assess the accuracy of the FeedSmart app, we utilised data from trials primary designed to examine the effects of gestational under-nutrition. Therefore, we can, with confidence, discus the performance of the app in relation to maintenance and sub-maintenance diets only. The trial requirement for daily allocations, resulted in extraordinarily high stocking rates ranging from 1500 to 7000 stock units/ha. The calculations underlying the FeedSmart app provide an estimate of nutritional energy requirements (Nicol & Brookes 2007). It is suggested to include a safety margin of 10% to prevent underfeeding of some individuals. In the current trials, to assess the accuracy, no safety margins were included.

In 2017, the app provided an accurate estimate of feed allocation to achieve the desired energy requirements at a maintenance level. Weight changes in both M and R groups were not significantly different from expected (paired t-test, P= 0.49 for M, P=0.32 for R), though liveweight loss in the R group was predicted from previous experimentation, rather than derived from the FeedSmart app. Similarly, in 2018, weight changes in HM and HR groups were not significantly different from expected (paired t-test, P=0.31 for HM, P=0.25 for HR). However, in both the LM and LR groups, weight changes were significantly different from expected (paired t-test, P<0.01 for both groups). The LM group, predicted to maintain weight, gained 1.8 kg, while the LR group, predicted to lose 4.4 kg, lost only 3.1 kg.

In relation to the mix-aged ewes, the two-tooths did not perform as well, e.g., in the LM group the two-tooths lost 0.5%, whereas the mixed-age gained 3.9%. Similarly, in the LR group the two-tooths lost 9.0% whereas the mixed-age lost 2.8%. Interestingly the predicted feed requirements in the FeedSmart app for a two-tooth is less than that for a mixed-age ewe, yet the two-tooths performed poorly compared to the mixed-aged ewes. We would hypothesis that the two-tooths were unable to compete with the mixed-age ewes for feed in a high-stocking environment; this hypothesis is supported in the literature (NRC 1985). This may offer some explanation for the significant deviation from the expected weight changes in both the LM and LR groups. While the two-tooth ewes in the 2018 heavy mob also performed differently from the mixed-age ewes, this did not impact significantly on the overall result, likely due to the lower proportion of two-tooth ewes in the heavy mobs. Additionally, there is evidence to suggest that the nutritional requirements of two-tooth ewes differ from those of mixed-age ewes (Stevens & Young 2013; Corbett & Ball 2002).

When feeding animals diets for growth or maintenance, the larger the weight range in the mob the more variable the outcome (Treacher 1983). There are potentially several reasons which contribute to outliers, for example, individual variations in energy requirements (Nicol & Brookes 2007; Corbett & Ball 2002), the ability to compete for feed, and animal health status. Certainly, within these trials, in a high-stocking-rate environment we observed animals which were consistently first to populate new breaks and dominated the grazing resource in specific areas, while others were slow to the new breaks, subsequently struggling to compete for feed. Comparison of these animals with their live weights suggests that this was a major contributor to the outliers of these studies.

Feed requirements can be underestimated in the sheep and beef industry due to incorrectly accounting for pasture wastage (utilisation) (Johns et al. 2016). The FeedSmart app automatically accepts that 10% of the feed will be lost due to pasture wastage. This can be changed in the app dependant on conditions. In the current study, the pasture wastage was set to 0. This primarily was based on the extraordinarily high stocking rate (between 1.4 m^2 and 6.6 m^2 per ewe). Further evidence supporting this was recorded in times of heavy rainfall. As part of ethics requirements for these trials, animals were housed under cover overnight in times of heavy rainfall. In these instances, animals were on their allocated area between four to six hours. The residuals recorded in these instances did not differ from those recorded after 24 hours, suggesting that the available pasture was predominantly eaten within the first four to six hours each day. In a normal farm situation, having a high stocking rate and intense grazing, especially during high-rainfall periods, will decrease animal production, as the amount of feed utilised is reduced significantly (Valentine et al. 2009). Animals cause treading and pugging to the pasture, triggering slow regrowth and a loss of dry matter of pasture, reducing feed intake (Nie et al. 2001; Valentine et al. 2009). This can be accounted for in the app by altering the utilisation settings.

An accurate nutrition program to feed animals throughout the year is imperative for the animal's health and production to be successful and can be achieved through feed management. Regular recording of live weights and BCS of animals is important to monitor the animals' requirements and the success of the farm's feed management. It is important to understand that BCS also plays a role in assessing an animal's response to nutrition (Freer et al. 1997) particularly over extended periods of restricted nutrition. For example, in the 2017 R group 7 of 24 animals had lost at least 0.5 of a BCS by day 25. It is interesting to note from day 25 to 55 when the weights stabilised an additional 15 animals lost at least 0.5 of a BCS. An additional value of BCS is apparent during mid to late pregnancy as live weights can be complicated to monitor effective feed budgeting. Arguably, one deficiency in the app is that it does not utilise BCS information. Users should be aware of the BCS of the animals when determining target weights in the app.

Research by Corner-Thomas and colleagues evaluated the use of specific tools by New Zealand sheep farmers (Corner-Thomas et al. 2019). The study showed 1220 respondents had used one or more tools for herbage measurement (rising-plate meter, pasture probe, sward stick or C-DAX pasture meter). On average 39.1% of farmers found them to be "not relevant", while 22.1% said there were "unwanted complications" and 13.8% found "benefits not apparent". Research also discloses widespread farmer use of visual pasture cover for herbage measurements. Therefore, the perception amongst farmers may be that cost, complexity, accuracy, and relevance can be discouraging factors to the uptake of apps such as FeedSmart. Tools for feed planning and management to farmers can improve prediction of pasture shortages and surplus throughout the year, and vital feeding assessments and decisions (Gray et al. 2003). The resistance to use the available tools can lead to poor feed planning, loss of profitability and productivity (Wall et al. 2012). In these trials we found the app to be versatile, easy to use, and accurate.

Conclusions

While intense stocking rates in these trials were not likely to be utilised in most farming situations, this approach provided a unique opportunity to assess the accuracy of the FeedSmart app. In these trials where maintenance and below-maintenance diets were used, we found the app to be accurate, quick, and easy to use in the field. In our trials, a high level of precision was required when determining pasture allocation and nutritional energy requirements, and we found the FeedSmart app provided us with the levels of accuracy required.

Acknowledgments

This research received funding from the AgResearch Strategic Science Investment Fund. The assistance of the farm staff at AgResearch Invermay was much appreciated.

References

Beef and Lamb NZ 2008. Feedsmart 2. 2 edition. Beef and Lamb.

Beukes PC, Mccarthy S, Wims CM, Romera AJ 2015. Regular estimates of paddock pasture mass can improve profitability on New Zealand dairy farms. Journal of New Zealand Grasslands Association 77: 29-34.

Clark DA, Woodward SL 2007. Supplementation of dairy cows, beef cattle and sheep grazing pasture. In: Rattray PV, Brookes IM & Nicol AM eds. Pasture and supplements for grazing animals, New Zealand Society of Animal Production. Hamilton New Zealand Occasional Publication 14. pp 117-131.

Coop IE, Clark VR, Claro D 1972. Nutrition of the ewe in early lactation. New Zealand Journal of Agricultural Research 15: 203-208.

Corbett J, Ball A 2002. Nutrition for maintenance. In: Sheep Nutrition M. Freer and H. Dove eds. CAB International, Wallingford, Oxon, UK. pp 143-163.

Corner-Thomas RA, Kenyon PR, Morris ST, Ridler AL, Hickson RE, Greer AW, Logan CM, Blair HT 2019. Farm-management tools: reasons for the decision not to use specific tools. Proceedings of the New Zealand Society of Animal Production 79: 125-130.

Densley RJ, Williams ID, Miller DB, Densley DW, Sapienza DA 2004. Advances in maize silage genetics and crop management and the impact on the NZ dairy industry. Proceedings of the New Zealand Society of Animal Production 64: 105-109.

Everett-Hincks JM, Blair HT, Stafford KJ, Lopez-Villalobos N 2004. The effect of ewe nutrition on maternal behaviour score and litter survival. Proceedings of the New Zealand Society of Animal Production 64: 188-191.

Freer M, Moore AD, Donnelly JR 1997. GRAZPLAN: Decision Support Systems for Australian Grazing Enterprises-II. The Animal Biology Model for Feed Intake, Production and Reproduction and the GrazFeed DSS. In Agricultural Systems 54. pp 77-126.

Gray DI, Parker WJ, Kemp EA, Kemp PD, Brookes IM, Horne D, Kenyon PR, Matthew C, Morris ST, Reid JI, Valentine I 2003. Feed planning- alternative approaches used by farmers. Proceeding of the New Zealand Grassland Association 65: 211-217.

Johns C, Johns J, Stevens DR 2016. Increasing lamb survival and lamb weaning weight through feeding high fecundity crossbred sheep. Hill Country-Grassland Research and Practice 16: 323-328.

Litherland A, Lambert M, Mclaren P 1999. Effects of herbage mass and ewe condition score at lambing on lamb survival and liveweight gain. Proceedings of the New Zealand Society of Animal Production 59: 104-107.

Nicol AM, Brookes IM 2007. The metabolisable energy requirements of grazing livestock. In: Rattray PV, Brookes IM & Nicol AM eds. Pasture and supplements for grazing animals, New Zealand Society of Animal Production. Hamilton New Zealand Occasional Publication 14. pp 151- 172.

Nie ZN, Ward GN, Michael AT 2001. Impact of pugging by dairy cows on pastures and indicators of pugging damage to pasture soil in south-western Victoria. Australian Journal of Agricultural Research 52: 37-43.

NRC 1985. Nutrient requirements of sheep. 6th Edition, National Academy of Sciences, National Research Council, Washington, D.C. pp 1-7.

Smith P 2017. Effects of gestational nutrition on post-natal fertility in sheep. Doctor of Philosophy thesis, University of Otago.

Smith P, Stanton J-A L, Quirke L & Juengel J L 2019 Gestational nutrition 1: alterations to gestational nutrition can increase indicators of fertility in sheep. Reproduction 157: 199-313.

Stevens DR 1999 Ewe nutrition: decisions to be made with scanning information. Proceedings of the New Zealand Society of Animal Production 59: 93-94.

Stevens DR, Young G 2013. Using data from large scale farming operations to understand the importance of feeding the ewe to improve whole farm performance. New Zealand Grassland Association Tauranga: 97-102.

Thompson BR, Stevens DR, Scobie DR, O'Connell D 2016. The impact of lamb growth rate pre- and post-weaning on farm profitability in three geo-climatic regions. Proceedings of the New Zealand Society of Animal Production 76: 132-136.

Treacher TT 1983. Nutrient requirements for lactation in the ewe. In: Haresign W ed. Sheep Production. Butterworths, London, pp. 133–153.

Valentine S, Lewis P, Cowan RT, DeFaveri J 2009. The effects of high stocking rates on milk production from dryland and irrigated Mediterranean pastures. Animal Production Science 49: 100-111.

Wall A J, Stevens D R, Thompson B R & Goulter C L 2012 Winter management practices to optimise early spring pasture production: a review. Proceeding of the New Zealand Grassland Association 74 85-90

Wu G, Bazer FW, Cudd TA, Meininger CJ, Spencer TE 2004. Maternal Nutrition and Foetal Development. American Society for Nutritional Sciences 134(9): 2169-2172.

Can GreenFeed automated emissions monitoring systems measure methane output from young red deer grazing pasture?

CP Bennett[a]*, JF Ward[a] and A Jonker[b]

[a]*AgResearch Invermay Agricultural Centre, Private Bag 50034 Mosgiel 9053, New Zealand;* [b]*AgResearch Grasslands, Private Bag 11008, Palmerston North 4442, New Zealand*

**Corresponding author.* Email: charlie.bennett@agresearch.co.nz

Abstract

There are limited data on the methane output of pasture-grazed farmed red deer (Cervus elaphus) in New Zealand. Identification and validation of methane-mitigation strategies for farmed deer requires a technology that can be deployed under grazing conditions. The GreenFeed automated emissions monitoring system (GAEM) integrated with an automated concentrate feeder has been successfully used to determine methane output from cattle but has not been tested with deer. The objective of this study was to evaluate large and small ruminant GAEM units with rising-yearling red deer, to determine if they would access the feed lure, and remain for sufficient duration in the hood of the unit to collect representative measurements of methane outputs. Groups of 14 mixed-sex red deer were used in three trials during spring of 2020. In each trial, approximately a third of individuals accessed the GAEM units after being habituated to the lucerne pellet lure, more than 85% of visits were < 2 min. With further refinement of deer-training protocols and configuration of the GAEM, pasture-based methane outputs should be able to be measured for a proportion of any mob with access to the unit.

Keywords: GreenFeed; red deer; methane

Introduction

New Zealand's agricultural sector is estimated to contribute to 47.8% of the country's total gross emissions (MfE 2020). The main source (74.1%) of agricultural emissions is methane (CH_4) produced from enteric fermentation in grazing ruminant livestock, making up 35.4% of New Zealand's gross emissions (MfE 2020). Obligations to mitigate emissions and meet the Kyoto and Paris Agreement targets require regular greenhouse gas inventories to accurately estimate the methane output of the ruminant livestock classes farmed in New Zealand (Climate Change Response (Zero Carbon) Amendment Act 2019).

There has been a large amount of data collected on the methane production of dairy cattle, beef cattle and sheep, but there is little methane-production data for farmed deer. A few comparable methods used to measure the methane output from ruminant livestock include respiration chambers (RC), the sulphur hexafluoride tracer technique (SF_6) and the GreenFeed automated-emissions-monitoring system (C-Lock Inc., Rapid City, USA). The development of these tools and methods to measure ruminant methane production have been focussed on cattle and sheep but might not be directly applicable to deer due to behavioural differences among livestock classes.

Respiration chambers are the most accurate method for directly measuring the total methane emissions of an individual. This method requires animals to be placed in a closed chamber, therefore removing them from the grazing environment. A long period of deer training is required before they can be measured in chambers (Pérez-Barbería et al. 2020) and this system cannot simulate the grazing situation. Swainson (2004) and Swainson et al. (2007) successfully used the SF_6 tracer technique to measure the methane output of young deer, both indoors and grazing pasture. This was an expensive and labour-intensive method, requiring training of animals to wear specialised measurement equipment. The GreenFeed emissions-monitoring system is a portable, stand-alone unit used to measure multiple short-term (up to about 5 minutes) emissions of methane from animals in confinement or at pasture. An automated feeder system uses a small lure of pelleted feed to attract animals to a hood where airflow drawn past the animal's head during feeding is sampled, described further by Hammond et al (2015a) and Waghorn et al (2016). When used at pasture GAEM are not labour intensive and require minimal human-animal interaction to collect data. This makes it potentially applicable for use with deer. The GreenFeed system has been extensively used with cattle, and in a limited number of studies with sheep, but has never been validated for deer.

There are GAEM units designed for small and large ruminants (see Fig. 1). The small-ruminant GAEM unit has a greater sensitivity to capture emissions from animals with low feed intake and methane emission (<150g/day) as is the case for deer, however, the hood designed for a sheep may be too small for a deer head, due to their larger ears and pedicles (males). The large-ruminant GAEM has a large hood suitable for cattle and, therefore, it may be more suitable for capturing methane production in deer. To generate accurate methane-production estimates, animals need to visit the GAEM at least 20 times (i.e., spot-samples) of a minimum of 2 min in duration and at least one visit per day. The aim of this trial was to test the feasibility of both the large and small ruminant GAEM to capture methane emissions from red deer, and to evaluate how they interact and visit the novel GAEM unit at pasture.

Materials and methods

A series of three deer pilot trials (1a, 1b and 2) were

performed at two different locations at the AgResearch Invermay Agricultural Research Centre (Mosgiel, New Zealand 45°58′S, 170°04′E) between August and December 2020. The animal experiments conducted adhered to the guidelines of the AgResearch Code of Ethical Conduct and were approved by the AgResearch Animal Ethics committee (approval number 15104).

A total of 28 mixed-sex rising yearling (R1) red deer (*Cervus elaphus*) between 8 and 12 months of age were used in the trials. The deer were randomly selected from a larger group of 56 and stratified by sex into two groups consisting of 14 animals each (seven females and seven males). On 26 August 2020, the 28 trial animals were weighed and drafted into Group 1 (n=14) and Group 2 (n=14).

In consecutive periods, these two groups were introduced to the large-ruminant GAEM and had visitation monitored (Trial 1a). The trial period went for 21 days for Group 1 and five days for Group 2. The most frequent visitors to the GAEM were then combined to form a new group of 14 animals (four female and ten males) and methane emission measurements were performed over a period of nine days using the large-ruminant GAEM (Trial 1b), followed by a period (18 days) of measurements using the small ruminant GAEM (Trial 2).

Equipment setup and configuration

Trials 1a, 1b and 2 were carried out in paddocks of 0.7 to 1.7 ha on both the flat and rolling hill area of Invermay farm. After finishing with Group 1 in Trial 1a, the GAEM unit was moved to a less-disturbed area of the farm for the remainder of Trial 1a with Group 2, and the following trials;1b and 2. No formal measurement of pasture quality and quantity was undertaken. The feed available to the animals was growing spring pasture, the visually estimated pre-grazing pasture cover offered was between 2,500 and 3,500 kg DM/ha in all paddocks, and pastures were not grazed below 1800 kg DM/ha which suggests that animal feed requirements were not limited. The pasture sward in these paddocks was perennial ryegrass (*Lolium perenne*) dominant and included other species such as white clover (*Trifolium repens*), cocksfoot (*Dactylis glomerata)*, sweet vernal (*Anthoxanthum odoratum)* and Yorkshire fog (*Holcus lanatus)*.

Detailed implementation of the large-ruminant GAEM unit with grazing dairy cows has been described previously by Jonker et al. (2019) and Waghorn et al. (2016). The small ruminant GAEM uses the same measurement system (but has a smaller pipe and lower airflow rate to enable capture of low methane emissions) and was setup and deployed in the same way. The GAEM were powered by mains power. Deer were excluded from all but the feed hood of the GAEM using deer gates (Fig. 1). Carbon dioxide recovery and gas calibration tests were performed at the start of each trial. The auto-chime (audio cue before a feed lure is dispensed) and auto-fan (switches on when head enters feeder) function were initially used then turned off for all trials. The fan was either turned off or left continuously running (when performing gas emission measurements). The large-ruminant GAEM unit was initially configured to drop lucerne pellets at 15 s intervals and five drops per visit (~35g/drop) with a waiting time of 2 h between visits before the same animal could receive another feed drop. The small

Figure 1 The large-ruminant (left) and small-ruminant (right) Greenfeed automated emissions-monitoring systems (GAEM) set up in Trial 1a and 2, respectively.

ruminant GAEM used the same initial settings, except that number of feed drops was increased to 10 drops per visit (~9 g/drop). In an attempt to increase animal visitation time, the number of feeds drops per visit and interval time were altered for individual animals using the C-Lock web interface during the course of Trial 2.

Data collected by the GAEM system included the number of animal visits (electronic tag reads), number of feed drops, time and duration of each individual's feeding bout and the proximity of the head in the feeder hood, to be used as an informative monitoring tool during the trials. The CH_4 and CO_2 concentrations of each visit were sampled when the fan system in the unit was turned on (i.e., Trials 1b and 2).

Animals

Animal groups were weighed before being introduced to the GAEM in each trial. The average live weight of animals was 69.8±11.9 kg and 78.3±11.2 kg in Trial 1a (Group 1 and 2, respectively), 82.9±13.8 kg in Trial 1b and 98.6±14.0 kg in Trial 2.

The large mob of 56 animals was acclimated to lucerne pellets trailed along the pasture daily for seven days prior to the start of Trial 1a. This was maintained for both groups of animals during Trial 1, in which the pellet trail was fed incrementally closer to the GAEM at a similar time of day until the trail met the feeder face.

Animals were observed immediately after they were released into the paddock with the GAEM and at least once a day for the following 14 days. After that, observations were kept to a minimum to reduce disturbance to the animals.

No formal statistical analysis was carried out because this was a pilot study without treatment comparisons.

Results

Training deer to use the large and small ruminant GAEM

Initial observations of Group 1 in the paddock with the large-ruminant GAEM showed signs that animals were unsettled and disturbed by the novel object in the paddock. Three animals in the group exhibited stress by excessively pacing the fence line, causing the remainder of the group to be in a constantly alert state. Animals would cease grazing and mob together after any disturbance such as noise from aircraft, heavy machinery and human/vehicle activity near the paddock. All animals in the group consumed pellets trailed onto the ground during the initial five days but would not graze within 15 m of the GAEM. The first animal visit occurred on Day 5 for a few seconds and it was assumed the automated chime and start-up of the fan frightened the animal as it did not return within 48 h. The auto-chime function was then switched off for the duration of Trial 1a to prevent the issue re-occurring. Pellets were eventually trailed out from the feeder face and by Day 9 some animals in the group were eating them within a metre of the GAEM feeder. The second animal tag read was on the Day 10 for a brief visit and this animal returned the next day for a longer period (four feed drops). A new animal tag read was recorded on Day 11, however, its head did not enter the hood far enough to trigger a feed drop. The feeding of pellets around the GAEM was stopped on Day 14 to see if inquisitive animals would search at the feeder. Feeder use increased in the following week (Week 3), with the original animal visiting the feeder multiple times for extended periods of time and on Day 18 received its maximum number (n=30) of feed drops in a day. A new animal received five feed drops and another one drop during this period. A previous animal again registered a tag read, but did not trigger a feed drop so the head proximity setting of the GAEM was altered (to dispense pellets when the head was further out of the hood). An adjacent paddock to the paddock with the GAEM was opened towards the end of Week 3 to increase pasture availability, which reduced animal GAEM visits/feed drops, however, one animal still returned to the unit regularly. In the final week of the Trial 1a, two animals were regularly receiving feed drops and another animal's tag was read, but did not receive a feed drop.

For the second group of animals of Trial 1a, the GAEM was moved to a new location in a less-disturbed area of the farm. Rolling topography allowed the animals to hide from disturbances as they chose, which was mainly passing vehicles and nearby stock movements. Group 2 was introduced to the paddock with the GAEM and the unit was turned off to avoid frightening animals. Pellets were trailed on the ground around the feeder and in the feeder hood to encourage new animals of Group 2. On Day 2 the animals seemed at ease and bolder animals approached the GAEM during human presence while calibrating the unit. The pellets around the unit and in the hood had been eaten while the unit was turned off. The GAEM was activated and pellets were trailed out for a further three days, on Day 3 the pellets had all been eaten but no tag reads had been registered. Feeder activity can be remotely monitored on a web-based interface, which helped to identify an issue with pellets jamming in the feeder opening. On the following day (Day 4) two animals had recorded tag reads and received pellet drops. The same two animals regularly returned to the feeder receiving multiple feed drops until the end of Trial 1a; a further two tags were read during this period but the animals didn't receive feed drops.

Trial 1b commenced with the recorded visitors (n=8) to the GAEM from Group 1 and 2 combined into a single group totalling 14 animals, with six other animals randomly selected from Group 2, that had not visited the GAEM. For this trial, the gas-sampling system (fan) of the large-ruminant GAEM was activated to enable methane emission measurements. Competition at the feeder meant there were only short individual methane measurements and frequently with multiple animals at once. One animal had over 100 tag reads in one day and was present at every visit by other animals in the group. Wooden wings were installed on the hood of the unit and a race (designed for cattle) leading up to the feeder was attached to restrict access to a single

Table 1 The number, duration (minute:second) and between-visit coefficient of variation (CV) of visits by individual rising-yearling red deer to the large-ruminant and small-ruminant Greenfeed automated emissions-monitoring system (GAEM) by the same four deer in Trial 1b and 2, respectively.

				Visit duration (mm:ss)			
Animal	Total visits	Visits/day	Total visits > 2 min	Average	CV (%)	Max	Min
Trial 1b, large-ruminant GreenFeed (9 measurement days)							
INV164	13	1.4	0	00.25	74	01.19	00.10
INV404	34	3.8	1	00.41	83	02.46	00.11
INV488	5	0.6	0	00.42	33	01.00	00.27
INV530	23	2.6	1	00.26	131	03.02	00.10
Trial 2, small-ruminant GreenFeed (18 measurement days)							
INV164	3	0.2	0	00.17	26	00.22	00.14
INV404	8	0.4	0	00.40	50	01.07	00.14
INV488	108	6.0	4	00.39	100	04.03	00.10
INV530	158	8.8	33	01.18	106	06.16	00.10

animal at a time (Garnett 2012). Following this, no animals accessed the feeder for 24 hours, and so only the wings were removed. Only one animal accessed the feeder with the race attachment in the following three days, until two new animals accessed the large-ruminant GAEM in the final two days of Trial 1b.

Commencing on 16 November 2021, Trial 2 used the same 14 animals and paddock from Trial 1b with the small-ruminant GAEM (see Fig. 1). By Day 2, three animals that had previously visited the large-ruminant GAEM in Trial 1b were regularly visiting and receiving pellets drops from the small-ruminant GAEM. Initial feeding periods of individual animals occurred without interruption from other animals, so feed-drop number and interval were adjusted for two of the animals in an attempt to achieve longer feeding bouts (>2 mins). One frequent-visiting animal from Trial 1b did not receive feed drops in Trial 2: this animal was large and had growing antler (velvet). After Day 5, animals were visiting the feeder at the same time with multiple animals competing for the feed. No animals fed for significant periods (>2 mins) before being interrupted by another individual.

Animal visitation of the large and small ruminant GAEM

In Trial 1a, the total number of animals registering tag reads by the large-ruminant GAEM was eight out of the 28 animals (29%), four deer from Group 1 and four from Group 2. Only four out of the 28 animals (14%) regularly visited the feeder for multiple feed drops, two deer from Group 1 and two from Group 2.

During Trial 1b and 2, the average number of daily visits to the GAEM by frequent users (i.e., four animals) were 2.1 and 3.9 visits/day respectively. Only two animals visited the small-ruminant GAEM more frequently in Trial 2. One (INV488) increased from 0.6 to 6 visits per day and the other (INV530) increased from 2.6 to 8.8 visits per day (Table 1). The duration (minute:second) of all visits ranged between 00:10 and 06:16, with a mean duration of 00:34 in Trial 1b compared to 00:43 in Trial 2. Approximately 13% of visits to the small-ruminant GAEM in Trial 2 were > 2 minutes (Table 1).

Table 2 Effect of retaining only visitation lengths of 0.5 and 2.9 minutes on the methane output (grams/day) and between-visit coefficient of variation (CV (%)) of rising-yearling red deer with the large-ruminant and small-ruminant Greenfeed automated emissions-monitoring system (GAEM) in Trial 1b and 2, respectively.

Visit duration	Number of visits	Mean (g/d)	CV (%)
Trial 1b, large-ruminant GreenFeed (9 measurement days)			
All	77	72.1	93.9
> 0.5 min	28	54.4	73.6
> 1 min	11	41.0	54.3
Trial 2, small-ruminant GreenFeed (18 measurement days)			
All	324	61.1	81.7
> 0.5 min	176	59.8	61.9
> 1 min	90	55.9	41.8
> 1.5 min	57	53.1	27.3
> 2 min	41	52.1	26.5
> 3 min	25	53.3	23.3

Methane emissions of large and small ruminant Greenfeed

The mean methane output estimate of deer in Trial 1b was 72.1 g/d (n=77) compared to 61.1 g/d (n=324) in Trial 2. The coefficient of variation (CV) and CH_4 output estimates in Trial 1b and 2 decreased with increasing duration of animal visits (Fig. 2). This was most notable in Trial 2, excluding all data shorter than 1.5 mins stabilised the CV and CH_4 production estimate (Table 2). The average CH_4 output from deer visiting the feeder for longer than 1.5 minutes was 53.1 g/d. There were only three visits longer than 1.5 minutes in Trial 1b, so methane estimates from this trial are not discussed further.

Discussion

The main finding of the current pilot trials was that it is possible to get direct methane measurements from deer at pasture using both the large- and small-ruminant GAEM. Animals that accessed the feeder and returned regularly for multiple feed drops did so when high-quality pasture was available. However, only a small proportion

Figure 2 Relationship between sampling duration (minutes) and estimated methane outputs (grams/day) in rising-yearling deer visiting the small-ruminant Greenfeed automated emissions-monitoring system (GAEM) in Trial 2.

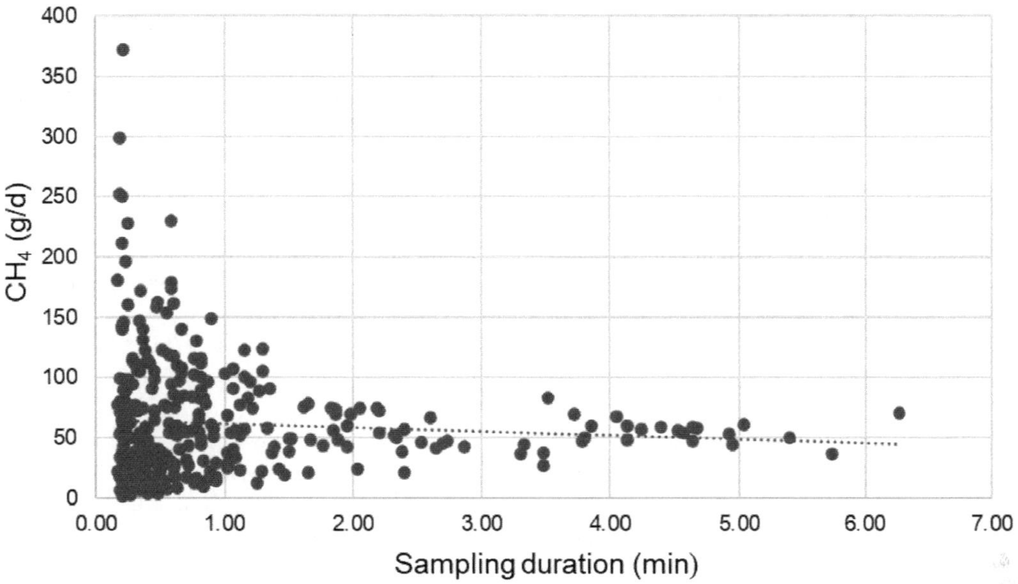

(15%) of the mob used the feeder regularly enough to get reliable measurements, described by Hammond et al. (2015) as multiple daily visits by an individual animal. The daily animal visitations during Trial 1b and 2 by frequent visitors were in a similar range to that reported for cattle (Hammond et al. 2016) and suitable for accurate methane estimates. Overall GAEM use by deer at pasture was lower than reported results (50-72%) in dairy cattle (Waghorn et al. 2016; Garnett 2012). Unlike typical deer-production systems, dairy cattle are exposed to yarding, disturbances, and novel objects daily, which may explain higher visitation rates in dairy cattle. Further trials with a larger group of deer and multiple feeders to reduce competition and improve animal GAEM access opportunities may increase the total animal visitation. Garnett (2012) showed targeted training of non-user dairy cattle to the feeder in confinement can increase individual animal visitation to the GAEM. Pre-training animals to use the system in confinement before pasture is recommended by GAEM manufacturers (Zimmerman 2020) to improve animal visitation. However, Gunter and Beck (2018) found this could take between 4-8 weeks in beef cattle in extensive grazing systems. Pre-training to increase GAEM visitation could be investigated with deer, but this may require cohorts of regularly handled, bold deer that are habituated to disturbances. The increased time and labour associated with training may limit the GAEM use to measure methane outputs of deer on a larger scale.

Determining the best GAEM configuration to keep animals at the feeder for adequate lengths of time proved difficult in Trial 2. Without the race attached, competition interrupted animal feed bouts, resulting in >85% of visits being shorter than 2 minutes. Arthur et al. (2017) found the accuracy of the Greenfeed system to estimate CH_4 production of cattle was significantly lower when visit duration records of <2 min were included. The CH_4 estimate results in Trial 2 (Table 2) reflect this, hence, these data (visits <2 min) are usually excluded before analysis by C-Lock (Mike Billars, C-Lock Inc; pers. comm.). The race attachment is key to capturing accurate data and should be attached when animals are habituated to the GAEM, restricting access to the hood to a single animal at any one time. Modifying the GAEM hood of either size unit with a design for deer may also help improve the methane measurements (i.e., the small-ruminant GAEM hood was too small and limited feeder access for larger animals or those growing antler).

The methane-output estimate for visits longer than 1.5 min was approximately 53 g/d, slightly higher than estimates in NZ farmed deer using the SF_6 tracer technique (24.2 to 40.1 g/d with increasing age from 4.5 to 11.5 months) (Swainson et al. 2007). In forage-fed cattle the SF_6 technique was found to result in lower CH_4 emission estimates than measurements made with respiration chambers and GAEM (Jonker et al. 2020). However, when expressed per unit of body weight (BW), CH_4 production was approximately 0.51 g/kg BW in the current study, which is in a similar range (0.48 to 0.56 g/kg BW) to that found by Swainson et al. (2007).

Implementing the GreenFeed system with pasture-grazing deer in the future may require longer training periods with animals initially in confinement, feeder-hood modification, and the use of multiple GAEM units in a paddock to produce increased animal visits and provide more accurate estimates of methane output.

Acknowledgements

This research received funding from the Pastoral Greenhouse Gas Research Consortium (PGgRc.co.nz). The remote help during the trial with initial setup, feeder configuration, interface use and unit repairs from those at C-Lock Inc (especially Mike Billars) and Anna Taylor at AgResearch was much appreciated.

References

Arthur PF, Barchia IM, Weber C, Bird-Gardiner T, Donoghue KA, Herd RM, Hegarty RS 2017. Optimizing test procedures for estimating daily methane and carbon dioxide emissions in cattle using short-term breath measures. Journal of Animal Science 95: 645-656.

Climate Change Response (Zero Carbon) Amendment Act 2019. https://www.legislation.govt.nz/act/public/2019/0061/latest/LMS183736.htm (assessed 21 April 2021).

Garnett EJ 2012. Evaluation of the Greenfeed system for methane estimation from grazing dairy cows. Masters Thesis, Massey University, Palmerston North, New Zealand.

Gunter SA, Beck MR 2018. Measuring the respiratory gas exchange by grazing cattle using an automated, open-circuit gas quantification system. Translational Animal Science 2: 11-18.

Hammond KJ, Humphries DJ, Crompton LA, Green C, Reynolds CK 2015a. Methane emissions from cattle: Estimates from short-term measurements using a GreenFeed system compared with measurements obtained using respiration chambers or sulphur hexafluoride tracer. Animal Feed Science and Technology 203: 41-52.

Hammond KJ, Waghorn GC, Hegarty RS 2016. The GreenFeed system for measurement of enteric methane emission from cattle. Animal Production Science 56: 181-189.

Jonker A, Farrell L, Scobie D, Dynes R, Edwards G, Hague H, McAuliffe R, Taylor A, Knight T, Waghorn G 2019. Methane and carbon dioxide emissions from lactating dairy cows grazing mature ryegrass/white clover or a diverse pasture comprising ryegrass, legumes and herbs. Animal Production Science 59: 1063-1069.

Jonker A, Green P, Waghorn G, van der Weerden T, Pacheco D, de Klein C 2020. A meta-analysis comparing four measurement methods to determine the relationship between methane emissions and dry-matter intake in New Zealand dairy cattle. Animal Production Science 60: 96-101.

Ministry for the Environment (MfE), 2020. New Zealand GreenHouse Gas Inventory 1990 – 2018. Volume 1, chapters 1.15.

Pérez-Barbería FJ, Mayes RW, Giráldez J, Sánchez-Pérez D 2020. Ericaceous species reduce methane emissions in sheep and red deer: Respiration chamber measurements and predictions at the scale of European heathlands. Science of the Total Environment 714.

Swainson NM 2004. Methane emissions from farmed red deer. Unpublished MSc thesis, Massey University, Palmerston North, New Zealand.

Swainson NM, Hoskin SO, Clark H, Lopez-Villalobos N 2007. The effect of age on methane emissions from young, weaned red deer (*Cervus elaphus*) stags grazing perennial ryegrass (*Lolium perenne*)-based pasture. New Zealand Journal of Agricultural Research 50: 407-416.

Waghorn GC, Jonker A, Macdonald KA 2016. Measuring methane from grazing dairy cows using GreenFeed. Animal Production Science 56: 252-257.

Zimmerman S 2020. GreenFeed Webinar Part 2 – Experimental Design and Applications. Retrieved from https://www.clockinc.com/blog/greenfeed-webinar-part-2-experimental-design-and-applications. (assessed 13 April 2021).

Effect of concurrent pregnancy on the persistency of lactation in dairy cows milked once a day

Y Fu*, SW Peterson and N Lopez-Villalobos

School of Agriculture and Environment, Massey University, PB11222, Palmerston North, New Zealand.

Corresponding author: Email: cdfyx11@126.com

Abstract

Persistency of lactation is a trait of economic importance in any dairy production system. The objective was to investigate the effect of concurrent pregnancy on the persistency of lactation in a commercial herd of dairy cows milked once a day. The data set included 1962 monthly herd-test records (total 8-9 tests per cow) of milk, fat, protein and lactose yield from 219 pregnant cows and 23 non-pregnant cows of different breeds and lactation number. The records were used to model lactation curves for pregnant and non-pregnant cows using random regression with a linear model (Wilmink function). This was not an experiment initially designed to examine the effects of pregnancy on persistency of lactation in dairy cows. Rather, in the course of normal farm operations, some cows became pregnant during lactation and others did not. Pregnancy did not affect the persistency of lactation of cows milked OAD. After conception, there was no difference in total yields between the two groups, but there was a trend that pregnant cows produced lower yields of fat (P=0.0519) and lactose (P=0.0884). This difference between our results and those of previous studies was probably because our cows were only milked once a day or because the data set was not suitable for testing the hypothesis. Further study is required to investigate persistence of lactation in pregnant and non-pregnant cows milked OAD.

Keywords: pregnancy; lactation persistency; cows milked once a day

Introduction

The majority (95%) of milk produced in New Zealand is based on a spring-calving, pasture-based dairy system (Blackwell et al. 2010; Lembeye et al. 2016). Dairy cows are pregnant for much of their lactation in this seasonal-calving system (Roche 2003; Penasa et al. 2016). The gestation length of a cow is approximately 280 days, which leaves a maximum of 85 days for the cow to resume ovarian cycling, be mated and conceive, if the cow is not to calve later each year (Blackwell et al. 2010; Yamazaki et al. 2016).

Although most dairy farmers milk their cows twice a day (TAD), milking cows once a day (OAD) for the entire season has been adopted by many farmers in New Zealand as a strategy to reduce labour requirements, improve animal welfare, and reduce milking costs (Tong et al. 2002; Lembeye et al. 2016). However, compared with TAD milking, the major disadvantage of OAD milking is 15 to 35% lower milk production in a full lactation (Hickson et al. 2006; Clark et al. 2006; Lembeye et al. 2016). Since milk yield is dependent on the persistency of lactation in dairy cows (Stefanon et al. 2002), increasing persistency of lactation is a way to reduce production losses in OAD systems by decreasing the rate of production decline after the peak, or extending the length of lactation (Harding 2002).

Lactation persistency is defined as the ability of an animal to maintain milk yield at a high level after the peak yield (Strapáková et al. 2016; Hermiz & Ahmed 2020). Concurrent pregnancy has been reported to have a negative effect on persistency of lactation of dairy cows milked TAD due to hormonal changes (Stefanon et al. 2002; Salama et al. 2005; Yart et al. 2012), causing regression of the mammary secretory tissue (Akers et al. 2006; Nørgaard et al. 2008), and also due to the nutrient requirements of the developing conceptus, significantly in the last stage of the pregnancy period, reducing nutrients available for milk production (Bell et al. 1995; Pesántez-Pacheco et al. 2019).

The combined exponential and linear model of Wilmink (1987) is one of the mathematical functions used to model the lactation curve. The Wilmink function consists of three terms that are combined additively, enhancing flexibility (Macciotta et al. 2011). Also, the Wilmink function can be easily linearized by setting the "k" parameter to a suitable fixed value and its parameters still maintain a relationship with shape of the lactation curve (Brotherstone et al. 2000). This model has been successfully used to fit individual lactation curves and was implemented in the earlier versions of random models (Olori et al. 1999; Silvestre et al. 2006).

Pregnancy affecting the persistency of lactation in dairy cows milked TAD has been reported in many previous studies, but no study has reported the effects of pregnancy on persistency of lactation in dairy cows milked OAD. With the possible introduction of extended lactation (no mating and no dry period), the effects of pregnancy, or lack of it, on lactation takes on a new importance. Therefore, the objective was to investigate the effect of concurrent pregnancy on persistency of lactation in a herd of commercial dairy cows milked OAD, modelled using a linear regression model with the Wilmink function.

Materials and methods

Data

The data set included 1962 monthly herd-test records for daily milk yield (MY), fat yield (FY), protein yield (PY) and lactose yield (LY) obtained from 242 cows milked OAD at Massey University Dairy 1 during the 2018-19 season. The herd tests began on 29 August 2018, and were completed on 23 April 2019. The herd included 219 pregnant cows and 23 non-pregnant cows.

Animal management and feeding system

All cows were milked OAD at 6:30 AM. From August to April, MY was determined at each monthly herd test throughout the lactation, using mechanical milk meters (Tru-Test Field Collection meter WB HI / Pullout) provided by Livestock Improvement Corporation (Hamilton, New Zealand). Percentages of fat (FP), lactose (LP) and protein (PP) were determined at each herd test using a Fossomatic™ MilkoScan FT 120 instrument (Foss Electric, Hillerød, Denmark). There were only morning samples, since cows were milked OAD. Yields of fat (FY), lactose (LY) and protein (PY) were calculated using daily MY along with FP, LP and PP, respectively.

Calving spread was from 8 July to 2 October 2018. Artificial insemination was from 20 October 2018 to 28 December 2018, and all cows were dried off before May 2019. Cows which failed to conceive (empty), or which were culled due to some other reason, such as mastitis, leg problems or temperament, were defined as non-pregnant cows (NONPREG) in this data set. Cows that conceived were classified as pregnant cows (PREG).

All cows were grazed together on fresh perennial ryegrass *(Lolium perenne)* and white clover *(Trifolium repens)* pasture, though the composition of the diet differed among months during the whole production season. Cows had 100% fresh perennial ryegrass *(Lolium perenne)* and white clover *(Trifolium repens)* only, during September and December. In October, November and February, mixtures of plantain *(Plantago lanceolata)* and chicory *(Cichorium intybus)* were offered to the cows, accounting for 13%, 33% and 23% of the diet, respectively. In January, all cows had access to turnips (*Brassica campestris ssp.* rapifera) which accounted for 63% of the diet. Since a severe shortage of pasture occurred, in February, March and April, pasture silage was offered directly on the paddock at a percentage of 34%, 67% and 14% of the diet, respectively. Similarly, maize silage was offered at 25% and 32% of the diet in August and April, respectively. No estimates of pasture intake were made.

Modelling lactation curves

Lactation curves for the population, and for each cow, were modelled with a linear random regression model based on the Wilmink function (Wilmink 1987) using the MIXED procedure of SAS version 9.4 (SAS Institute Inc.), which was represented as,

$$y_{ti} = (A + Be^{kt} + Ct) + (a_i + b_i e^{kt} + c_i t) + e_{ti}$$

where y is the yield of the cow (i) at day t, parameter A, B and C are the fixed regression coefficients for the population, parameter a_i, b_i, c_i are the random regression coefficients as a deviation from the population for each cow, k is a fixed value and e_{ti} is the random error.

Total MY, FY, PY and LY were estimated from the herd-test values using the predicted daily yields for both pregnant and non-pregnant cows from the linear random regression model with the Wilmink function.

Persistency of lactation

Parameter ($C + c_i$) of the Wilmink function was chosen to measure persistency of lactation, because of its interpretation as the slope of the lactation curve after peak milk yield.

Statistical analyses

This was not an experiment initially designed to examine the effects of pregnancy on milk yield in dairy cows. Rather it was a retrospective analysis of data from a herd in which most of the cows became pregnant during their lactation and some did not do so. The mean conception date of those that conceived was day 88 of the lactation, and the whole lactation length was 205 days (the mean lactation length of NONPREG cows), Accordingly, the statistical analysis was divided into two parts for modeling.

Comparison made over the whole lactation

Estimates of random regression coefficients for each cow obtained from the random regression linear model and measures of persistency of lactation and total yields for MY, FY, LY and PY over the whole lactation were analysed with a mixed linear model with the MIXED procedure of SAS version 9.4. The linear model was the following:

$$y_{ijmk} = \mu + B_i + L_j + R_m + \beta_1 d_{ijmk} + e_{ijmk}$$

where y_{ijmx} is the dependent variable measured in cow k, μ is the population mean; B_i is the fixed effect of breed i; L_j is the fixed effect of parity j; R_m is the fixed effect of reproductive status m, β_1 is the regression coefficient of the dependent variable on deviation from median herd calving date of cow k; e_{ijmk} is the residual random error associated with the observation y.

Comparison made after the mean conception date

Yields from day 88 to 205 days were analysed with the same mixed linear model, as described above, including yield from day 1 to 87 as covariate.

For both models, least-squares means and standard errors of the mean (SEM) for the fixed effects were obtained and used for multiple mean comparisons using Fisher's Least Significant Difference as implemented in the LSMEANS option of the MIXED procedure. Significant differences between means were declared at $P<0.05$, and trends declared at $0.05<P<0.10$.

Results

Parameters "b" of MY, FY and LY for PREG cows were larger than those for NONPREG cows (Table 1), indicating that PREG cows increased to peak yields of milk, protein, fat and lactose more quickly than did NONPREG cows (Fig. 1). However, reproductive status did not affect parameters "a" and "c" for the lactation curves for MY, FY, PY and LY of cows milked OAD (Table 1). Since parameter "c" is related to lactation persistency (Mikhchi et al. 2015), persistency of lactation for MY, PY, FY and LY were similar in PREG and NONPREG cows (Fig. 1).

Table 1 Parameters (a, b and c) used in the model for the lactation curves of total yields of milk, protein, fat and lactose, using linear random regression with the Wilmink function fitted to PREG and NONPREG cows, and significance of reproductive status on each of parameters a, b and c. Values presented are the least-squares means ± SEM and significance of difference in each parameter.

Yield	Reproductive status	Curve Parameters		
		a	b	c
Milk	PREG	22.2556 ± 0.26	-6.6665^{a} ± 0.34	-0.05129 ± 0.00
	NONPREG	21.7086 ± 0.74	-8.6407^{b} ± 0.95	-0.04865 ± 0.00
	P value	0.4709	0.0435	0.3037
Protein	PREG	0.7906 ± 0.01	-0.1308 ± 0.01	-0.00139 ± 0.00
	NONPREG	0.7943 ± 0.02	-0.1822 ± 0.04	-0.00138 ± 0.00
	P value	0.8790	0.1638	0.9656
Fat	PREG	1.0896 ± 0.01	-0.1960^{a} ± 0.02	-0.00227 ± 0.00
	NONPREG	1.1066 ± 0.03	-0.3442^{b} ± 0.06	-0.00232 ± 0.00
	P value	0.6344	0.0172	0.6508
Lactose	PREG	1.1414 ± 0.01	-0.3763^{b} ± 0.02	-0.00281 ± 0.00
	NONPREG	1.1047 ± 0.04	-0.5257^{a} ± 0.05	-0.00263 ± 0.00
	P value	0.3600	0.0072	0.2058

^{a,b}Least-squares means with different superscripts in each column are significantly different ($P<0.05$). PREG: pregnant cows milked once a day, NONPREG: non-pregnant cows milked once a day.

Figure 1 Lactation curves for milk yield (a), protein yield (b), fat yield (c) and lactose yield (d) in pregnant (PREG: dash line) and non-pregnant (NONPREG: solid line) cows milked OAD, modelled with a linear random regression model using the Wilmink function. Day 88 of the lactation (the mean time of the pregnant cows from calving date to conception date) was assumed as the first day of pregnancy for PREG cows. Since parameters "b" of MY, FY and LY for PREG cows were larger than those for NONPREG cows, PREG cows increased to the peak yields of milk, protein, fat and lactose more quickly than did NONPREG cows. However, persistency of lactation between two groups were nearly the same. Data include 23 NONPREG cows milked OAD and 219 PREG cows milked OAD.

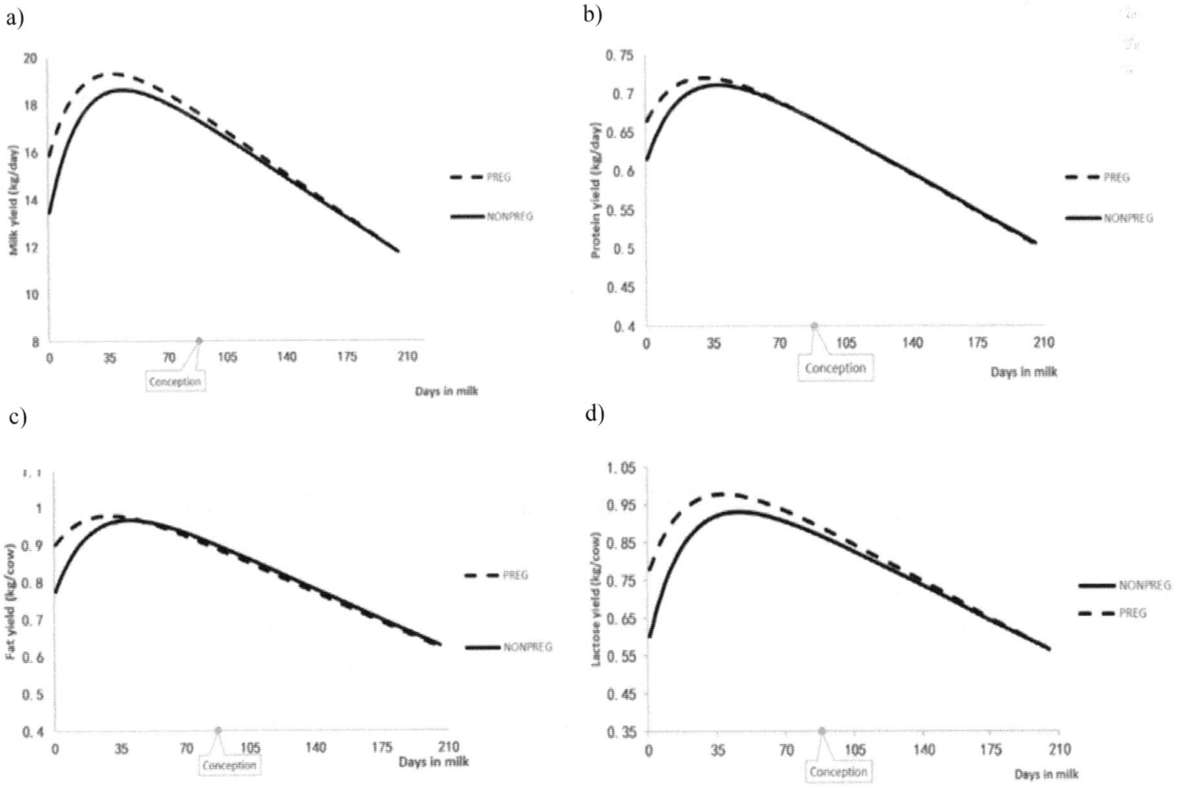

Table 2 Effect of reproductive status on total yields (kg/cow) of milk, fat, lactose and protein over the actual whole lactation, or at the same lactation length (205 days), or including only data from the start of conception (day 88 of the lactation length), modelled using linear random regression with the Wilmink function. Values presented are the least-square means±SEM and significance of each factor. The cows milked OAD were in a commercial herd at Massey University Dairy 1 during the 2018-19 season.

Period	Reproductive status	Number of cows	Total yields			
			Milk	Fat	Lactose	Protein
Whole lactation (actual days)	PREG	219	3798.78^{b}±46.09	196.43^{b} ± 2.23	188.68^{b}±2.33	150.58^{b} ± 1.72
	NONPREG	23	3211.85^{a}±130.54	168.34^{a} ± 6.31	158.64^{a}±6.60	128.17^{a} ± 4.88
	P value		<0.0001	<0.0001	<0.0001	<0.0001
Same lactation length (205 days)	PREG	219	3349.30±36.67	171.69±1.68	167.35±1.90	130.21±1.29
	NONPREG	23	3254.53±103.86	171.20±4.75	160.63±5.37	130.05±3.65
	P value		0.3736	0.9199	0.2225	0.9665
After conception (117 days)	PREG	219	1815.15±9.73	94.38±0.45	89.65±0.53	72.50±0.34
	NONPREG	23	1851.88±26.19	96.64±1.16	92.11±1.44	73.37±0.88
	P value		0.1632	0.0519	0.0884	0.3322

^{a,b} Least-squares means with different superscripts within each time are significantly different (P<0.05). PREG: pregnant cows, NONPREG: non-pregnant cows.

Table 3 Effects of reproductive status, lactation number and breed on lactation length (days) modelled using linear random regression with the Wilmink function. Values presented are the least-squares means±SEM and significance of each factor. The cows milked OAD were in a commercial herd at Massey University Dairy 1 during the 2018-19 season.

Factor		Number of cows	Lactation length	P value
Reproductive status	NONPREG	23	204.05a ± 4.24	<.0001
	PREG	219	248.82b ± 1.50	
Lactation number	1	58	228.43a ± 2.32	0.2158
	2	45	228.63a ± 2.55	
	3	139	224.92a ± 1.60	
Breed	F	54	229.13a ± 2.30	0.2388
	F×J	49	228.26a ± 1.86	
	J	115	224.60a ± 2.29	

^{a,b} Least-squares means with different superscripts within each effect are significantly different (P<0.05). PREG: pregnant cows, NONPREG: non-pregnant cows. Lactation number 1: cows with only 1 previous lactation, 2: cows with two previous lactations, 3: cows with three or more previous lactations. F: Holstein-Friesian cows, J: Jersey cows. F×J: crossbreed cows of Holstein-Friesian and Jersey.

Over the whole lactation, PREG cows had higher total MY, FY, LY and PY than did NONPREG cows (Table 2). However, PREG cows had longer lactation length than did NONPREG cows (Table 3), and when all cows were compared at the same lactation length (205 days), reproductive status did not affect total yields (Table 2). When including only data collected after the mean date of conception, pregnancy did not affect total yields of milk and protein (Table 2). However, PREG cows tended to produce lower yields of fat (P=0.0519) and lactose (P=0.0884).

Discussion

Lactation curves for MY, LY, PY and FY modelled with the Wilmink function were of standard shape for both PREG and NONPREG cows milked OAD, and were consistent with previous studies in which cows were milked TAD (Macciotta et al. 2005; Mikhchi et al. 2015).

Parameter "a" of the Wilmink function is associated with the level of milk production at the start of the lactation (day 1 of the lactation). The higher the value of "a", the higher the level of milk production at the start of the lactation. Parameter "b" is associated with the rate of variation of milk yield in the first part of the lactation curve (higher absolute values of parameter "b" indicate faster increasing values). Parameter "c" is mainly related to the rate of decline after peak yield (the lower the value of parameter "c", the lower the persistency of lactation) (Macciotta et al. 2005; Mikhchi et al. 2015; Otwinowska-Mindur et al. 2016). In the current study, there was no difference in parameter "c" of MY, FY, PY and LY between PREG and NONPREG cows (Table1), and as it is presented in Figure 1, PREG cows had similar persistency of MY, PY, FY and LY. PREG cows increased to the peak yields of milk, protein, fat and lactose more quickly than did NONPREG cows (Fig. 1), as indicated by parameters "b" of lactation curves for MY, FY, PY and LY of PREG cows were greater than those NONPREG cows (Table 1). This was probably because PREG cows had better body condition. This is consistent with previous studies in which parameter "b" was associated with the rate of variation of milk yield in the first part of the lactation curve of cows milked TAD (Macciotta et al. 2005; Mikhchi et al. 2015; Otwinowska-Mindur et al. 2016).

Most previous studies indicated that pregnant cows had lower total yields of milk, fat, protein and lactose (Olori et al. 1997; Brotherstone et al. 2004; Penasa et al. 2016), although all of those experiments used cows milked TAD. Unlike the previous studies, in the present study, over the whole lactation, PREG cows had higher total MY, FY, LY and PY than did NONPREG cows. A likely reason was that the non-pregnant cows had shorter lactation lengths because they were culled. The mean date of removal of non-pregnant cows was 26 February 2019, which was nearly 60 days earlier than the mean dry-off date of the herd. Hence, we needed to exclude the possibility that pregnant cows tended to have higher total milk yield simply because they had longer lactation length.

To determine whether total MY, FY, LY and PY of cows was dependent on the lactation length of cows, we analysed the effects of reproductive status, lactation number and breed on the lactation length, corrected for the deviation of median calving date (Table 3). Only reproductive status significantly affected lactation length, pregnant cows having longer lactation length. Thus, longer lactation length most likely resulted in the higher total yields in PREG cows.

To determine whether the higher total yields in PREG cows were due only to the longer lactation length, we re-analysed the data, assuming that PREG cows and NONPREG cows had the same lactation length (205 days: the mean lactation lengths of NONPREG cows). There were no significant differences in total yields between PREG and NONPREG cows when compared at the same lactation length (Table 3). Thus, a major reason why pregnant cows had higher total milk yield was due to the longer lactation lengths; non-pregnant cows had lower total milk yield because they were culled before the pregnant cows ceased lactation.

Since the cows became pregnant after approximately 88 days of the lactation, assuming the lactation length was 205 days and excluding the effect of previous total yields of cows (day 1 to 87 of the lactation), we re-analysed the effect of concurrent pregnancy on total MY, FY, LY and PY of cows milked OAD after day 88 of the lactation, modelled using the Wilmink function. There were still no differences in total yields of cows milked OAD (Table 2), but there was a trend that PREG cows produced lower FY and LY than did NONPREG cows. One likely reason why there was no difference in total MY is that cows were only milked OAD in this study, and pregnant cows were perhaps better able to maintain yields relative to non-pregnant cows when the metabolic challenge was lower than that imposed by TAD milking and associated higher yields. Previous studies that demonstrated differences in persistency between pregnant and non-pregnant cows all used cows milked TAD (Olori et al. 1997; Brotherstone et al. 2004; Penasa et al. 2016). Another possibility is that the data set in this study is unsuitable for testing the hypothesis of this study, since there was no randomized allocation of cows to treatment groups. Also, the number of cows in this study was small, including only 23 NONPREG cows and 219 PREG cows, limiting the statistical power of the study.

According to the hormonal explanation, pregnant cows should have lower total milk yield due to increasing concentrations of estrogen and progesterone which interferes with prolactin binding to the receptors on the mammary epithelial cell (Stefanon et al. 2002; Yart et al. 2012). However, the concentration of estrogen remains low during the first two-thirds of pregnancy and rises during the third trimester (Henricks et al. 1972; Davis 2017). Since PREG cows in this study were mated at day 88 of the lactation and the lactation length considered was limited to 205 days to avoid inducing a large error by extrapolating far beyond the lactation length of NONPREG cows, the third trimester of gestation was not included in the lactation curves, which could partly explain why the PREG cows had similar total yields to NONPREG cows after the mean mating date in this study. Also, since NONPREG cows in this study were those which had failed to conceive, lower body condition of NONPREG cows was probably one of the factors which caused them not to conceive and also prevented them from producing well (Domecq et al. 1997; Haresign & Cole 2013), and could also explain why there were no differences of total yields between PREG and NONPREG cows from day 88 to day 205 of the lactation. The non-pregnant cows may not have had sufficient body condition to maintain yields at higher values than those of the pregnant cows.

Conclusion

The objectives of this study were to investigate the effect of concurrent pregnancy on persistency of lactation in dairy cows milked OAD modelled using a linear regression model with the Wilmink function. There was no significant difference in persistency of lactation for milk, fat, lactose and protein between PREG and NONPREG cows milked OAD. After conception, there were no differences in total yields between the two groups, but pregnant cows tended to produce lower yields of fat ($P=0.0519$) and lactose ($P=0.0884$). This difference between our results and those of previous studies, was probably because our cows were only milked once a day, or because the data set was not suitable for testing the hypothesis. Further study is required to investigate persistence of lactation in pregnant and non-pregnant cows milked OAD.

Acknowledgments

The authors are grateful to the staff at the Massey University Dairy 1 farm for support and collecting data.

References

Akers RM, Capuco AV, Keys JE 2006. Mammary histology and alveolar cell differentiation during late gestation and early lactation in mammary tissue of beef and dairy heifers. Livestock Science 105: 44-49.

Bell AW, Slepetis R, Ehrhardt RA 1995. Growth and accretion of energy and protein in the gravid uterus during late pregnancy in Holstein cows. Journal of Dairy Science 78: 1954-1961.

Blackwell MB, Burke CR, Verkerk GA 2010. Reproductive management practices in New Zealand dairy farms: what will the future hold in a consumer-focused, export-driven marketplace? Society of Reproduction and Fertility supplement 67: 341-355.

Brotherstone S, Thompson R, White IMS 2004. Effects of pregnancy on daily milk yield of Holstein–Friesian dairy cattle. Livestock Production Science 87: 265-269.

Brotherstone S, White IMS, Meyer K 2000. Genetic modelling of daily milk yield using orthogonal polynomials and parametric curves. Journal of Animal Science 70: 407-415.

Clark DA, Phyn CVC, Tong MJ, Collis SJ, Dalley DE 2006. A systems comparison of once-versus twice-daily milking of pastured dairy cows. Journal of Dairy Science 89: 1854-1862.

Davis SR 2017. Triennial lactation symposium/BOLFA: mammary growth during pregnancy and lactation and its relationship with milk yield. Journal of Animal Science 95: 5675-5688.

Domecq JJ, Skidmore AL, Lloyd JW, Kaneene JB 1997. Relationship between body condition scores and milk yield in a large dairy herd of high yielding Holstein cows. Journal of Dairy Science 80: 101-112.

Harding L 2002. Milking cows once daily since 1985: Our reasons and experiences. Massey University Dairy Farming Annual 54: 85-87.

Haresign W, Cole DJA 2013. Body condition, milk yield and reproduction in cattle. Recent Advances in Animal Nutrition–1979: Studies in the Agricultural and Food Sciences 107.

Henricks DM, Dickey JF, Hill JR, Johnston WE 1972. Plasma estrogen and progesterone levels after mating, and during late pregnancy and postpartum in cows. Endocrinology 90: 1336-1342.

Hermiz H, Ahmed J 2020. Factors affecting persistency and repeatability in several breeds of dairy cattle. Plant Archives 20: 9-12.

Hickson RE, Lopez-Villalobos N, Dalley DE, Clark DA, Holmes CW 2006. Yields and persistency of lactation in Friesian and Jersey cows milked once daily. Journal of Dairy Science 89: 2017-2024.

Lembeye F, Lopez-Villalobos N, Burke JL, Davis SR 2016. Milk production of Holstein-Friesian, Jersey and crossbred cows milked once-a-day or twice-a-day in New Zealand. New Zealand Journal of Agricultural Research 59: 50-64.

Macciotta NPP, Dimauro C, Rassu SPG, Steri R, Pulina G 2011. The mathematical description of lactation curves in dairy cattle. Italian Journal of Animal Science 10: 4.

Macciotta NPP, Vicario D, Cappio-Borlino A 2005. Detection of different shapes of lactation curve for milk yield in dairy cattle by empirical mathematical models. Journal of Dairy Science 88: 1178-1191.

Mikhchi A, Mashhadi MH, Jafarabadi GA 2015. Estimation of lactation curve parameters and its correlation with udder type traits in Iranian Holstein dairy cattle using Wilmink's function. The Indian Journal of Animal Sciences 85: 165-178.

Nørgaard JV, Sørensen MT, Theil PK, Sehested Jakob, Sejrsen K 2008. Effect of pregnancy and feeding level on cell turnover and expression of related genes in the mammary tissue of lactating dairy cows. Animal 2: 588-594.

Olori VE, Brotherstone S, Hill WG, Mcguirk BJ 1997. Effect of gestation stage on milk yield and composition in Holstein Friesian dairy cattle. Livestock Production Science 52: 167-176.

Olori VE, Brotherstone S, Hill WG, Mcguirk BJ 1999. Fit of standard models of the lactation curve to weekly records of milk production of cows in a single herd. Livestock Production Science 58: 55-63.

Otwinowska-Mindur A, Ptak E 2016. Factors affecting the shape of lactation curves in Polish Holstein-Friesian cows. Animal Science Papers and Reports 34: 373-386.

Penasa M, De Marchi M, Cassandro M 2016. Effects of pregnancy on milk yield, composition traits, and coagulation properties of Holstein cows. Journal of Dairy Science 99: 4864-4869.

Pesántez-Pacheco J, Heras-Molina A, Torres-Rovira L, Sanz-Fernández MV, García-Contreras C, Vázquez-Gómez M, Feyjoo P, Cáceres E, Frías-MM, Hernández F 2019. Maternal metabolic demands caused by pregnancy and lactation: association with productivity and offspring phenotype in high-yielding dairy ewes. Animals 9: 295.

Roche JR 2003. Effect of pregnancy on milk production and bodyweight from identical twin study. Journal of Dairy Science 86: 777-783.

Salama-Ahmed AK, Caja G, Such X, Casals R, Albanell E 2005. Effect of pregnancy and extended lactation on milk production in dairy goats milked once daily. Journal of Dairy Science 88: 3894-3904.

Silvestre AM, Petim-Batista F, Colaco J 2006. The accuracy of seven mathematical functions in modeling dairy cattle lactation curves based on test-day records from varying sample schemes. Journal of Dairy Science 89: 1813-1821.

Stefanon B, Colitti M, Gabai G, Knight C H, Wilde C J 2002. Mammary apoptosis and lactation persistency in dairy animals. Journal of Dairy Research 69: 37-52.

Strapáková E, Candrák J, Strapák P 2016. Genetic relationship of lactation persistency with milk yield, somatic cell score, reproductive traits, and longevity in Slovak Holstein cattle. Archives Animal Breeding 59: 329-335.

Tong MJ, Clark D, Cooper C 2002. Once-a-day milking: Possible and profitable? Proceedings of the New Zealand Grassland Association 64: 33-37.

Wilmink JBM 1987. Adjustment of test-day milk, fat and protein yield for age, season and stage of lactation. Livestock Production Science 16: 335-348.

Yamazaki T, Hagiya K, Takeda H, Osawa T, Yamaguchi S, Nagamine Y 2016. Effects of stage of pregnancy on variance components, daily milk yields and 305-day milk yield in Holstein cows, as estimated by using a test-day model. Animal 10: 1263-1270.

Yart L, Dessauge F, Finot L, Barbey S, Marnet P, Lollivier V 2012. Ovariectomy improves lactation persistency in dairy cows. Journal of Dairy Science 95: 3794-3802.

Genomic evaluations of sheep in New Zealand

MA Lee[a,b*], S-AN Newman[c], KG Dodds[c], MA Nilforooshan[b,e], B Auvray[d], S McIntyre[a] and D Campbell[a]

[a]*Beef+Lamb NZ Genetics, PO Box 5501, Dunedin 9054, New Zealand;* [b]*Department of Mathematics and Statistics, University of Otago, PO Box 56, Dunedin, New Zealand;* [c]*AgResearch, Invermay Agricultural Centre, Private Bag 50034, Mosgiel, New Zealand;* [d]*Iris Data Science, Level 2, 7 Bond St, Dunedin, New Zealand; Livestock Improvement Limited, 605 Ruakura Road, Newstead 3286, New Zealand*

[*]*Corresponding author:* Email: michael.lee@otago.ac.nz

Abstract

Genetic improvement of farmed livestock in New Zealand is achieved through selective breeding using predictions of an animal's genetic merit called 'breeding values'. Inclusion of genomic data in genetic evaluations (known as 'genomic selection') has increased the accuracy of breeding values compared to best linear unbiased prediction (BLUP) methods based on relationships from known pedigree; particularly for young animals with limited recording. The additional accuracy gained from genotype information is most valuable for traits that are: hard to measure; sex limited; measured late in life; and/or expensive to measure. The 'New Zealand Genetic Evaluation' (NZGE) for sheep has used single-step genomic BLUP (ssGBLUP) since late 2018, which provides the basis for genomic selection in the NZ sheep industry. This method allowed pedigree, performance records and genomic data to be included in the same analysis to predict breeding values in contrast to performing a separate pedigree and genomic analysis and then combining these results (multi-step genomic BLUP; msGBLUP). ssGBLUP is a better alternative for genomic evaluations compared to msGBLUP from a perspective of bias, best use of information and computational-ease. This paper describes the NZGE genomic analysis and details some of the current challenges relating to optimising accuracy and minimising bias in the evaluation. Genomic selection is a useful additional tool for breeders. However, it is important to be mindful that data recording is still the foundation for any genetic improvement program, as is allowing breeders to best leverage genomic selection.

Keywords: sheep; genetics; genomics; breeding

Introduction

The best linear unbiased prediction (BLUP) statistical method (Henderson et al. 1959; Henderson 1975) has been used to predict breeding values in sheep in New Zealand for decades (Newman et al. 2000). In BLUP (equation 1), the breeding values (**u**) are predicted from the information from phenotypes (**y**) and fixed effects (**b**), where **X** and **Z** are design matrices and **e** is a vector of residual errors.

Equation 1

$$\mathbf{y} = \mathbf{Xb} + \mathbf{Zu} + \mathbf{e}$$

In order to predict breeding values or random effects ($\hat{u}$) from equation 1, the mixed-model equations (equation 2; MME in matrix notation) described by Henderson 1975, where λ is the ratio of the variances of the residual additive effects over the variances of the additive genetic effects assuming an animal model are solved. The model assumes homogeneous variances for the residual errors (e) that are independent of the random effects ($\hat{u}$). This also implies that population genetic parameters are known (e.g., genetic and residual additive variances), requiring pre-estimation of parameters for the population to be evaluated.

BLUP assumes an infinitesimal or polygenic model (Falconer 1996) – in that a given quantitative trait is explained by an infinite number of genes, each having a small effect.

The relationship among the breeding values is given by the covariance matrix (**G**) that is estimated from pedigree for BLUP, so that the numerator relationship matrix (A) is used for G in the MME. Increasingly, genotype data are used to estimate the relationship matrix. For example, for genomic BLUP (GBLUP) the genomic relationship (**G**), calculated from genotype data, is used in the MME.

Equation 2

$$\begin{pmatrix} X'X & X'Z \\ Z'X & Z'Z + G^{-1}\lambda \end{pmatrix} \begin{pmatrix} \hat{b} \\ \hat{u} \end{pmatrix} = \begin{pmatrix} X'y \\ Z'y \end{pmatrix}$$

A national across-flock and -breed genetic evaluation has been available to the NZ sheep industry through Sheep Improvement Limited (SIL) called the New Zealand Genetic Evaluation (NZGE). Genomic data have been incorporated into the NZGE, but the models and genetic parameters that have been determined over the past decades and used in SIL BLUP analyses still form the basis for NZGE, irrespective of the integration of genomics. Most the traits are analysed in multi-trait models that use genetically correlated traits to improve accuracy and account for selection bias. NZGE is computed using cloud-based high performance computing infrastructure and reported weekly to suit the structure of sheep seed-stock breeders, which comprises many different breeders needing evaluation results at various times.

The number of pedigreed animals available for evaluations has increased from thousands of animals in the 1980s to a total of about 10.75 million in 2020; whilst the number of animals with genotypes (over 5,000 markers) has increased to over 300,000 (Figure 1).

Figure 1 Cumulative number of sheep evaluated in NZGE with pedigree (in millions (Mil)) and genotypes

Figure 1 Cumulative number of sheep evaluated in NZGE with pedigree (in millions (Mil)) and genotypes from 1985-2020 calculated by the year of birth of animals.

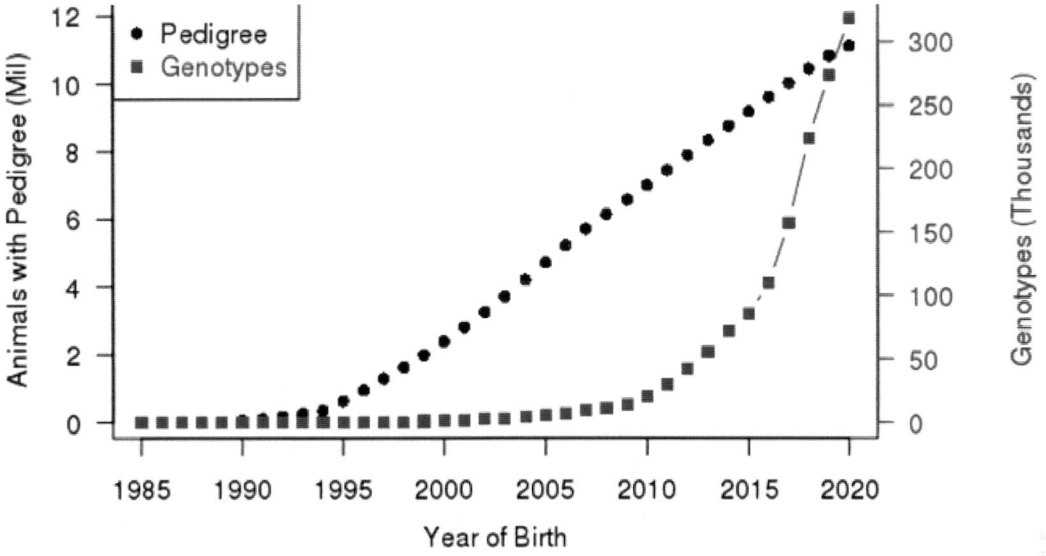

from 1985-2020 calculated by the year of birth of animals.

This paper describes the NZGE with a focus on the implementation of the genomic component and highlights some of the challenges.

Earlier genomic predictions via GBLUP

The basis of genomic prediction for sheep was made possible after the development of the Illumina Ovine 50K single nucleotide polymorphism (SNP) chip as part of the International Sheep Genomics Consortium in 2009, comprising AgResearch, Baylor College of Medicine (Houston, TX), the Commonwealth Scientific and Industrial Research Organisation (Australia), and the United States Department of Agriculture. Since then, a number of SNP chips have been developed and used. The density ranged from an early chip of about 5K to a high density 600K chip. Significant numbers of animals were only genotyped for genomic prediction once cost-effective, lower density, SNP chips (~5K) were available.

The first genomic predictions in 2013 were based on Genomic BLUP (GBLUP) (Meuwissen 2009), where the genotypes were used to create a genomic relationship matrix (**G**) that is used in the MME instead of the pedigree relationship matrix (**A**) (equation 2). The process was described by Auvray et al. 2014 and predictions were restricted to the maternal breeds Romney, Coopworth, Perendale and composites of these breeds.

A problem with GBLUP is that breeding values are predicted only for genotyped animals and most of the animals to be evaluated from a NZGE perspective have only pedigree data (see Figure 1). Consequently, the strategy for genomic prediction was to prioritise sires for genotyping and to optimally use all information by calculating a pseudo-phenotype (e.g. deregressed breeding value with parent average removed) for these animals. These animals formed the reference population which were used to calibrate the genomic predictions. However, as these animals also had a breeding value from BLUP (estimated breeding value; EBV) there was a need to combine or blend the predictions from GBLUP (molecular breeding value; MBV) with EBV (equation 3). The blending function was based on a selection index approach (VanRaden 2008; Harris & Johnson 2010), where the blended genomic breeding value ($\hat{g}$) is calculated from three different evaluations namely, the GBLUP and two pedigree BLUP evaluations, consisting of a "national" evaluation and a subset evaluation of only genotyped animals. The reliabilities from these three evaluations were, respectively R_G, R_N and R_A and the breeding values to be blended, respectively, value $\hat{a}_G$, $\hat{a}_N$ and $\hat{a}_A$. Importantly, the $\hat{a}_G$ and $\hat{a}_N$ are not assumed to be independent and their assumed covariance is calculated from reliabilities of the breeding values by equation 4.

Equation 3 Blending of EBV and GBV

$$\hat{g} = \frac{(1-R_S)(1-R_G)\hat{u_N} - 1 - R_N 1 - R_G \hat{u_S} + (1-R_N)(1-R_S)\hat{u_G}}{1 - 2R_S - R_N R_G + R_S R_G + R_N R_S}$$

Equation 4 Covariance of the national and molecular breeding values

$$cov(\hat{u_G}, \hat{u_N}) = R_S + \frac{(R_G - R_S)(R_N - R_S)}{1 - R_S}$$

The number of genotyped animals was modest enough to estimate the accuracies of the breeding values from the solutions of the left hand side of the MMEs. This allowed accuracies to be calculated for each animal to be evaluated and used for blending in equation 3 as a reliability. Similarly, when the software ASREML (Gilmore et al. 2009) was used for BLUP the accuracies could also be calculated. The goal of the blending procedure was to allow, as much as possible, breeders to have one breeding value that was comparable across groups of animals, such as flocks.

However, this multistep evaluation was problematic as recalibrations (i.e. including new information into the components needed to run GBLUP) were time-consuming and the use of all available information was not optimal, and with each calibration there was significant re-ranking amongst animals.

The New Zealand Genetic Evaluation

Single-step GBLUP

NZGE with ssGBLUP (Aguilar et al. 2011) was implemented in 2018 to allow for a single national evaluation that included genomic information and in which most animals still have only pedigree without any genotype data. This replaced the pedigree BLUP NZGE that was the main national evaluation for sheep in NZ at that time.

ssGBLUP was attractive as the models (mostly multi-trait), genetic parameters and much of the infrastructure, developed over a number of decades, to adjust and quality control the data could be re-used and the transition would allow for minimal changes (re-ranking), particularly for those animals without genomic information.

In principle, ssGBLUP is simple in that the relationship matrix used in the MME is replaced by a matrix constructed from pedigree and genotypes. The pedigree matrix (**A**) can be partitioned as in equation 5, where the partition $\mathbf{A}_{22}$ corresponds to animals with genotype and $\mathbf{A}_{11}$, $\mathbf{A}_{12}$ **and** $\mathbf{A}_{21}$ are those without genotypes. The combined pedigree and genomic relationship matrix (**H**) used in ssGBLUP is given in equation 6. Importantly, **H** gives the covariances of all animals to be evaluated using both genotype and pedigree information.

Equation 5

$$A = \begin{pmatrix} A_{11} & A_{12} \\ A_{21} & A_{22} \end{pmatrix}$$

Equation 6

$$H = \begin{pmatrix} A_{11} - A_{12}A_{22}^{-1}A_{21} + A_{12}A_{22}^{-1}GA_{22}^{-1}A_{21} & A_{12}A_{22}^{-1}G \\ GA_{22}^{-1}A_{21} & G \end{pmatrix}$$

In practise, the $\mathbf{H}^{-1}$ is calculated directly as in equation 7 and used in the MME ($\mathbf{G}^{-1}$ in equation 2).

Equation 7 Calculation of the H inverse.

$$H^{-1} = A^{-1} + \begin{pmatrix} 0 & 0 \\ 0 & G^{-1} - A_{22}^{-1} \end{pmatrix}$$

G generally needs to be regularised (Mäntysaari et al. 2017) before solving and in NZGE it is weighted with $\mathbf{A}_{22}$ (**A** in equation 8) by a value of α to give G_w. However, the value of α is also optimised within trait by validation. An interpretation of α is the proportion of additive genetic variance that is not explained by **G** and in NZGE this can vary from 0.05-1.

Equation 8 Weighting of the **G** matrix with **A** ($\mathbf{A}_{22}$).

$$G_w = (1 - \alpha)G + \alpha A$$

As an example, a validation for fleece-weight at 12-months of age (FW12), within a multivariate model that included weigh and lamb fleece weight, using ~10 million pedigreed animals and 50,630 genotyped animals each with 41,958 SNPs to make **G**. The genotypes were deduced from a 600K, 50K or lower than 50K chip. The lower density genotypes were imputed to 41,958 SNPs using FImpute2 (Sargolzaei et al. 2014), where only genotypes that imputed with an accuracy >97% concordance were used. The accuracy of imputation was determined by validation within the chip used.

In order to estimate an optimal α, 810 progeny-tested sires (born from 2013-17) were used as validation animals. The mean number of progeny with phenotype for the sires was 38.6 (range 21-69). A reduced dataset was constructed by removing all the phenotypes from the sires and all their descendants. Deregressed breeding values with parent average removed (dBV) (Garrick et al. 2009) that were calculated from a BLUP using the full dataset with 2,016,945 phenotypes were used to compare against predicted breeding values with the reduced dataset to assess the prediction accuracy of ssGBLUP and pedigree BLUP. The results, analysed within and across breed groups, as correlation and the slope from regressing dBV on predicted breeding values are given in Table 1. The results from a Romney flock that was sizeable and well-recorded with a significant amount of genotyping are also given in Table 1, where 67 of the 810 sires used in validation were from this flock.

In general, ssGBLUP correlations were higher than for BLUP correlation across all α-values and groups. The slopes from regression also suggested ssGBLUP was also less biased than BLUP. The correlation was 0.26 for BLUP and at best 0.44 for ssGBLUP which was also less biased. In NZGE an α-value of 0.25 is used. The aim is to use a value that optimises accuracy and minimises bias across all breeds. A similar procedure is used to estimate α across all other NZGE traits.

Imputation

Most of the genotypes submitted for genomic predictions are from lower density chips than 50K, for economic reasons. These genotypes are imputed to a density of about 42K and used for ssGBLUP. Imputation refers to the inference of unobserved genotypes (e.g., inferring the genotypes that are on the 42K panel of genotypes that are not on the lower density 15K panel). We recommend that breeders routinely genotype the sires that they use for mating with a higher-density SNP chip (e.g., 600K) to update imputation reference panels.

Imputation inaccuracy reduced prediction accuracy and could result in bias. To demonstrate this, we used a prediction dataset of 1,301 Romney animals that were genotyped with an Ovine 15K SNP chip and 15,885

Table 1 Correlations (r) and slopes from a validation for ssGBLUP with FW12. The slopes were the dBVs regressed onto the predictions for the animals used in validation.

	ssGBLUP alpha value used									
	1 (BLUP)		0.75		0.5		0.25		0.05	
Group	r	slope	r	slope	r	slope	r	slope	r	slope
All	0.49	0.88	0.53	0.88	0.54	0.85	0.54	0.8	0.5	0.7
Romney	0.43	0.85	0.46	0.89	0.47	0.84	0.46	0.76	0.41	0.62
Coopworth	0.31	0.6	0.36	0.67	0.4	0.7	0.43	0.7	0.41	0.63
Perendale	0.38	0.74	0.42	0.77	0.44	0.76	0.44	0.74	0.43	0.68
Composites	0.54	0.59	0.57	0.76	0.58	0.88	0.57	0.86	0.52	0.42
Romnney flock	0.26	0.59	0.38	0.76	0.42	0.76	0.44	0.7	0.42	0.59

suitable training animals genotyped with a 50K chip. The total number of pedigreed animals in the evaluation was 9,513,189. Different imputation reference panels, generated by random sampling from available reference genotypes, were used to impute the genotypes of the prediction dataset and the accuracy of imputation, for each animal for a given imputation reference panel, was estimated by validation.

An example for the trait number of lambs born is given in Figure 2, where the prediction dataset had an imputation accuracy of 89% (panel A; GBV89) and 70% (panel B; GBV70). The breeding values of animals imputed with an accuracy of 89% were indistinguishable from those that were imputed to a high average accuracy (96%) with a correlation of >0.99. Whereas those imputed to an average accuracy of 70% were less accurate, with a correlation of 0.9 compared to those imputed at 96%. They were also biased (in this case upwards), such that there was a shift in the mean for these animals. From the perspective of ram breeders, this bias might result in sub-optimal predictions across groups (e.g., flocks). Poor convergence in solving the ssGBLUP equations was another result of increased numbers of poorly imputed genotypes in the evaluations. We observed similar results for other traits (results not shown).

In practise, with NZGE the imputation accuracy of each animal is estimated by validation and only animals that reach a threshold of imputation accuracy (90%) are included. This is also a useful quality control check to ensure that only genotypes from animals that are likely to obtain useful predictions are included (i.e. correct breed and relatives of previously genotyped and phenotyped individuals). This is because the reference panel comprises animals of correct breed that are generally sires from flocks using genomic selection. Therefore, animals that are of the wrong breed type and unrelated will generally be poorly imputed.

Traits with low levels of recording

There are some traits in NZGE that have low levels of recording - often traits for which phenotypes are difficult and/or expensive to measure and where breeders are generally unwilling to record. As such, they are also the traits for which genomic predictions are most useful (e.g., faecal egg count (parasite resistance) and green-house gas emission (Pickering et al. 2015)), where the ratio of animals with phenotype to those with genotype is low.

As the phenotyping to genotyping ratio decreased for a trait, the convergence for ssGBLUP takes more iterations to

Figure 2 Graph of breeding values for the trait number of lambs born from a ssGBLUP with prediction dataset genotypes imputed to an average accuracy of 89% (GBV89) and 70% (GBV70) given in red compared to when imputed to an accuracy of 96% (GBV96) . Sheep that were not imputed (training animals) are given in black.

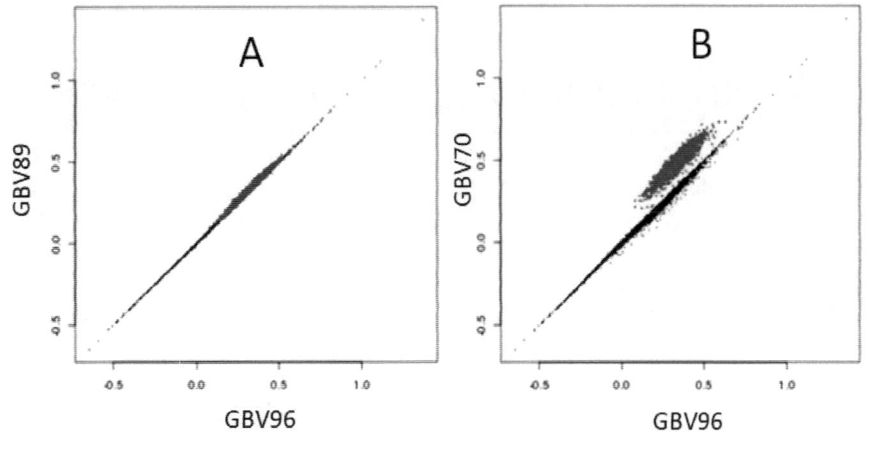

a point where there is a failure to converge. Consequently, in NZGE for a given trait, genotypes from animals that meet all the following criteria: are not sires, have no phenotype; were culled or deceased and with a recorded year of birth greater than two years in the current evaluation year are not included. These animals do not offer any useful information and are not of interest to breeders, but they remain in the evaluation as a pedigreed animal.

A longer-term solution to cater for breeding schemes and scenarios for groups that undertake limited or no phenotyping or pedigree-recording and rely on genomic predictions (e.g., multiplier flocks) is to not include these sheep in NZGE, but make predictions only based on their genotypes using separate analyses.

Dealing with a high number of genotyped animals

So that the evaluation can still be run when the number of genotypes animals exceeds about 150,000, other strategies to estimate the **G** and to invert or solve it may be needed. This is because calculating and inverting of **G** is computationally demanding, where the computations increase about cubically with the number of genotypes. Here we consider some future computing options.

We investigated the algorithm for proven and young animals (APY)(Misztal et al. 2014) (Fragomeni et al. 2015), that gives an approximation to $\mathbf{G}^{-1}$. In general, the results were satisfactory with a core of animals selected randomly (Nilforooshan & Lee 2019). However, the APY was not used in the NZGE, because breeding values from some flocks re-ranked between successive evaluations with only slight changes in the core. It was problematic to resolve this even with a tailored core (results not included). It was concluded that the population being evaluated was too heterogenous for APY to be reliably used. Focus was therefore moved to exact methods rather than approximations.

In exact ssGBLUP (Fernando et al. 2016), the genotyped animals were partitioned into core and noncore animals. The MME used were constructed with only the breeding values of the core animals and breeding values of the non-core animals were estimated as a linear function of the estimates of the breeding values of the core animals. This gives exact solutions for all animals. It is considered that this method should not be any more difficult to implement than APY, however, this was not proven for a large dataset and would require significant changes to the NZGE system.

Another method ssGTBLUP (Mäntysaari et al. 2017), uses a T-matrix and is equivalent to ssGBLUP, but is computationally less demanding. Furthermore, the T-matrix has dimensions $m \times n$, where m is the number of markers and n the number of genotyped animals. This is in contrast to the H-inverse that has dimensions $n \times n$. This becomes important with large numbers of genotypes in the evaluation as the memory usage to store these matrices becomes large. ssGTBLUP is being further investigated as a longer term solution for NZGE.

An alternative strategy is to use brute force for matrix inversion. For example, a **G** of 2,097,152 × 2,097,152 matrix (of size 32TB) could be inverted in 6.4 hours on 2,048 compute nodes (49,152 cores) (Gray, 2016). However, from a practical and economic perspective this approach is not preferred. Another way of overcoming the bottleneck due to large numbers of genotypes is to construct and invert **G** in partitions. This uses mathematical operations for block matrices and recognises that a square matrix can be inverted by partitioning, where the largest block to solve is n divided by the number of partitions. **G** can be constructed and solved for 200,000 genotyped animals partitioned into 4×4 blocks on a modest machine (results not shown).

Hardware and analysis

The NZGE is run on cloud-based computer services and is automated. This is facilitated using docker containers which are orchestrated for weekly evaluations. The use of containers adds portability and ease of automation, where in principle, it can be used on any server or operating system. The use of cloud-based compute infrastructure has allowed immediate and seamlessly change in machine types and number of machines. The NZGE uses about 20 high-performance computing machines (typically 128-512 GB of RAM and 16-48 processors). An evaluation with about 150,000 genotyped animals and about 10 million animals running up to 20 different goal trait groups models will typically take about 20 hours to run, currently, including pre-processing of data.

Conclusion

An overview of the current NZGE system and some of the issues in developing and maintaining this evaluation system, particularly regarding when genotype data is included. The population evaluated in NZGE is heterogeneous and includes different breeds and stabilised crosses of these breeds. The major proportion are Romney-based and genomic prediction for these breeds has been integrated into NZGE. The foundation for NZGE is decades of data recording and models and parameter estimation by many researchers. The implementation of genomic prediction has allowed more utility to be gained from this foundation.

In the future, new and/or significantly more data may need to be evaluated such as data from whole genome sequencing, markers associated with QTL and data from non-seedstock farms and processors (e.g. from image analysis using artificial intelligence). A significant challenge remains to include disparate breeds into a single genetic evaluation with genomic predictions.

While genomic selection is a useful tool for breeders, it is important to be mindful that data recording is still the foundation for genetic improvement which will then allow breeders to best leverage genomic technologies for their selection decisions.

Acknowledgements

We acknowledge the commitment of many individuals and organisations involved in data recording, research and development and for financial investment that forms the basis of NZGE. Meat levy payers through Beef+Lamb New Zealand have supported and funded development and routine NZGE. Supported by the Ministry of Business, Innovation and Employment (NZ) is also acknowledged. We thank Dr Neville Jopson for useful comments related to this paper.

References

Aguilar I, Misztal I, Legarra A, Tsuruta S 2011. Efficient computation of the genomic relationship matrix and other matrices used in single-step evaluation. Journal of Animal Breeding and Genetics 128: 422-428.

Auvray B, McEwan JC, Newman SA, Lee MA, Dodds KG 2014. Genomic prediction of breeding values in the New Zealand sheep industry using a 50K SNP chip. Journal of Animal Science 92: 4375-4389.

Falconer DS 1996. Introduction to Quantitative Genetics. Pearson Education.

Fernando RL, Cheng H, Garrick DJ 2016. An efficient exact method to obtain GBLUP and single-step GBLUP when the genomic relationship matrix is singular. Genetics Selection Evolution 48: 80.

Fragomeni BO et al. 2015. Hot topic: Use of genomic recursions in single-step genomic best linear unbiased predictor (BLUP) with a large number of genotypes. Journal of Dairy Science 98: 4090-4094.

Garrick DJ, Taylor JF, Fernando RL 2009. Deregressing estimated breeding values and weighting information for genomic regression analyses. Genetics Selection Evolution 41: 55.

Gray A 2016. Invertastic: Large-scale Dense Matrix Inversion, EPPC, The University of Edinburgh. June, 2016. https://www.archer.ac.uk/documentation/white-papers/invertastic/invertasticGray.pdf

Gilmore A, Gogel BJ, Cullis BR, Thompson R, Butler D 2009. ASReml User Guide Release 3.0. In: V. I. Ltd (ed.), Hemel Hempstead, United Kingdom.

Harris BL, Johnson DL 2010. Genomic predictions for New Zealand dairy bulls and integration with national genetic evaluation. Journal of Dairy Science 93: 1243-1252.

Henderson CR 1975. Best linear unbiased estimation and prediction under a selection model. Biometrics 31: 423-447.

Henderson CR, Kempthorne O, Searle SR, von Krosigk CM 1959. The Estimation of Environmental and Genetic Trends from Records Subject to Culling. Biometrics 15: 192-218.

Mäntysaari EA, Evans RD, Strandén I 2017. Efficient single-step genomic evaluation for a multibreed beef cattle population having many genotyped animals. Journal of Animal Science 95: 4728-4737.

Meuwissen TH 2009. Accuracy of breeding values of 'unrelated' individuals predicted by dense SNP genotyping. Genetics Selection Evolution 41: 35.

Misztal I, Legarra A, Aguilar I 2014. Using recursion to compute the inverse of the genomic relationship matrix. Journal of Dairy Science 97: 3943-3952.

Newman SA, Dodds KG, Clark JN, Garrick DJ, McEwan JC 2000. The Sheep Improvement Limited (SIL) genetic engine. Proceedings of the New Zealand Society of Animal Production. 60: 195-197.

Nilforooshan MA, Lee M 2019. The quality of the algorithm for proven and young with various sets of core animals in a multibreed sheep population. Journal of Animal Science 97: 1090-1100.

Pickering NK, Oddy VH, Basarab J, Cammack K, Hayes B, Hegarty RS, Lassen J, McEwan JC, Miller S, Pinares-Patiño CS, de Haas Y 2015. Animal board invited review: genetic possibilities to reduce enteric methane emissions from ruminants. Animal 9: 1431-1440.

Sargolzaei M, Chesnais JP, Schenkel FS 2014. A new approach for efficient genotype imputation using information from relatives. BMC genomics 15: 478-478.

VanRaden PM 2008. Efficient methods to compute genomic predictions. Journal of Dairy Science 91: 4414-4423.

Dominant mutation(s) for low fibre curvature in the wool of sheep

DR Scobie*, C van Koten, M Richena, JE Plowman and DP Harland

AgResearch Lincoln, Private Bag 4749, Christchurch, 8140

**Corresponding author:* Email: scobie@agresearch.co.nz

Abstract

Two sires judged to have phenotypically lustrous wool as neonates were mated to ewes that did not exhibit lustre in their fleece as adults. Around half their progeny were lustrous as neonates. Wool characteristics of these progeny were measured on an optically based fibre diameter analyser (OFDA) when they were weaned and again as yearlings. There was a profound difference in mean fibre curvature ($P<0.001$) such that the lustrous phenotype had less curvature in both sire groups (22.8, 24.5°/mm) compared with the non-lustrous (67.5, 56.2°/mm) as lambs, and 29.8, 28.1°/mm compared with 74.1, 55.0°/mm as yearlings. Curvature of the sire groups was significantly different in concert with breed mix, Merino in one sire and Romney in the other, but only when their fleece was not lustrous. The greasy fleece of the lustrous yearlings was significantly lighter (1.8 *vs* 3.0 kg, $P<0.05$) than their non-lustrous contemporaries, while there was no significant difference in fibre diameter between the two phenotypes as lambs or yearlings. Fibre curvature is inexpensive to measure and well-defined, whereas lustre is a qualitatively assessed trait. These findings suggest "low fibre curvature" would be a more appropriate name for the trait than "lustre".

Keywords: lustre mutant; wool fibre; crimp; curvature

Introduction

Perhaps the first scientifically documented sheep that resemble those we discuss here were reported by Short (1958) and were called "felting lustre" mutants. A very similar phenotype had been observed in America in the 1930s but was called "Silky", and breeding experiments were reported much later (Warwick et al. 1960). These experiments concluded that the phenotype was due to a single dominant gene, and both were uncovered in Merino sheep. McGuirk and Short (1967) reported breeding experiments conducted following the original discovery and found that slightly less than a 1:1 ratio of normal to lustre phenotype were born. *In utero* survival of lustre embryos may have been impaired. The lustre phenotype was more common in those lambs found dead at birth and in those which died within four days after birth. Warwick et al. (1960) reported similar poor viability of lambs of the "Silky" phenotype.

Additional reports have occurred more recently in Australia (Li et al. 2009) and New Zealand (Blair 1990; Plowman et al. 2019). The appearance of an unusual coat is most often noted in Merinos, perhaps because the lustrous phenotype is so different to the normal lamb coat. Blair (1990) reported a lustre mutant in a Border Leicester X Romney where the wool of the wild-type has a lustrous appearance, and the animals might stand out less at first glance. We have reported the phenotype in Perendale and Merino breeds (Plowman et al. 2019) and also observed it in Suffolk (DR Scobie, Unpublished data) where they were particularly noticeable.

The sheep we have observed certainly have a lustrous appearance at birth, but do not, in all cases, maintain that lustrous appearance as they age. Lustre is a visual effect that is judged subjectively. The following experiment was undertaken to test the hypothesis that the lustre gene affects fibre crimp, which can be measured cheaply.

Methods

Two sires were obtained that had a lustrous coat at birth. One was born on the AgResearch Lincoln research farm in 2016 at 175 Boundary Road, Lincoln (-43.628, 172.471), while the other was born in 2017 at 1624 Branch Creek Road, Wanaka (-44.833, 169.052) and obtained by AgResearch at approximately 140 days of age. Both these sires were a combination of breeds that are common in New Zealand. One was a Romney-type (named "Dazzle", born at Lincoln), while the other was a Merino-type (named "Razzle", born at Wanaka). Such is the state of composite breeds in New Zealand and on the two farms where these lambs were bred that it is not possible to describe their breed composition accurately. However, there was no contribution from the known lustre breeds (English Leicester, Border Leicester or Lincoln) in their background. These sires were each mated to 12 ewes that were also of unknown composite breeding but were not lustre breeds nor did they exhibit lustrous wool.

Mating was timed such that lambs would be born in late spring. During the period of lambing, the ewes were monitored every six hours. Lambs were tagged at birth and identified to dam and sire. Lambs that were phenotypically lustrous were fitted with a woollen cover to prevent hypothermia (Woolover Limited, Christchurch, New Zealand). The entire flock was moved into a shed during an episode of rain and strong cold wind.

Lambs were tail docked and vaccinated at around 21 days of age, in two separate groups based on date of birth. All male lambs were castrated so that both sexes could be pastured together and experience identical environments. The lambs were weaned at an average of 120 days of age. Wool samples were collected from all using a conventional shearing handpiece at weaning. The lambs were subsequently shorn. Fleece weight was not recorded because the lambs were a range of ages, and a range of weaning weights that are known to influence fleece weight. Shearing as a lamb resets fleece growth and makes the measurement of fleece weight at yearling more comparable (Rogan et al. 1995). The live weight of the lambs was recorded after shearing.

All animals were run together as a single mob from weaning until they were shorn as yearlings at 443 days of age, whereupon a further wool sample was collected from the right mid-side during shearing. Fleece removal was conducted by a professional shearer and the weight recorded. This weight included the fleece without skirting but did not include belly wool. Live weight was recorded after shearing.

Lustrous fleeces often "felt", "matt" or "cott" which is largely due to fibre entanglement. This explains the use of "felting" in the nomenclature of some mutants (Short 1958). A few fleeces of lambs and yearlings of the lustrous phenotype were so profoundly cotted that the entire fleece was taken to the laboratory, where the mid-side was identified and removed. It was also necessary to tease these samples apart to obtain fibre measurements.

The wool samples were measured at a commercial wool-testing laboratory on an optically-based fibre diameter analyser (OFDA) (BSC Electronics, 13 Willcock Street, Ardross, Western Australia 6153, Australia) for average fibre curvature, average fibre diameter and variation of fibre diameter. These measurements were undertaken in laboratory conditions of 20°C and 65%RH.

Each variable of interest was analysed using analysis of variance (ANOVA), consisting of a full factorial combination of three factors: sire, progeny type (lustrous or wild-type) and sex – i.e., a 2-by-2-by-2 factorial analysis. Effects of age, birth rank and rearing rank covariates were also examined for each variable, and included in the ANOVA if any of these covariates had a statistically significant effect on the variable. The age covariate (days) was calculated from each progeny's birth date. All statistical analyses were carried out with SAS 9.4 software.

Results

Lambs (n=46) were born during a 42-day period. Four neonatal mortalities were diagnosed to have perished as a result of dystocia upon autopsy by a professional veterinarian. Two orphan lambs, one of each phenotype, were artificially reared but not returned to the flock after weaning (n=40). With a number of triplet births in this small flock, some triplets were supplemented with formula (AnLamb, NZAgbiz Ltd, PO Box 844, Hamilton 3240, New Zealand) for up to a month after birth and remained with their dam. A short while after weaning, one lamb that was not of the lustre phenotype died from unidentified causes, and the remainder (n=39) were run together until yearling shearing.

Live weight is affected by age, birth rank and rearing rank, and these were examined as potential covariates for the traits analysed here. In turn, live weight is known to affect fleece weight and fibre diameter. Such effects were evident in lamb live weight at weaning, but only age and rearing rank showed significant relationships to justify inclusion in an analysis of covariance. Lambs that were older were significantly heavier at weaning (P=0.005) and lambs that were reared as multiples were significantly lighter than singletons (P=0.023). No significant relationship was uncovered for age, birth rank or rearing rank with any of the other traits discussed here (Table 1). These traits included average fibre curvature, average fibre diameter and fibre-diameter variation in fleece samples taken at weaning and yearling shearing, and fleece weight and live weight at yearling shearing.

Interesting differences in live weight were evident between the sire groups as yearlings when there was a significant difference between males that were judged to be lustrous at birth (P=0.002) and females that were judged to be wild-type at birth (P=0.019), and also between the sexes of the progeny of Razzle (P=0.010). However, the yearlings were generally more than 50 kg and the small differences between groups were not sufficient to suggest poor health (Table 1).

Lambs that had been identified as having a lustrous phenotype at birth had a slightly lower fleece weight when they were shorn as yearlings (Fig. 1) from Razzle (P=0.044 for females and P< 0.001 for castrate males) or Dazzle for females (P< 0.001), but not castrate males (P=0.072). There was no evidence of a difference in fleece weight between the two sire groups.

When the wool samples of weaned lambs were measured for fibre curvature, there was a highly significant difference between wool phenotypes judged at birth. Those judged to have lustrous wool were significantly lower in fibre curvature (P< 0.001), regardless of sire group or sex. As yearlings, this significant difference prevailed in all comparisons, although the magnitude of the significance had declined in the male progeny of Dazzle (P=0.044) compared to the other three groupings (P< 0.001). Despite the Romney in the background of Dazzle and Merino in Razzle, the fibre-curvature values of those in the two sire groups judged to have a lustrous phenotype at birth were not different. However, the greater fibre curvature of the Merino genotype was evident in those that were judged to have a wild-type fleece phenotype at birth whether female (P=0.016) or male (P=0.002) as lambs, and again as female (P=0.005) or male (P=0.002) as yearlings.

Fibre diameter was not significantly different between those judged to have lustrous and wild-type fleeces except in the female progeny of Dazzle as lambs (P=0.030), but this did not persist in yearling fleece samples where there were no significant differences at all. Comparing between the sire groups did reveal that, in mean fibre diameter of lamb wool samples, the progeny of Dazzle had coarser fibres than those of Razzle, but only for those lambs (female (P=0.003) or male (P=0.002)) that were not judged to have lustrous wool. However, when sampled again as yearlings the sire groups were found to have significantly different mean fibre diameter (lustrous male (P=0.004), wild-type

Table 1 Means of live weight, fibre curvature, fibre diameter and fibre diameter variation at weaning and yearling shearing, and fleece weight at yearling shearing of the progeny of two lustrous sires. Each progeny was judged to be lustrous or non-lustrous (wild-type) at birth. Means sharing a common alphabetical letter were not different at 5% significance level within each row, followed by the standard error of mean (SEMs) in brackets, with NA indicating only one progeny in the subset.

Characteristic	Sampling time	Number of samples	Merino-type sire "Razzle"			
			Lustrous		Wild type	
			Female	Male	Female	Male
Number of sheep	Weaning	24	1	8	10	5
	Yearling shearing	23	1	8	9	5
Live weight (kg)	Weaning	24	30.0ª (NA)	29.1ª (1.5)	25.0ª (1.6)	31.2ª (1.9)
	Yearling shearing	23	59.6ᵃᵇᶜ (NA)	64.9ª (1.9)	54.5ᵇ (1.6)	62.4ᵃᶜ (2.8)
Fleece weight (kg)	Weaning	0	-	-	-	-
	Yearling shearing	23	1.90ᵃᵇ (NA)	1.88ᵃᵇ (0.12)	2.82ᶜ (0.13)	2.84ᶜᵈ (0.20)
Mean fibre curvature (°/mm)	Weaning	24	23.7ª (NA)	22.5ª (0.9)	66.7ᵇ (0.9)	68.2ᵇ (4.0)
	Yearling shearing	23	28.4ª (NA)	26.9ª (1.9)	76.5ᵇ (3.9)	72.0ᵇᶜ (7.6)
Mean fibre diameter (μm)	Weaning	24	25.9ᵃᵇᶜ (NA)	24.5ᵃᵇ (1.0)	24.3ª (0.7)	23.4ª (0.7)
	Yearling shearing	23	24.3ª (NA)	25.5ª (0.6)	27.3ª (0.9)	26.7ª (1.1)
Fibre diameter variation (μm)	Weaning	24	7.5ᵃᵇ (NA)	6.5ª (0.6)	5.6ª (0.3)	5.4ª (0.4)
	Yearling shearing	23	4.8ª (NA)	5.7ª (0.4)	5.2ª (0.4)	5.5ª (0.6)
			Romney-type sire "Dazzle"			
			Lustrous		Wild type	
			Female	Male	Female	Male
Number of sheep	Weaning	16	3	1	5	7
	Yearling shearing	16	3	1	5	7
Live weight (kg)	Weaning	16	27.8ª (1.0)	21.5ᵇ (NA)	33.0ª (1.0)	33.7ª (2.0)
	Yearling shearing	16	56.4ᵇᶜ (0.6)	45.9ᵇ (NA)	61.6ᵃᶜ (1.8)	63.9ª (2.3)
Fleece weight (kg)	Weaning	0	-	-	-	-
	Yearling shearing	16	1.43ª (0.12)	2.50ᵇᶜᵈ (NA)	3.06ᶜᵈ (0.30)	3.33ᵈ (0.13)
Mean fibre curvature (°/mm)	Weaning	16	22.9ª (1.2)	29.1ª (NA)	57.6ᶜ (2.8)	55.1ᶜ (1.7)
	Yearling shearing	16	27.7ª (1.8)	29.4ª (NA)	59.5ᶜᵈ (4.5)	51.8ᵈ (2.2)
Mean fibre diameter (μm)	Weaning	16	24.6ᵃᵇ (1.0)	29.1ᵇᶜ (NA)	28.2ᶜ (0.9)	27.8ᶜ (0.7)
	Yearling shearing	16	28.6ᵃᵇ (1.1)	34.1ᵇᶜ (NA)	31.1ᵇᶜ (1.0)	32.6ᶜ (1.4)
Fibre diameter variation (μm)	Weaning	16	6.5ª (0.6)	9.1ᵇ (NA)	6.5ª (0.4)	6.1ª (0.4)
	Yearling shearing	16	6.5ᵃᵇ (0.3)	8.2ᵇ (NA)	6.1ᵃᵇ (0.5)	6.4ᵃᵇ (0.4)

Figure 1 A pair of yearlings born as twins and images of their wool sampled at shearing. The sheep with normal wool is on the left and its sibling carrying the mutant gene for lustrous wool, with low curvature is on the right.

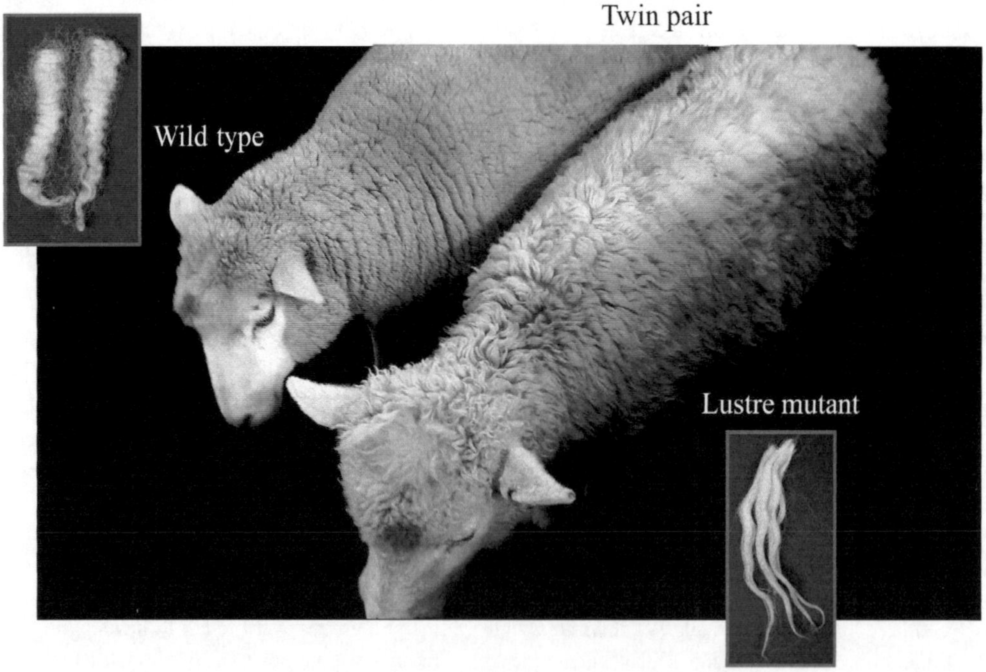

female (P=0.014) and wild-type male (P<0.001)) except for females which had been judged lustrous at birth.

The standard deviation of mean fibre diameter average values are summarised in Table 1. With respect to variation in diameter among fibres within a sample, there was a significant difference for male lambs that were the progeny of Dazzle between those having lustrous and non-lustrous fleeces at birth (P=0.017), but this was not evident in females from that sire and either sex sired by Razzle. This difference had disappeared at the time of yearling wool samples. The lustrous male progeny were significantly different between sire groups in standard deviation of fibre diameter in their lamb fleece (P=0.036) and this difference persisted as yearlings (P=0.033), but this was not evident in other such subsets.

Discussion

Lambs that were identified as phenotypically lustrous at birth exhibited much lower fibre curvature when shorn as lambs and again as yearlings when they were also found to express lower fleece weight. Lambs that were identified as lustrous at birth were generally distinguishable by their wool type at weaning, but did not necessarily exhibit a lustrous fleece as yearlings, though they continued to exhibit much lower fibre curvature than did their siblings. This has been noted in a small number of other sheep that we have observed as adults, including the sires of the groups reported here. Immediately after shearing, these sheep can exhibit a shiny appearance that might be regarded as lustrous, but a disorganised and matted fleece with a somewhat chalky appearance can rapidly appear in the following few weeks.

McGuirk and Short (1967) reported that fleece weight of the Merino lustre phenotype was around half that of those with a normal fleece. Blair (1990) reported a similar difference in fleece weight between the phenotypes in the Border Leicester X Romney for fleece weight at 5 months, 12 months and 18 months of age. The progeny of Dazzle (the Romney-type sire) observed here, exhibited a similar difference in fleece weight between phenotypes, but the difference was smaller between the different phenotypes among the progeny of Razzle (the Merino-type sire used here).

The wool of lambs and yearlings that were phenotypically lustrous at birth was not significantly finer in diameter than the wool of wild-type animals observed here. Blair (1990) found no difference between lustrous and non-lustrous wools in fibre diameter. McGuirk and Short (1967) reported fibre diameter was very slightly greater in the fleeces of ewes of the lustre phenotype but not significantly so. The greater detail in the work of McGuirk and Short (1967) showed no significant difference in the fibres grown from secondary wool follicles, but fibres growing from primary follicles were 2 to 4 µm finer in skin samples from the wild-type phenotype. The primary wool fibres were outnumbered by the secondary follicles by between 16 and 19-to-one, such that fibre diameter of the whole sample was not significantly different. Despite a discernible difference in fibre diameter between the wild-type progeny of the two sire groups observed here, the wool of the lustrous phenotype was quite similar in diameter in both sire groups.

A report from Li et al. (2009) noted that the wool of the lustre mutants they observed was fine (15.9, 17.8 and 19.0 µm), and like those measured here, was low in fibre curvature (59 to 68°/mm). However, the three non-lustrous sheep they observed were randomly obtained from an unrelated neighbouring flock that was pastured under very different conditions and not bred specifically to test the hypothesis, as we have done here. Li et al. (2009) suggested that differences in the protein composition of the lustrous phenotype account for the physical properties of the wool fibres. Alternatively, we suggest that felting is largely due to lower fibre curvature, as shown by Sumner et al. (2013).

The measurement of lustre in wool (Nicol 1991) and felting in Romney and Merino X Romney wools (Kenyon et al. 1998) and of Merino wools (Greeff & Schlink 2002) has been investigated and reported. The lustrous fleeces observed here have a strong propensity to form cotts. Lustrous wool lacks loose wool bulk (Elliot 1986) and in some respects, bulkiness is the opposite of lustre, although lustre is generally associated with a shiny appearance. However, the testing methods for felting and bulk are time consuming and expensive while the determination of lustre is qualitative. Measuring fibre curvature is inexpensive and can be obtained along with fibre diameter and other parameters, and we suggest that discrimination of these phenotypes can be made using fibre curvature measurements alone.

Although there was a wide range in live weight, there was no consistent trend for those with a lustrous phenotype to be different in live weight at weaning or as yearlings. Not surprisingly, the sire group with a part-Merino genotype exhibited finer wool, as lambs and yearlings, than did those with a sire influenced by Romney, but the lustrous phenotype did not seem to affect fibre diameter. However, with respect to curvature, those phenotypically judged as lustrous as neonates in the current experiment exhibited dramatically lower curvature as lambs and yearlings across both sire groups and sexes. We would like to suggest that what has been called a lustrous phenotype should perhaps more appropriately have been judged a phenotype low in curvature and, therefore, crimp. The felting or cotting that has sometimes been associated with lustre, in the "felting lustre mutant" is likely a consequence of the low curvature. Without doubt, the lustrous appearance of the lamb is easily judged from photographs in historical reports and nomenclature such as "silky" and our own observations concur, but as the animals age, their fleece becomes deranged and fails to appear lustrous.

As a research tool, the difference in fibre curvature is profound, without differences in fibre diameter or fibre diameter variability. These animals provide a good model for establishing differences in protein composition and perhaps to find a gene locus that influences crimp in wool.

Conclusions

In many respects the lustrous phenotype observed here resembles those described in other studies where they were described as 'lustre', 'felting-lustre' or the more poetic 'silky'. The phenotype appears lustrous at birth but that may not persist, while the low fibre curvature persists in yearlings. Fibre curvature is cheap to measure and is a well-defined and readily quantified measurement. "Low fibre curvature" appears to be an appropriate name for the trait segregating in these two sire groups.

References

Blair HT 1990. Inheritance of a major gene for excessively lustrous wool in sheep. Journal of Heredity 81: 220-222.

Elliott KH 1986. The ability of the WRONZ Automatic Bulkometer to rank wools for lustre. WRONZ Report No R135.

Greeff JC, Schlink AC 2002. The inheritance of felting of Merino wool. Wool Technology and Sheep Breeding 50: 6-10.

Kenyon PR, Wickham GA, Blair HT 1998. The assessment of natural means of controlling loose wool feltability and yarn shrinkage. Proceedings of the New Zealand Society of Animal Production 58: 291-293.

Li SW, Ouyang HS, Rogers GE, Bawden C 2009. Characterization of the structural and molecular defects in fibres and follicles of the merino felting lustre mutant. Experimental Dermatology 18: 134-142.

McGuirk BJ, Short BF 1967. The "Lustre" mutation in Merino sheep. Australian Journal of Experimental Agriculture and Animal Husbandry 7: 296-303.

Nicol JJ 1991. Within flock ranking of loose wool bulk and fibre diameter to predict wool lustre. Proceedings of the New Zealand Society of Animal Production 51: 319-321.

Plowman JE, Harland DP, Scobie DR, O'Connell D, Thomas A, Brorens PH, Richena M, Meenken E, Phillips AJ, Vernon JA, Clerens S 2019. Differences between ultrastructure and protein composition in straight hair fibres. Zoology 133: 40-53.

Rogan IM, Atkins KD, Coelli KA 1995. Shearing Merino ram weaners – Is it worth it? Wool Technology and Sheep Breeding 43: 56-68.

Short BF 1958. A dominant felting lustre mutant fleece-type in the Australian merino sheep. Nature 181: 1414-1415

Sumner RMW, Cuthbertson IM, Upsdell MP 2013. Evaluation of the relative significance of fibre diameter and fibre curvature when processing. New Zealand Romcross type wool. Journal of The Textile Institute 104: 1195-1205.

Warwick BL, David SP, Berry RO, Patterson RE 1960. "Silky" a dominant mutation in sheep. Journal of Heredity 51: 39-42.

Optimising a panel of single nucleotide polymorphisms for genomic prediction in New Zealand sheep

MA Lee[ab*], KG Dodds[c], S-AN Newman[c], AS Hess[c], D Campbell[b] and SM Clarke[c]

[a]*Beef+Lamb NZ Genetics, PO Box 5501, Dunedin 9054, New Zealand;* [b]*Department of Mathematics and Statistics, University of Otago, PO Box 56, Dunedin, New Zealand;* [c]*AgResearch, Invermay Agricultural Centre, Private Bag 5004, Mosgiel, New Zealand*

[*]*Corresponding author:* Email: michael.lee@otago.ac.nz

Abstract

A national genetic evaluation called the New Zealand Genetic Evaluation (NZGE) is undertaken weekly to support the genetic improvement of sheep via selective breeding in New Zealand (NZ). This is based on single step genomic best linear unbiased predictions (ssGBLUP), which allows animals that have genotypes and pedigree, or just pedigree, to be included in the same analysis (or evaluation). In theory, an optimal strategy is that the density and quality of SNPs used should be high enough to track all the haplotypes segregating in the population to be evaluated. Since genomic predictions have been used in NZ sheep, the content of the SNP panel used has decreased to a panel of 41K; despite that there are new SNP panels with more content available.. The effect of using this additional content and additional SNPs from an Ovine 600K chip was investigated using the trait fleece weight at 12-months. We found that a high density panel provided more accurate predictions. However, a SNP panel based on the 50K SNPs plus a relatively small subset of additional SNPs (e.g. from the 600K Chip), were as predictive as using the high density panel. We conclude that increased SNP density, beyond the 41K selected from the Ovine 50K SNP, chip will improve prediction accuracy. Our results suggest that the inclusion of SNPs from putative quantitative trait loci (QTL) may also improve prediction accuracy. However, further analyses will be helpful to distinguish the relative benefit to accuracy of increased SNP density across the genome and/or inclusion of SNPs associated with QTL.

Keywords: sheep, genetics, genomics, breeding, ssGBLUP, genomic prediction

Introduction

The process of genetic evaluation aims to take data (e.g. pedigree, phenotypes and fixed effects) to predicts breeding values. More-often the statistical method best linear unbiased prediction (BLUP) (Henderson et al. 1959; Henderson 1975) is used. In BLUP (Equation 1), the breeding values (**u**) are predicted from the information from phenotypes (**y**), fixed effects (**b**) and relationships between individuals, where **X** and **Z** are design matrices and **e** residual errors, assumed independent of the random effects, with variance proportion to I (identity matrix) and the variance of **u** is proportional to the animal relationship matrix.

Equation 1 BLUP

$$y = Xb + Zu + e$$

In genomic BLUP (GBLUP) the genotypes are used to create a genomic relationship matrix (GRM) in contrast to a pedigree relationship matrix (or numerator relationship matrix; NRM) for standard pedigree BLUP. In single step GBLUP (ssGBLUP) the NRM and GRM are combined to give matrix **H** that is used in the BLUP mixed model equations described by Henderson (Henderson 1975) to predict breeding values.

It is the inverse of the relationship matrices that are used in the mixed model equations (MME). For example, for ssGBLUP (Legarra et al. 2009), the inverse of **H** ($\mathbf{H}^{-1}$) is used in the MME (Equation 2) to obtain solutions for fixed effects ($\hat{b}$) and random effects or breeding values ($\hat{u}$), where λ is the ratio of the variance of residual additive effects over the variance of additive genetic effects. This implies that these parameters are known.

Equation 2 Mixed model equation described by Henderson

$$\begin{pmatrix} X'X & X'Z \\ Z'X & Z'Z + H^{-1}\lambda \end{pmatrix} \begin{pmatrix} \hat{b} \\ \hat{u} \end{pmatrix} = \begin{pmatrix} X'y \\ Z'y \end{pmatrix}$$

Currently, NZGE for maternal breeds (*i.e.* Romney, Coopworth, Perendale and Composites of these breeds) uses single nucleotide polymorphisms (SNPs) based on the Illumina Ovine 50K SNP chip (Anonymous 2015). In practice, most animals are genotyped using a lower density chip and imputed up to 50K. The density of these lower density chips has increased with time from about 5,000 SNPs to 18,000 SNPs. Many of these additional SNPs are not on the 50K SNP chip, but are present on a higher density 600K chip.

Information from data such as whole-genome sequence (WGS) is increasingly being applied into genomic predictions, offering a potential for increased prediction accuracy by including causal mutations or single-nucleotide polymorphisms (SNPs) in strong linkage disequilibrium (LD) with causal mutations affecting the trait of interest. Research to include information from WGS into genomic predictions to increase prediction accuracy is a focus in many species. The hypothesis is that causal mutations or SNPs in strong linkage disequilibrium with causal mutations affecting the trait may increase prediction accuracy, particularly across breeds. It is impractical to use all markers available for routine NZGE; therefore a subset of those available is sought. The selection of an "ideal" marker set is a challenging task. The longer term aim of this study was to understand the effect of using SNPs additional to those used routinely in NZGE and in particular those associated with quantitative trait loci (QTL). The longer term goal is to use this information to improve predictions and hence the profitability in farming. Here we report some initial results for the trait wool fleece weight at 12-months of age (FW12). Future work will investigate and contrast other traits with higher marker densities.

Materials and methods

Data

This study used historic data collected by NZ sheep farmers and researchers and prior information (e.g. models and parameters) provided by Beef+Lamb New Zealand Genetics (B+LNZG). The number of pedigreed animals in the evaluation was 10,751,607.

Genotype data from 6,706 animals that had both phenotype and/or progeny with phenotypes (FW12) were included in the evaluation. The number of animals available were low as there were much lower numbers of animals with 600K SNP panel data compared to 50K data. This genotype data included terminal sire breeds, which were sheep breed to produce lambs for meat (e.g. Sufolks, Texels and Sufteexs) and maternal sire breeds from which ewe replacements are taken from (e.g. sheep of breed Romney, Coopworth, Perendale or composites of these breeds). The SNP panels used and treatments are described in Table 1. The genotypes were from an Ovine 600K SNP chip (n=2,243) or imputed from lower density chips (n=4,463) using a reference panel from B+LNZG and using the software FImpute2 (Sargolzaei et al. 2014) or Beagle 5.0 (Browning et al. 2018). The mean accuracy of the imputed animals was 0.99 calculated as an allelic concordance. The accuracy of imputation for each animal was estimated by validation, where markers were removed from the animals genotyped and after imputation the removed markers compared to those imputed.

In a previous study using genotype data from 131,916 animals each with 13,209 SNPs we undertook a GWAS across 40 traits in NZ sheep, with sample sizes ranging from 1,316 to 105,248 animals - only 24 traits had significant QTL (data not shown). The goal of this GWAS was to use non-imputed SNPs to identify large QTL segregating in the population. We used the results from this study and data assembled from a number of more recent ovine SNP chips to investigate the effect of increased SNP content on prediction accuracy compared to a SNP chip based solely on the initial Ovine 50K chip.

The results for the SNP panels of 41K and 46K all used 5% of NRM and 95% GRM. The 41K panel is derived from the 50K Ovine SNP chip and the 46K panel is the 41K panel with the additional SNPs from the HD SNP Chip that are also on the more recent SNPs of density >15K used in NZ. The panel 41KKMM (Table 2) is the 46K panel with 5,292 SNPs of the most significant SNPs from the GWAS described above removed. These SNPs are putatively associated with QTL for FW12. The panel 45KMF is based on the 46K panel; except that the significant SNPs (Figure 1) and those flanking these SNPs within a genetic distance of 2,500,000 base pairs were removed. Consequently, 188, 210 and 185 SNPs were removed, respectively, on chromosomes 1, 3 and 8 (see Figure 1).

Computing and Software

Breeding values were predicted with BLUP and ssGBLUP by using preconditioned conjugate gradient in Mix99 (Strandén and Lidauer 1999). Other data handling used Linux shell commands or R (The_R_Development_Core_Team 2011) built with the Intel® Math Kernel Library.

Assessing Prediction accuracy of genomic prediction

The accuracy of evaluations was assessed by validation with 185 recent sires that had a mean (standard deviation) number of progeny with phenotypes of 36.26 (24). The year of birth of these sires ranged from 2014 to 2017. In the reduced dataset all of the phenotypes from the sires and their descendants were removed (*i.e.* un-recorded). The number of FW12 phenotypes in the full and reduced data set was, respectively, 2,022,154 and 1,946,506. The accuracy using the reduced data was compared as the correlation between the deregressed breeding values with the parent average removed (Garrick et al. 2009) calculated from a BLUP that included all of the data. A higher correlation for a given prediction with the reduced dataset implies a higher accuracy of prediction.

The percentage of NRM relative to the GRM used to calculate the H ranged from 5-75% for ssGBLUP evaluations with the highest density SNP panel (n=564,998) . Equation 3 and Equation 4 explain how this weighting is done, where the weighted GRM (G_w) is calculated from the GRM (G) and the NRM (A) depending on α (proportion of the NRM to be used).

Equation 3 Weighted G matrix

$$G_w = (1-\alpha)G + \alpha A$$

Equation 4 The H-inverse matrix used in the mixed model equations for ssGBLUP

$$H^{-1} = A^{-1} + \begin{pmatrix} 0 & 0 \\ 0 & G_w^{-1} - A_{22}^{-1} \end{pmatrix}$$

A description of the treatments investigated is given in Table 1. The model used for BLUP and ssGBLUP was identical except for the inverse of the covariance matrix (NRM versus **H**). The model and assumed genetic parameters used were proprietary to B+LNZG.

Results

A plot of the results from the genome wide association study (GWAS), described in materials and methods, based on 26,024 animals is given in Figure 1 for the trait FW12. There were three putative QTL regions that met a threshold of 0.05 divided by the product of the number of traits by the number of markers.

Table 1 Treatments and SNP panels used in this study. The treatments used for the different evaluations were: BLUP, pedigree only; ss0.75, ssGBLUP with a GRM calculated from 564,998 SNPs; ss0.5, ssGBLUP with a GRM calculated from 564,998 SNPs; ss0.25, ssGBLUP with a GRM calculated from 564,998 SNPs; ss0.05, ssGBLUP with a GRM calculated from 564,998 SNPs; ss0.05_41K, ssGBLUP with a GRM calculated from 40,881 SNPs; ss0.05_46K, ssGBLUP with a GRM calculated from 46,173 SNPs that consisted the same SNPs from the 41K panel and additional SNPs from the 564,998 panel; ss0.05_41KMM, 5,292 SNPs removed from the 46K panel that were the most significant SNPs from GWAS; ss0.05_45MF, the SNPs removed from the 46K panel that flanked three putative QTL detected by GWAS.

Treatment	SNP panel	Comment
BLUP	Non applicable	uses just pedigree
ss0.75	564998 SNPs from 600K SNP chip	H-matrix uses 75% A, 25% G
ss0.5		H-matrix uses 50% A, 50% G
ss0.25		H-matrix uses 25% A, 75% G
ss0.05		H-matrix uses 5% A, 95% G
ss0.05_41K	40,881 SNPs from 50K Chip	H-matrix uses 5% A, 95% G
ss0.05_46K	46,173 SNPs updated from HD and 50K	H-matrix uses 5% A, 95% G
ss0.05_41KMM	removed 5,292 GWAS SNPs from the 46,173 panel giving 40,881 SNPs	H-matrix uses 5% A, 95% G
ss0.05_45KMF	removed SNPs flanking 3 putative QTL	H-matrix uses 5% A, 95% G

Table 2 Validation results comparing different SNP panels and BLUP. Maternal Animals consisted of animals of breed Romney, Coopworth, Perendale and composites of these and Terminal Animals consisted of those of mainly Suffolk, Texel and composites of these. The Combined group included animals from both Maternal and Terminal breeds. See Table 1 for abbreviations.

Group	n	BLUP	ss0.75	ss0.5	ss0.25	ss0.05	ss0.05_41K	ss0.05_46K	ss0.05_41KMM	ss0.05_45KMF
Maternal Animals	96	0.23	0.27	0.28	0.3	0.31	0.29	0.31	0.29	0.3
Terminal Animals	89	0.56	0.57	0.56	0.55	0.5	0.5	0.5	0.49	0.5
Combined	185	0.56	0.57	0.58	0.57	0.55	0.54	0.54	0.53	0.54

Figure 1 Manhattan plot of GWAS results for the trait FW12. The horizontal line is a Bonferroni significance threshold of 0.05.

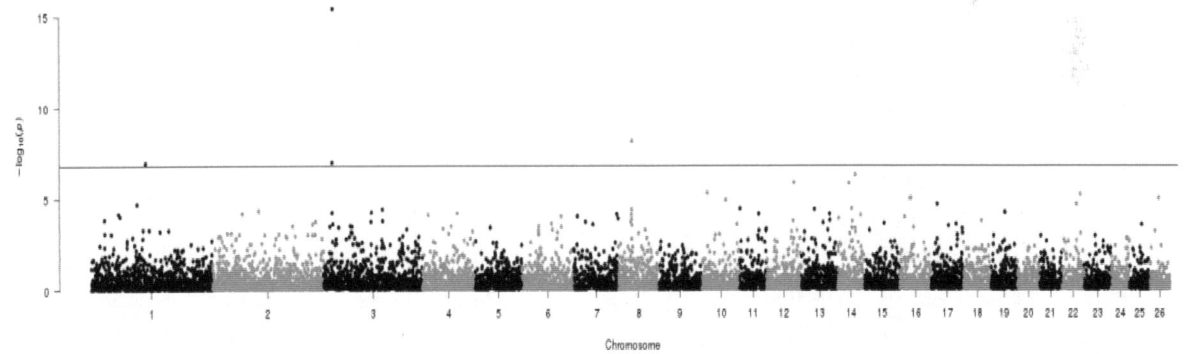

The validation results as correlations from comparing different treatments (see Table 1) are given in Table 2 for the full validation set and for the validation set split by breed type.

Discussion

Previous GWAS undertaken suggested for most current economically important traits there is not an abundance of large QTL, such as the myostatin mutation that affects muscling in sheep (Clop et al. 2006), segregating in the NZ sheep population (data not shown). This is relevant because an ongoing goal of research is to improve prediction accuracy by better using all data that is available. This might include WGS, prior knowledge on causal mutations etc.

Unsurprisingly, we found that a higher density panel gives more accurate prediction than a lower density (treatment ss0.05 was at least as good as any other method with 0.05 NRM weighting). The best correlation, 0.31, for maternal animals was seen using an α-value of 0.05 for the high density panel, whereas, for the lower density 41K chip based on the 50K Ovine SNP chip the correlation was 0.29. This was also shown in dairy cattle when the reliability of genomic breeding value was 0.5% and 1.0% higher using a 777K SNP chip compared to a 54K SNP chip respectively in Holstein and Red Dairy cattle (Su et al. 2012). The use of whole genome sequence for predictions, however, showed only a small increase in accuracy (1%) over a 60K panel in chickens (Heidaritabar et al. 2016). Practical and/or economic considerations would probably preclude the use of such high marker density in routine genetic evaluation.

Prediction accuracy in diverse sheep populations increased when using variants selected from WGS data compared to standard 50K genotypes using GBLUP and Bayesian methods (Moghaddar et al. 2019). In the context of ssGBLUP the inclusion of additional SNPs, associated with QTL, from sequencing data could improve the accuracy for some, but not all, milk traits (Liu et al. 2020). They concluded that two more sophisticated ssGBLUP methods (e.g. weighted ssGBLUP) offered no significant advantage over standard ssGBLUP. In French dairy goats, weighted ssGBLUP was between 2 and 14% more accurate when a QTL was known to be segregating in the population, but otherwise less accurate or as accurate as, standard ssGBLUP (Teissier et al. 2019). In the context of NZGE, it would be difficult to implement weighted ssGBLUP, given the complexity of the population to be evaluated. Instead, the incorporation of markers associated with QTL may provide a better strategy for this population.

In relation to this we expected that the removal of the SNPs detected by GWAS would decrease prediction accuracy. In a comparison of ss0.05_41K with ss0.05_41KMM there was no difference for maternal animals and a modest (1%) decrease for terminal animals, when comparing the terminal animals with an α-value of 0.05. There was a 1% decrease in correlation between ss0.05_46K and ss0.05_45KMF for the maternal animals. This result is consistent with our expectation, but further analyses will need to be done to substantiate them.

In NZGE genotypes from maternal animals are evaluated separately of those from terminal animals. In this study, we combined the genotypes into one evaluation and analysed the results as correlations across and within these two groups. The results highlight that for ssGBLUP by combining animals from disparate breeds a trade-off needs to be made to optimally predict across all breeds. For example, the optimal percentage of the NRM to include in making **H** was 0.05 for maternal animals (see Table 2), whereas, for terminal animals this percentage (α in Equation 3) the correlation was worse than using BLUP (*i.e.* pedigree only). Using a higher percentage of NRM (less GRM) would be necessary to provide predictions for both the maternal and terminal animals that are more accurate than BLUP.

The inclusion of SNPs from GWAS was indicative of improving prediction accuracy and warrants further investigation (e.g. investigation across other traits). We were restricted by low sample size in this study as the number of progeny tested sires that had 600K genotypes was low. This meant that it was hard to interpret bias, where this is commonly analysed by comparing the slope from regressing the deregressed breeding value on the breeding value predicted, from the results so these results were omitted from this study. However, the results suggested that increased density might also decrease bias (data not shown).

We conclude that for the population and trait being evaluated the use of an increased SNP density, beyond the 41K selected from the Ovine 50K SNP chip, will improve prediction accuracy and that further investigation into including QTL effects into evaluations is warranted.

Acknowledgements

The authors gratefully acknowledge Beef + Lamb New Zealand Genetics, Levy paying farmers and Sheep breeders for contributing to this work. The research was also, in part, funded by Genomics Aotearoa.

References

Anonymous 2015. OvineSNP50 Genotyping BeadChip, In: Data Sheet: Agrigenomics.

Browning BL, Zhou Y, Browning SR 2018. A One-Penny Imputed Genome from Next-Generation Reference Panels. The American Journal of Human Genetics 103: 338-348.

Clop A, Marcq F, Takeda H, Pirottin D, Tordoir X, Bibe B, Bouix J, Caiment F, Elsen JM, Eychenne F, Larzul C, Laville E, Meish F, Milenkovic D, Tobin J, Charlier C, Georges M 2006. A mutation creating a potential illegitimate microRNA target site in the myostatin gene affects muscularity in sheep. Nature Genetics 38: 813-818.

Garrick DJ, Taylor JF, Fernando RL, 2009. Deregressing estimated breeding values and weighting information for genomic regression analyses. Genetics Selection Evolution 41: 55.

Heidaritabar M, Calus MP, Megens HJ, Vereijken A, Groenen MA, Bastiaansen JW 2016. Accuracy of genomic prediction using imputed whole-genome sequence data in white layers. Journal of Animal Breeding and Genetics 133: 167-179.

Henderson CR 1975. Best linear unbiased estimation and prediction under a selection model. Biometrics 31: 423-447.

Henderson CR, Kempthorne O, Searle SR, von Krosigk CM, 1959. The estimation of environmental and genetic trends from records subject to culling. Biometrics 15: 192-218.

Legarra A, Aguilar I, Misztal I 2009. A relationship matrix including full pedigree and genomic information. Journal of Dairy Science 92: 4656-4663.

Liu A, Lund MS, Boichard D, Karaman E, Guldbrandtsen B, Fritz S, Aamand GP, Nielsen US, Sahana G, Wang Y, Su G 2020. Weighted single-step genomic best linear unbiased prediction integrating variants selected from sequencing data by association and bioinformatics analyses. Genetics Selection Evolution 52: 48.

Moghaddar N, Khansefid M, van der Werf JHJ, Bolormaa S, Duijvesteijn N, Clark SA, Swan AA, Daetwyler HD, MacLeod IM 2019. Genomic prediction based on selected variants from imputed whole-genome sequence data in Australian sheep populations. Genetics Selection Evolution 51: 72.

Sargolzaei M, Chesnais JP, Schenkel FS 2014. A new approach for efficient genotype imputation using information from relatives. BMC genomics 15: 478-478.

Strandén I, Lidauer M 1999. Solving large mixed linear models using preconditioned conjugate gradient iteration. Journal of Dairy Science 82: 2779-2787.

Su G, Brøndum RF, Ma P, Guldbrandtsen B, Aamand GP, Lund MS 2012. Comparison of genomic predictions using medium-density (~54,000) and high-density (~777,000) single nucleotide polymorphism marker panels in Nordic Holstein and Red Dairy Cattle populations. Journal of Dairy Science 95: 4657-4665.

Teissier M, Larroque H, Robert-Granie C, 2019. Accuracy of genomic evaluation with weighted single-step genomic best linear unbiased prediction for milk production traits, udder type traits, and somatic cell scores in French dairy goats. Journal of Dairy Science 102: 3142-3154.

The R Development Core Team 2011. R: A language and environment for statistical computing. R Foundation for Statistical Computing, Vienna, Austria.

Do cows with facial eczema feel pain?

M Sutherland[a]*, F Huddart[a], NR Cox[b], E Cuttance[c] and K Schütz[a]

[a]*AgResearch Ltd., Ruakura Research Centre, Hamilton 3216, New Zealand;* [b]*NeilStat Ltd, 9 Ngaere Ave, Hamilton 3210, New Zealand;* [c]*VetEnt Research, Te Awamutu 3800, New Zealand*

**Corresponding author.* Email: mhairi.sutherland@beeflambnz.com

Abstract

Facial eczema (FE) negatively influences production and welfare. One consequence of FE is reduced feed intake, which could possibly be a consequence of cows experiencing nausea or pain or both. The aim of this study was to evaluate the association of cortisol and acute phase protein (APP) concentrations (indirect measures of pain) with gamma glutamyl transferase (GGT) and bile acid concentrations (indirect measures of liver damage) in dairy cattle. Blood samples were collected from 2,080 lactating dairy cattle from seven farms that were affected by FE and located in the North Island of New Zealand. Blood samples were analysed for GGT, bile acid, cortisol and APP (serum ferroxidase (SF), serum amyloid A and haptoglobin) concentrations. Data were analysed using best subsets linear regression including farm as a factor and various other prediction variables. There was a significant positive relationship between GGT and bile acid ($P<0.001$, Adjusted $R^2=37.3\%$), SF ($P<0.001$, Adjusted $R^2=27.4\%$), and haptoglobin ($P<0.001$, Adjusted $R^2=1.9\%$) concentrations, but not with cortisol. The positive association between GGT concentrations and APP (SF, haptoglobin) may partly reflect the degree of liver damage caused by sporidesmin intoxication, however, it could also suggest that animals with higher GGT levels are experiencing systemic inflammation and potentially pain.

Keywords: acute phase proteins; facial eczema; gamma-glutamyl transferase; inflammation; pain

Introduction

Facial eczema (FE) or pithomycotoxicosis is a disease caused by the ingestion of fungal spores produced by *Pithomyces chartarum* which contain the toxin sporidesmin. Sporidesmin causes occlusion of the bile ducts and liver damage (Smith & Towers 2002) which can then lead to photosensitisation (Clare 1944) in sheep, cattle, deer and goats (Di Menna et al. 2009). Facial eczema is generally not fatal but can negatively affect production and welfare of dairy cattle. One consequence of FE is a reduction in feed intake (Morris et al. 2004), however, the mechanism behind this response is not clear, it may be possible that animals are feeling pain or nausea or both and that this may result in behavioural changes. The pathology of this disease is well described in the literature (Laven et al. 2020; Munday et al. 2020; Fernández et al. 2021); however, we still know little about how animals experience having FE.

An important part of assessing welfare includes measuring affective state, which refers to how the animal is experiencing its life, e.g., free from negative affective states, such as pain or fear, or by experiencing positive affective states, such as pleasure (Fraser et al. 1997; Mellor et al. 2020). Clinical signs of FE, such as skin lesions, are likely to cause pain and discomfort, thereby resulting in negative affective states, however, this has not been well researched. Also, little is known about how cattle may perceive liver damage when clinical signs are not observed.

Facial eczema in ruminants can cause fibrosis which can lead to liver cirrhosis (Di Menna et al. 2009; Laven et al. 2020; Munday et al. 2020; Fernández et al. 2021). Hepatic fibrosis has been shown to develop in sheep after only a single exposure to sporidesmin intoxication (Munday et al. 2020). In humans, pain is commonly experienced by patients with liver cirrhosis (Rogal et al. 2013; Klinge et al. 2018), which may partially be due to systemic inflammation as pro-inflammatory biomarkers (e.g., Interleukin-6 and C-Reactive protein) have been associated with disease severity and pain in patients with cirrhosis (Rogal et al., 2015a; Costa et al. 2021). Therefore, dairy cattle may also experience pain in response to liver damage caused by FE and this could potentially be assessed by measuring pro-inflammatory biomarkers, such as acute phase proteins (APP).

It is difficult to directly assess affective states in animals, therefore, proxy measures are often used instead. Behaviour is commonly used to measure positive and negative affective states in animals, however, this can be time consuming and expensive. Alternatively, biomarkers in the blood can be used as proxy measures of stress and pain. Changes in cortisol concentrations are often used to measure pain and it has been shown to increase in response to painful husbandry procedures in cattle, such as dehorning and castration (Fisher et al. 1997; Ting et al. 2003; Sutherland et al. 2013). Moreover, APP can be used as a diagnostic tool in cattle (Eckersall 2000). The acute phase response refers to a wide range of neuroendocrinal, physiological and metabolic changes that occur in animals in response to infection, inflammation or trauma (Eckersall 2000). Part of this response involves the release of APP, such as haptoglobin, into the circulation. In cattle, APP have been shown to become elevated in response to acute painful procedures such as castration (Fisher et al. 1997; Ting et al. 2003; Ballou et al. 2013) and acute and chronic inflammatory conditions (Horadagoda et al. 1999). Therefore, in the present study cortisol and APP (serum ferroxidase, serum amyloid A and haptoglobin) were used as proxy measures of pain. The aim of this study was to evaluate the association of cortisol and APP concentrations with indirect measures of liver damage (gamma glutamyl transferase (GGT) and bile acids concentrations) in dairy cattle selected from herds with evidence of FE.

Materials and methods

Between March and April 2019 (Southern Hemisphere autumn) seven commercial dairy farms located in the North Island of New Zealand were selected based on these herds having evidence of FE. The methodology for identifying herds for the study is described by Cuttance et al. (2021). All procedures involving animals were approved by the Ruakura Animal Ethics Committee under the New Zealand Animal Welfare Act 1999.

Herd selection

This study was part of a larger research project designed to determine both the financial cost and animal welfare implications of sub-clinical facial eczema in dairy cattle. Briefly, a notice was sent to members of the Dairy Cattle Veterinarians branch of the New Zealand Veterinary Association and Livestock Improvement Corporation (LIC) asking whether they knew of any farmers who had farms where clinical or sub-clinical FE had been identified during the current season and who would be interested in being involved in a FE study. From this identification procedure, seven herds were identified as being suitable for enrolment in our study. These herds were selected based on at least one cow in the herd having clinical FE or the presence of serum GGT activities >300 IU/L (Cuttance et al. 2021). Blood samples were taken from all cows in these seven herds and then a herd test was conducted by LIC between -12 and 15 days in relation to blood sampling (mean -0.3 days). Herd-test data included milk solids, milk volume and somatic cell counts. Blood samples were taken from 2,080 cows in total. General information about the animals was also collected and used in the analysis, including calving date, age and breed.

Blood sampling

Blood samples were collected from the coccygeal (tail) vein of cattle into 10mL vacutainers (BD Diagnosis, Auckland NZ) that contained no anticoagulant. All blood samples (except for two missing samples) were analysed by Gribbles Veterinary pathology laboratory for GGT concentrations and a sub-set of samples were further analysed for bile acid, cortisol, serum ferroxidase (SF), serum amyloid A (SAA) and haptoglobin concentrations. The sub-set of samples was selected based on GGT concentrations (a representative spread of GGT concentrations), age, breed and farm. Number of blood samples analysed for each blood parameter are summarised in Table 1 and below.

Statistical analysis

Data were analysed using Genstat, version 19 (VSN International, Hemel Hempstead, UK) using best subsets linear regression including farm as a factor and various other prediction variables. Predictor variables included in the analysis were: log cortisol, log haptoglobin, log SF, log bile acids, calving date, milk total (kg), milk solids (kg), age and farm. The relationships among GGT (log transformed) and the range of explanatory variables measured were examined using linear regression. In addition, best subsets regression was used to determine the combination of variables that best correlated with GGT. The optimal prediction for the relationship between GGT and the other variables was calculated based on Mallows Cp, a criterion for selecting among alternative subset regressions (Thompson 1978). The regression model only used data where all variables were present for an individual animal.

The number of samples per variable used in the model are: GGT (n = 2078), bile acid (n = 851), cortisol (n = 806), SF (n = 851), haptoglobin (n = 852), total milk volume (n = 1931), total milk solids (n = 1929) and somatic cell counts (n = 1792).

Serum amyloid A concentrations were not statistically analysed as all animals tested had levels below the detectable limit (5 µg/mL), except for one animal that measured 8 µg/mL.

Results

Number of blood samples analysed and mean and range concentration for each blood parameter are summarised in Table 1. Herd testing was conducted on 1,931 cows. Milk production data are summarised in Table 2.

A linear regression model was fitted to log (GGT) against a range of explanatory variables (e.g., blood parameter, cow parity, calving date). When one variable at a time was fitted to the model, there were significant (linear) positive relationships between log GGT and log bile acids ($P<0.001$, Adjusted $R^2 = 37.3\%$; Figure 1), farm ($P<0.001$, Adjusted $R^2 = 29.1\%$), log SF ($P<0.001$, Adjusted $R^2 = 27.4\%$; Figure 2), milk solids ($P<0.001$, Adjusted $R^2 = 19.0\%$), milk total ($P<0.001$, Adjusted $R^2 = 14.9\%$), log haptoglobin ($P<0.001$, Adjusted $R^2 = 1.9\%$) and calving date ($P=0.002$, Adjusted $R^2 = 1.2\%$). The strongest relationship was between GGT and bile acid and the weakest was between GGT and calving date.

When two or more variables were fitted to the model, the optimal prediction (based on Mallows Cp) for the relationship between GGT (log) concentrations and the other variables involved six variables, which all contributed significantly: farm (factor), calving date, haptoglobin (log), SF (log), bile acids (log) and milk solids (kg). The adjusted R^2 for this model is 65.2% and the Mellows Cp is 12.9 with 12 df.

Discussion

In the present study, GGT was used as a proxy for liver damage caused by FE (ingestion of sporidesmin in spores of *P. chartarum*). Gamma glutamyl transferase is commonly used as an indirect measure of FE as several studies (in sheep and cattle) have shown a positive relationship between GGT concentrations and liver damage associated with sub-clinical and clinical FE (Blackshaw 1978; Towers

Table 1 Gamma glutamyl transferase (GGT), bile acid, cortisol, serum ferroxidase (SF), and haptoglobin concentrations of cows from seven dairy farms located in the North Island of New Zealand.

Farm		Age	GGT (IU/L)	Bile acids (µmol/L)	Cortisol (nmol/L)	SF (IU/L)	Haptoglobin (mg/dL)
1	Number	222	272	153	152	153	153
	Mean	4.6	1539.6	98.9	73.6	21.2	47.0
	Range	2.0 - 12.0	2.0 - 6337.0	13.2 - 784.0	8.1 - 226.0	10.4 - 37.8	28.0 - 273.2
2	Number	261	262	58	59	58	59
	Mean	4.4	763.9	135.1	64.4	23.2	40.3
	Range	2.0 - 11.0	3.0 - 6456.0	26.1 - 712.0	7.4 - 173.0	13.6 - 41.9	20.7 - 128.2
3	Number	216	216	93	94	93	93
	Mean	4.9	342.7	87.5	66.6	17.6	43.6
	Range	1.9 - 9.1	2.0 - 4704.0	8.5 - 718.0	6.6 - 156.0	5.9 - 31.2	22.9 - 127.2
4	Number	539	540	72	72	72	72
	Mean	4.6	88.8	62.8	58.3	15.8	46.2
	Range	1.8 - 14.9	2.0 - 5233.0	14.7 - 171.0	10.6 - 147.0	5.6 - 24.0	28.5 - 83.3
5	Number	226	225	118	118	118	118
	Mean	4.4	1196.0	126.1	83.7	17.5	51.0
	Range	2.0 - 10.0	2.0 - 6406.0	8.4 - 858.0	14.2 - 448.0	7.1 - 34.5	22.3 - 196.0
6	Number	398	399	296	250	296	296
	Mean	4.7	270.0	60.7	39.4	16.1	33.0
	Range	2.0 - 11.0	2.0 - 4032.0	7.4 - 662.0	5.1 - 138.0	3.6 - 37.8	3.2 - 300.7
7	Number	108	164	61	61	61	61
	Mean	5.4	1794.5	225.5	45.7	20.6	50.1
	Range	2.0 - 12.0	2.0 - 5300.0	29.0 - 699.0	9.7 - 132.0	12.8 - 33.0	0.0 - 184.7
All farms	Number	1970	2078	851	806	851	852
	Mean	4.7	856.5	113.8	61.7	18.9	44.5
	Range	1.8 - 14.9	2.0 - 6456.0	7.4 -858.0	5.1 - 448.0	3.6 - 41.9	0.0 - 300.7

Table 2 Selected herd-test data from seven dairy farms located in the North Island of New Zealand.

Farm		Age	Total milk volume (kg)*	Total milk solids (kg)	Somatic cell counts (kcells/mL)
1	Number	222	222	221	222
	Mean	4.6	5.0	0.5	231.5
	Range	2.0 - 12.0	0.2 - 12.1	0.0 - 1.3	0.0 - 2232.0
2	Number	261	259	259	132
	Mean	4.4	6.0	0.6	152.6
	Range	2.0 - 11.0	0.9 - 13.4	0.1 - 1.4	1.0 - 1524.0
3	Number	216	186	186	186
	Mean	4.9	11.0	1.2	191.8
	Range	1.9 - 9.1	4.9 - 20.7	0.6 - 2.3	0.0 - 9999.0
4	Number	539	536	536	535
	Mean	4.6	9.5	1.0	209.6
	Range	1.8 - 14.9	3.1 - 16.6	0.3 - 1.7	12.0 - 15182.0
5	Number	226	223	223	212
	Mean	4.4	9.3	1.0	224.0
	Range	2.0 - 10.0	0.4 - 17.6	0.1 - 1.8	11.0 - 7352.0
6	Number	398	398	397	398
	Mean	4.7	7.8	1.0	137.6
	Range	2.0 - 11.0	2.5 - 15.9	0.4 - 1.6	0.0 - 3440.0
7	Number	108	107	107	107
	Mean	5.4	7.3	0.6	228.0
	Range	2.0 - 12.0	1.3 - 12.4	0.1 - 1.1	0.0 - 1791.0
All farms	Number	1970	1931	1929	1792
	Mean	4.7	8.0	0.8	196.4
	Range	1.8 - 12.0	0.2 - 20.7	0.0 - 2.3	0.0 - 15182.0

*Some of these values are low as some of the animals were being milked once a day.

Figure 1 Linear relationships between log gamma-glutamyl transferase (GGT) and log bile acids concentrations in cattle (n = 851) on seven dairy farms in the North Island of New Zealand between March and April 2019. Different-coloured dots signify different farms

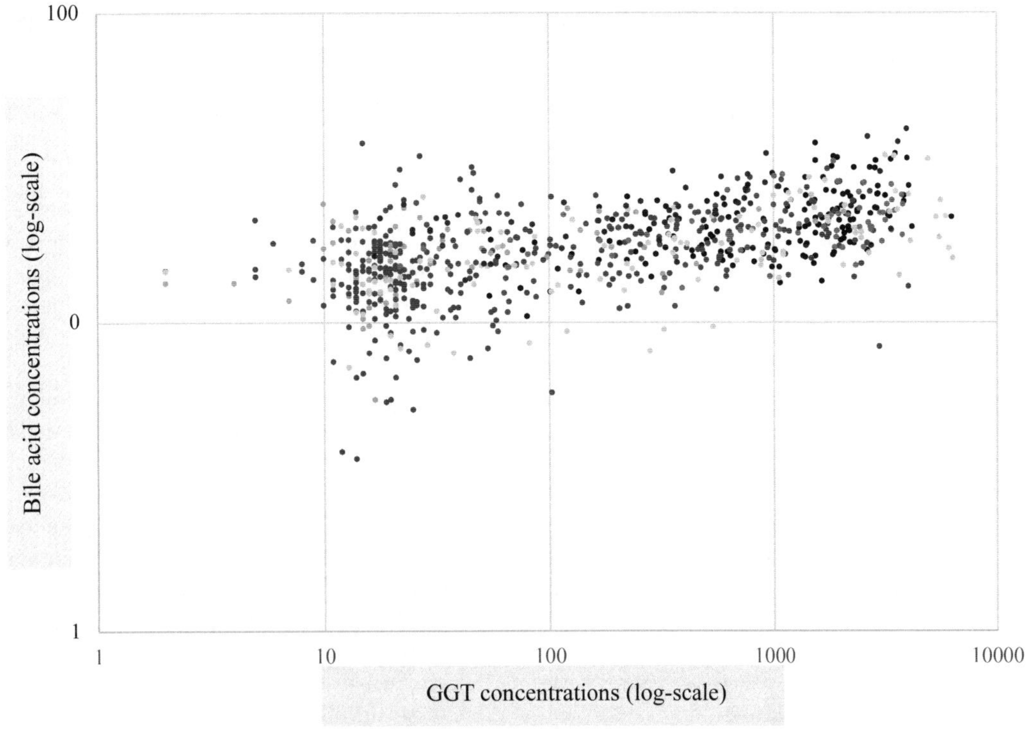

Figure 2 Linear relationships between log gamma-glutamyl transferase (GGT) and log serum ferroxidase concentrations in cattle (n = 851) on seven dairy farms in the North Island of New Zealand between March and April 2019. Different-coloured dots signify different farms.

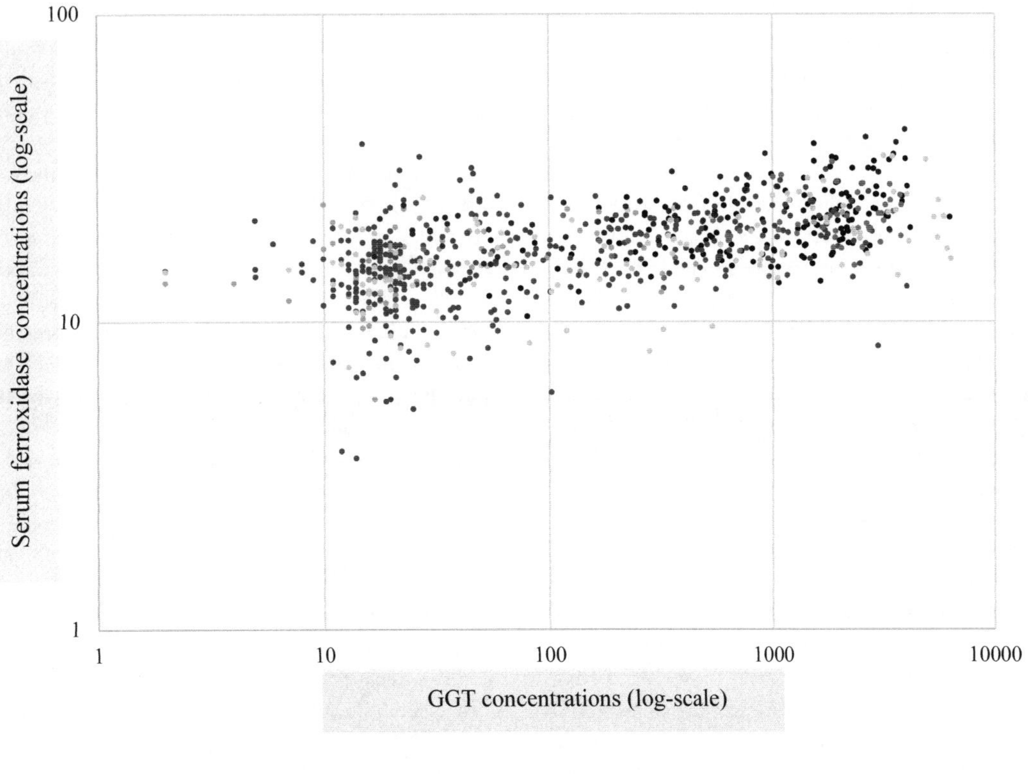

& Smith 1978; Towers & Stratton 1978). However, elevated GGT concentrations in the autumn could also have been caused by ingestion of *Brassica spp*., muscle damage, ragwort toxicity, fasciolosis, lipidosis or kidney damage (Blackshaw 1978; Cuttance et al. 2021). Cuttance et al. (2021) considered these possible alternative causes of liver damage in relation to the farms and animals enrolled in their study (and ours); given the time of year, region, lack of *Brassica spp*. being fed and low levels of liver fluke exposure they concluded that the liver damage (elevated GGT concentrations) seen in these herds was most likely due to ingestion of sporidesmin. Histopathology of the liver would be required for definitive diagnosis of FE, but this type of sample collection is invasive and is not practical on a large sample size.

Our results suggest that there is a positive relationship between GGT concentrations and the acute phase proteins haptoglobin and SF; however, SAA concentrations were below detectable levels for the majority of animals tested. The acute phase response is a systemic response to infection, inflammation or trauma and results in the increased production of acute phase proteins by the liver (Eckersall 2000). Haptoglobin, SAA and SF (also called ceruloplasmin) are APP that have been used in bovine medicine diagnostics. Haptoglobin and SAA have been shown to be elevated in response to a range of inflammatory conditions in cattle (Horadagoda et al. 1999) and ceruloplasmin concentrations have been shown to increase in response to conditions such as mastitis (Conner et al. 1986) and endometriosis (Kaya et al. 2016). Moreover, haptoglobin has been shown to increase in response to painful husbandry procedures, such as castration (e.g., Fisher et al. 1997; Ting et al. 2003; Ballou et al. 2013). Interestingly, providing animals with pain relief (a local anaesthetic and analgesic) prior to castration prevented the increase in haptoglobin concentrations that were observed in animals not provided with pain relief (Ballou et al. 2013). Therefore, it appears as though APP concentrations can increase in response to pain as well as disease in cattle, suggesting that animals with higher GGT levels may be experiencing systemic inflammation and potentially pain. It is unclear, why SAA was not elevated in response to FE. This could possibly be explained due to the short half-life of SAA and the timing of the initial liver damage in relation to when animals were sampled. Therefore, more research is needed to understand the relationship between FE and the APP response in cattle, including the relationship between the degree of liver damage and the APP response.

In humans, pain is commonly experienced by patients with liver disease (Rogal et al. 2013; Klinge et al. 2018) and pain has been reported in up to 80% of patients with cirrhosis caused by hepatitis C virus infection, non-alcoholic steatohepatitis or alcohol (Rogal et al. 2015a). Abdominal pain is reported by the majority of patients with liver disease (Klinge et al. 2018), however, a large proportion of patients also report widespread pain (Rogal et al. 2015a,b). Several factors are likely related to abdominal pain in patients with cirrhosis including ascites, hepatic capsular distension, and splenomegaly (Rogal et al. 2015a; Klinge et al. 2018). However, widespread pain may be due to systemic inflammation, as cirrhosis is a proinflammatory state and an increase in pro-inflammatory cytokines (e.g., IL-6: Interleukin-6) and acute phase proteins (e.g., CRP: C-Reactive Protein) have been associated with liver damage in humans (Rogal et al. 2015a). Moreover, IL-6 and CRP have been associated with the presence of abdominal pain and the severity of pain related to cirrhosis (Rogal et al. 2015a). These findings have led Rogal et al. (2015a) to suggest that increases in systemic inflammation may contribute to the prevalence of pain in patients with cirrhosis. Like CRP, Haptoglobin, SAA and SF are produced by the liver in response to pro-inflammatory cytokines (Eckersall 2000). Therefore, if increased IL-6 and CRP are a predictor of pain experienced by people with liver damage due to cirrhosis, then Haptoglobin and SF could potentially be a predictor of pain experienced by dairy cattle due to liver damage. Though, some histological features are similar between sporidesmin-induced liver damage in ruminants (e.g. sheep) and cirrhosis in humans, there are significant differences in the pathogenesis and clinical effects of these diseases (Munday et al. 2020). Therefore, more research is needed to investigate whether liver damage associated with FE causes pain in dairy cattle. If liver damage caused by FE results in animals experiencing chronic pain, then this has greater welfare implications for the disease than currently considered.

Cortisol concentration ranged from 5.1 to 448 nmol/L (mean 60 nmol/L) across the seven farms used in the present study; however, there was no relationship between cortisol and GGT concentrations. Cortisol concentrations are elevated in response to acute stress and pain in cattle (Fisher et al. 1997; Ting et al. 2003; Sutherland et al. 2013) and were elevated after the onset of photosensitisation in sporidesmin-dosed sheep (Smith & Payne 1991). Therefore, we predicted that any discomfort/pain caused by FE may result in elevated cortisol concentrations. However, it is difficult to interpret the stress response of an animal from a discrete sample as many factors can influence an animal's cortisol response, including handling, physical activity and the time of day. Therefore, the very large range in cortisol concentrations within and across farms in the present study could be due to multiple factors and may have hidden any potential effect of FE on the cortisol response in these animals. Alternatively, cortisol (measured in serum) may be a more sensitive measure of acute stress (e.g., castration, photosensitisation), but a less sensitive indicator of chronic stress caused by disease, such as liver damage. In sheep, cortisol concentrations in sporidesmin-dosed hoggets tended to increase more when photosensitisation developed rather than the onset of liver injury (Smith & Payne 1991). In the present study, cattle showing clinical signs of FE were not identified, so it is not possible to establish from this research if cattle with high cortisol concentrations had clinical FE.

In the present study there was a positive relationship between bile acid and GGT concentrations. In sheep dosed with sporidesmin, bile acid concentrations rose to very high levels during the biliary obstruction phase of the disease (Peters & Mortimer 1970). A positive correlation between bile acids and GGT activity in cattle has also been found, which the authors concluded could be considered a 'downstream' consequence of sporidesmin induced liver injury (Morris et al., 2004). Therefore, the positive relationship between GGT and bile acid concentrations in the present study further supports that GGT was associated with liver damage.

To increase our understanding of the impact of FE on the welfare of cattle, future studies should evaluate biomarkers of pain and disease obtained simultaneously, include other biomarkers of pain (e.g., substance-P) and increase the number of animals sampled. Moreover, it would be advantageous to confirm that animals with elevated GGT concentrations have FE and the degree of liver damage, however, currently the methodology to confirm liver damage in cattle is very invasive and not practical on a large sample size.

In summary, few studies evaluating the effects of FE on animals have focused on assessing the welfare of these animals beyond their health status. Further research focused on assessing affective state in animals with FE would help our understanding of the disease and its implications to animal welfare.

Acknowledgments

The authors gratefully acknowledge the technical assistance from AgResearch staff and the farmers who kindly agreed to participate in this study. This project was funded by the Ministry of Primary Industry Sustainable Farming Fund and by the Ministry of Business, Innovation and Employment (Wellington, New Zealand, contract no. C10X0813).

References

Ballou MA, Sutherland MA, Brooks TA, Hulbert LE, Davis BL, Cobb CJ 2013. Administration of anesthetic and analgesic prevent the suppression of many leukocyte responses following surgical castration and physical dehorning. Veterinary Immunology and Immunopathology 151: 285-293.

Blackshaw C 1978. Serum gamma-glutamyl transferase in the diagnosis of liver disease in cattle. New Zealand Veterinary Journal 26: 1-2.

Costa D, Simbrunner B, Jachs M, Hartl L, Bauer D, Paternostro R, Schwabl P, Scheiner B, Friedrich Stättermayer AF, Pinter M, Trauner M, Mandorfer M, Reiberger T 2021. Systemic inflammation increases across distinct stages of advanced chronic liver disease and correlates with decompensation and mortality. Journal of Hepatology 74: 819-828.

Clare NT 1944. Photosensitivity diseases in New Zealand. III The photosensitizing agent in facial eczema. The New Zealand Journal of Science and Technology 25: 202-220.

Conner JG, Eckersall PD, Doherty M, Douglas TA 1986. Acute phase response and mastitis in the cow. Research in Veterinary Science 40: 26-128.

Cuttance EL, Mason WA, Laven RA 2021. The association of milk-solid production during the current lactation with liver damage due to presumptive ingestion of spores from Pithomyces chatarum by dairy cattle. New Zealand Veterinary Journal (doi.org/10.1080/00480169.2020.1861570).

Di Menna ME, Smith BL, Miles CO 2009. A history of facial eczema (pithomycotoxicosis) research. New Zealand Journal of Agricultural Research 52: 345-376.

Eckersall 2000. Recent advances and future prospects for the use of acute phase proteins as markers of disease in animals. Revue de Medecine Veterinaire 151: 577-584.

Faull BW 1986. Facial eczema surveillance. Surveillance, New Zealand 13: 3-4.

Fernández M, Pérez V, Fuertes M, Benavides J, Espinosa J, Menéndez J, García-Pérez AL, Ferreras MC 2021. Pathological Study of Facial Eczema (Pithomycotoxicosis) in Sheep. Animals 11: 1070.

Fisher AD, Crowe MA, Ó'Nualláin EM, Monaghan ML, Larkin JA, O'Kiely P, Enright WJ 1997. Effects of cortisol on in vitro interferon-γ production, acute phase proteins, growth, and feed intake in a calf castration model. Journal of Animal Science 75: 1041-1047.

Fraser D, Weary DM, Pajor EA, Milligan BN 1997. A scientific conception of animal welfare that reflects ethical concerns. Animal Welfare 6: 187-205.

Horadagoda NU, Knox KMG, Gibbs HA, Reid SWJ, Horadagoda A, Edwards SER, Eckersall PD 1999. Acute phase proteins in cattle: discrimination between acute and chronic inflammation. Veterinary Record 144: 437-441.

Kaya S, Merhan O, Kacar C, Colak A, Bozukluhan K 2016. Determination of ceruloplasmin, some other acute phase proteins, and biochemical parameters in cows with endometritis. Veterinary World 10: 1056-1062.

Klinge M, Coppler T, Liebschutz JM, Dugum M, Wassan A, DiMartini A, Rogal S 2018. The assessment and management of pain in cirrhosis. Current Hepatology Reports 17: 42-51.

Laven RA, Cuttance EL, Yang DA, Mason WA 2021. The prevalence of gross pathological damage in the livers of dairy cattle at processing plants in autumn in the North Island of New Zealand and an assessment of the gross liver pathology score as a method for estimating the prevalence of facial eczema. New Zealand Veterinary Journal 69: 113-120.

Mellor DJ, Beausoleil NJ, Littlewood KE, McLean AN, McGreevy PD, Jones B, Wilkins C 2020. The 2020 Five Domains Model: Including Human–Animal Interactions in Assessments of Animal Welfare. Animals 10: 1870.

Morris CA, Towers NR, Hohenboken WD, Maqbool N, Smith BL, Phua SH 2004. Inheritance of resistance to facial eczema: a review of research findings from sheep and cattle in New Zealand. New Zealand Veterinary Journal. 52: 205-215.

Munday JS, Ridler A, Aberdein D, Thomson NA, Griffiths K 2021. Chronic facial eczema in sheep: description of gross and histological changes in the liver and association with serum gamma-glutamyltransferase activity at the time of sporidesmin intoxication. New Zealand Veterinary Journal 69: 104-112.

Peters JA, Mortimer PH 1970. Sporidesmin poisoning in sheep. The effects of lipid and bile acid metabolism as reflected in serum. Research in Veterinary Science 11: 183-188.

Rogal SS, Winger D, Bielefeldt K, Szigethy E 2013. Pain and opioid use in chronic liver disease. Digestive Diseases and Sciences 58: 2976-2985.

Rogal SS, Bielefeldt K, Wasan AD, Lotrich FE, Zickmund S, Szigethy E, DiMartini AF 2015a. Inflammation, psychiatric symptoms, and opioid use are associated with pain and disability in patients with cirrhosis. Clinical Gastroenterology and Hepatology 13: 1009-1016.

Rogal SS, Bielefeldt K, Wasan AD, Szigethy E, Lotrich FE, DiMartini AF 2015b. Fibromyalgia symptoms and cirrhosis. Digestive Diseases and Sciences 60: 1482-1489.

Smith BL, Payne E 1991. Adrenal-associated changes in experimental sporidesmin poisoning of sheep. New Zealand Veterinary Journal 39: 46-49.

Smith BL, Towers NR 2002. Mycotoxicosis of grazing animals in New Zealand. New Zealand Veterinary Journal 50: 28-345.

Sutherland MA, Ballou MA Davis BL, Brooks TA 2013. The effect of castration and dehorning singularly or combined on the behavior and physiology of Holstein calves. Journal of Animal Science 91: 935-942.

Ting STL, Earley B, Hughes JML, Crowe MA 2003. Effect of ketoprofen, lidocaine local anesthesia, and combined xylazine and lidocaine caudal epidural anesthesia during castration of beef cattle on stress responses, immunity, growth, and behavior. Journal of Animal Science 81: 1281-1293.

Thompson ML 1978. Selection of variables in multiple regression: Part I. A review and evaluation. International Statistical Review 46: 1-19.

Towers NR, Smith BL 1978. The protective effect of zinc sulphate in experimental sporidesmin intoxication of lactating dairy cows. New Zealand Veterinary Journal 26: 199-202.

Towers NR, Stratton G 1978. Serum gamma-glutamyl-transferase as a measure of sporidesmin-induced liver damage in sheep. New Zealand Veterinary Journal 26: 109-12.

The effects of out-of-season breeding on plasma prolactin concentrations, wool production and lamb birth weight

SW Peterson, DL Burnham and ST Morris

School of Agriculture and Environment, Massey University, PB11222, Palmerston North, New Zealand.

Corresponding author: Email: sampeterson90@outlook.com

Abstract

Performance of autumn- and spring-lambing ewes was compared in a year-long trial. Ewes were shorn in November and mated to lamb in May or August. Midside wool samples and plasma were collected at seven-week intervals to determine wool-growth rate, fleece characteristics, and prolactin concentrations. Autumn lambs were lighter ($P<0.001$) at birth and weaning (3.92±0.1; 15.6±0.7 kg) than spring lambs (5.27±0.1; 25.3±0.7 kg). Plasma prolactin concentrations were high (150 ng/ml) at mating (Dec) of autumn-lambing ewes, but low (50 ng/ml) at mating (March) of spring-lambing ewes. Both groups exhibited periparturient prolactin surges, but in autumn-lambing ewes this created high plasma prolactin concentrations during early pregnancy when plasma prolactin is usually very low. In autumn-lambing ewes, clean fleece weights (3.75±0.11 kg), midside wool growth over 345 d (374±12 mg/cm^2) and fibre diameter (39.8±0.5 μm) were significantly higher ($P<0.05$) than those of spring-lambing ewes (3.26±0.12 kg; 331±14 mg/cm^2; 38.2±0.5 μm respectively) whilst bulk and colour did not differ. Bulk was higher ($P<0.05$) in single-bearing ewes (21.1±0.4 cm^3/g) than in twin-bearing ewes (19.2±0.7 cm^3/g). It is postulated that elevated plasma prolactin concentrations (at key times) are related to seasonal differences in birth and fleece weights. Autumn lambing is beneficial for wool growth but detrimental for lamb birth weight and growth.

Keywords: out-of-season lambing; sheep; prolactin; wool growth; birth weight

Introduction

Autumn lambing may be practiced in areas with good autumn pasture supply to meet market requirements for the year-round supply of chilled lamb (Morris et al. 1994). However, autumn- or winter-born lambs are 0.4-1.0 kg lighter at birth than are spring-born lambs (Reid et al. 1988; Peterson et al. 1990; Morris et al. 1993). August-lambing ewes produced heavier lambs than did June-lambing ewes, even though groups had similar residual herbage mass (Morris et al. 1993). Subsequently, Jenkinson et al. (1994) confirmed that seasonal differences were evident in fetuses at day 140 of gestation (P140) and were associated with differences in placental development. Autumn-lambing ewes have the same number of caruncles, but a lower number of placentomes and lower total placental weight, compared with spring-lambing ewes (Jenkinson et al. 1994). Fetal:placental weight ratios were not different in the two seasonal groups which implies that the smaller placenta in the autumn-lambing ewes was the cause of their lower fetal growth rate. These differences occurred despite nutritional management that ensured that live weight (LWT) of the ewes in both groups did not differ at corresponding stages of gestation. (However, maternal nutrition may have had some influence, due to the pasture consumed by the ewes in the different seasons.) Furthermore, other trials have shown a greater mean fibre diameter (MFD) and staple strength (Morris et al. 1993) and that milk yield is lower (Peterson et al. 1990) in autumn-lambing ewes compared to spring-lambing ewes.

The New Zealand Romney exhibits an annual cycle of wool growth with concomitant changes in fibre length growth rate and MFD resulting in maximum fibre volume in summer and minimum fibre volume in winter. This cycle is entrained by changes in photoperiod via changes in circulating prolactin and is influenced by complex interactions among nutrient supply and reproductive hormones (see review by Kendall et al. 2006)

In sheep, circulating prolactin is strongly influenced by season, being highest in summer (Munro et al. 1980) (when formation of placenta is occurring in autumn-lambing ewes), and ambient photoperiod markedly influences plasma prolactin concentrations in the fetus (Bassett et al. 1988; Ebling et al 1989). Jenkinson et al. (1995) found a significant stage-of-gestation-by-season interaction such that maternal plasma prolactin concentrations were higher in December-mated ewes at P70 and in March-mated ewes at P100, but did not differ significantly between ewes mated in December and March at P140. Despite the lack of difference in maternal prolactin concentrations at P140, fetal plasma prolactin concentrations at P140 were higher in March-mated ewes than in December-mated ewes (Jenkinson 1994). Such results suggest that seasonal differences in birth weight of lambs may be a consequence of seasonal differences in circulating prolactin concentrations in the dam and/or fetus.

Seasonal differences in fetal weight were apparent at P84 and associated with small differences early in placental development (McCoard et al. 1996), suggesting that the seasonal effect occurs during that time. However, it is also possible that fetuses of ewes mated to lamb in the autumn or spring initially form the same number of placentomes but that the autumn lambs subsequently lose more placentomes. Jenkinson et al. (1995) demonstrated that the number of placentomes was lower in December-mated ewes. Data from other studies (SH Min, DDS Mackenzie and SN McCutcheon, unpublished results) indicate that something retarded the loss of placentomes in spring-lambing ewes. Growth hormone is unlikely to be the mediator since exogenous bGH treatment had little discernible effect on measures of placental size (Jenkinson et al. 1999; Gargosky et al. 1991). Insulin-like growth factor-I (IGF-1) may be the mediator since increasing maternal IGF-1 in mice and rats altered the pattern of fetal growth by removing maternal constraint (Gluckman et al. 1992).

There are major seasonal differences in the physiology of sheep for which we have been unable to ascertain the cause. Elucidating those mechanisms may enable manipulation of parameters of importance to animal industries and perhaps to human clinical medicine (Jenkinson et al. 1999; Gluckman et al. 1992). From another perspective, lambing in autumn has detrimental effects on lamb birth weight and lactation of ewes, but may have advantages in terms of wool-growth characteristics.

The objectives of the experiment were related to both wool growth and fetal growth. Our hypothesis was that wool-growth rate (WGR), fleece weight, lamb birth weight, ewe LWT and plasma prolactin concentrations differed between autumn- and spring-lambing ewes grazed at pasture, independently of seasonal nutritional differences.

We wished to evaluate wool-growth (quantity and quality) differences between ewes lambing at different times of year and to identify possible causes of the low wool-growth response observed in out-of-season-lambing ewes. Further, we wished to gain insight into a possible role of prolactin leading to the differences in wool growth, and in fetal growth and birth weight

Materials and methods

Animals and treatments

The trial was carried out at Keeble's farm, Massey University, and was approved by the University Animal Ethics Committee.

The trial began with 137 Romney x Border Leicester ewes but only ewes (n=47) that successfully birthed and reared one or two lambs were included in the analyses. Ewes were shorn in November and a 10x10 cm midside patch was cleared on each ewe on 15 Dec. Following synchronisation (using progesterone-impregnated CIDRs for 14 d) ewes were mated to lamb in either May or August. Ewes mated in December were treated with 500 i.u. PMSG as Folligon (Intervet International B.V., Boxmeer, The Netherlands) the day before CIDR removal, whilst ewes mated in March were not treated.

Rams were fitted with mating harnesses, and crayon marks on ewes were recorded daily. Following ultrasound pregnancy diagnosis at P50, groups were balanced for pregnancy rank, age and LWT. Ewes in each group were managed as separate mobs on pasture, mainly ryegrass (*Lolium perenne*) and white clover (*Trifolium repens*), following normal pasture management procedures and allowance for pregnancy. Both groups were flushed four weeks prior to mating on the best pasture available.

The midside patch was re-clipped approximately every seven weeks (Table 1) with clippers (Oster; Milwaukee, WI, USA) fitted with a No. 40 blade, until the sheep were shorn in November at the conclusion of the experiment. Patch wool samples were washed in warm water and detergent, dried, conditioned to 16% regain and weighed. The midside wool samples were used to determine growth rate and diameter. Mean fibre diameter was measured by OFDA (Edmunds 1995). Fleece weights were recorded after 360 days (November shearing) and sample were tested for bulk using a WRONZ bulkometer following the procedure of Bigham et al. (1984a) and were colour-tested using a Hunterlab D 25 D2M Colorimeter (Bigham et al. 1984b).

Plasma samples were collected at about seven-week intervals throughout the year for analysis of prolactin. Live weight of the ewes was recorded (Tru-Test Distributors, Auckland) at the same time. Birth weights and weaning weights of lambs were also recorded. The weaning dates were 12 Aug and 25 Nov (98 and 100 d respectively after first lamb birth) (Table 1).

Assays

The prolactin assay was a homologous double-antibody competitive-binding radioimmunoassay based upon the method of van Landeghem and van de Wiel (1978). Details of the assay were described previously (Peterson et al. 1994). Standards ranging from 1-1200 μg/l were prepared using biological grade ovine PRL (NIADDK-oPRL-18, AFP-82277E, 30 i.u./mg). The linear range was approximately 10-800 μg/l. The first antibody, rabbit anti-oPRL antiserum, was used at a working dilution of 1:50,000 (final dilution 1:550,000). The second antibody, goat anti-rabbit IgG (Lot No. 8103, Immuno-Chemical Products Limited, Auckland, New Zealand) was used at a working dilution of 1:40. Plasma samples were assayed in triplicate. The mean intra-assay coefficient of

Table 1 Comparison of stages of pregnancy and lactation in two groups of Romney x Border Leicester ewes lambing in either autumn or spring. On the dates indicated, plasma and midside wool samples were collected and animals were weighed. Dates of lambing were; autumn 6-18 May (mean 12 May) and spring 17 Aug-10 Sept (mean 5 Sep).

Date of wool and blood samples	Day of trial	Autumn-lambing ewes	Spring-lambing ewes
15 Dec	0	clear initial midside patch	
04 Feb	51	early pregnancy	pre-mating
25 Mar	100	mid pregnancy	post-mating
13 May	149	late pregnancy	early pregnancy
06 Jul	203	early lactation	mid pregnancy
12 Aug	240	late lactation lambs weaned	late pregnancy
07 Oct	296	post-weaning	early lactation
25 Nov	345	dry period	late lactation lambs weaned

Figure 1 (a) Live weight (kg) and **(b)** plasma prolactin concentration (μg/l) measured every seven weeks, of 25 autumn-lambing ewes (♦——♦) and 22 spring-lambing ewes (■------■) grazed at pasture and managed for normal commercial production. The February plasma samples were lost. Vertical bars represent the standard error of the mean.

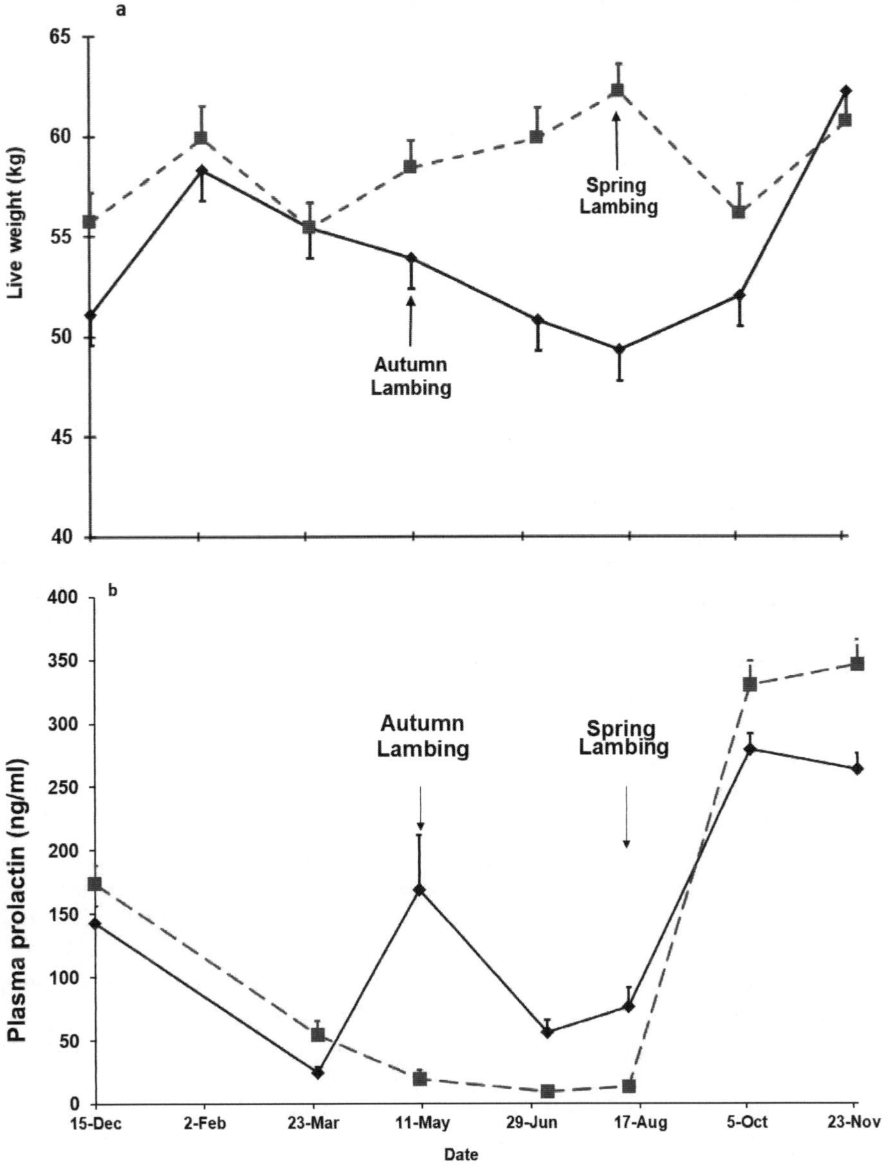

variation (CV) calculated over 12 assays was 9.7% and the mean inter-assay CV was 12.6% for three reference plasma samples corresponding with the linear portion of the standard curve.

Statistical analyses

Multivariate (repeated-measures) analyses of variance were used to analyse time-series data. Prolactin data were log transformed and were analysed using the computer statistical package REG (Gilmour 1990). Remaining data were analysed using SAS proc GLM. In addition to main effects (season), models included number and sex of lambs. Least-squares means (±SEM) are reported unless stated otherwise.

Results

Although both groups of ewes had similar LWT at the beginning of the trial, and identical LWT in March (55 kg) when the spring-lambing ewes were mated, the effect of pregnancy and lactation resulted in the autumn-lambing ewes being 10 kg lighter than the heavily pregnant spring-lambing ewes by the beginning of August (Fig 1a). However, following good spring feeding, there were no differences between the groups by November. Prior to birth, autumn-lambing ewes were 53.9±1.3 kg and declining in weight whilst the spring-laming ewes were 58.4±1.4 kg and increasing in weight.

Plasma prolactin profiles, for the most part, did

not differ between the two treatment groups with the exception of samples taken in May, when plasma prolactin concentrations were significantly higher in the autumn-lambing ewes than those in the spring-lambing ewes (Fig 1b).

Wool growth rate did not differ (P=0.3) between groups at the beginning of the trial but by March, WGR in the autumn-lambing ewes had declined significantly (P=0.0002) below that of the spring-lambing ewes (Fig. 2). Then in July and August, the autumn-lambing ewes had significantly higher WGR than did the spring-lambing ewes. Subsequently, WGR did not differ between the groups. Over the whole trial period, autumn-lambing ewes had higher (P<0.05) midside WGR and fleece weights than did spring-lambing ewes (Table 2) and their November fleeces (4.55 kg) were significantly (P<0.05) heavier than those of the spring-lambing group (4.01 kg), with slightly coarser (P<0.05) wool fibres. Bulk and colour did not differ between autumn- and spring-lambing ewes, however, bulk was greater (P<0.05) in ewes with single lambs than in those with twins, while no other fleece characteristics differed with birth rank.

Mean fibre diameter (MFD) differed significantly (P<0.01) between groups at the start of the trial (before treatments started) but by March and May, MFD did not differ (Fig. 2). Subsequently, MFD remained higher in autumn-lambing ewes for the rest of the trial (P<0.001).

Autumn lambs were on average 1.35 kg lighter at birth, they grew significantly slower and their weaning weights were 10 kg lighter than those born in spring (P<0.001) (Table 2). Autumn lambs were on average 92 days of age at weaning while the spring lambs were 88 days old.

Discussion

The objectives of the trial were to evaluate differences in the reproductive performance and wool growth characteristics of ewes lambing in autumn and those lambing in spring, with a view to identifying possible relationships among seasonal differences in plasma prolactin concentrations and the production differences observed.

Live weight

Live weight in each group followed the changes expected in pregnant ewes, and, since they conceived at different times, their LWT profile differed (Fig. 1). Autumn-lambing ewes were 4.5 kg lighter and declining in weight compared to spring-lambing ewes, which were increasing in weight. These small differences are unlikely to have resulted in the lower autumn lamb birth weight since large differences in maternal nutrition (*ad libitum* versus maintenance) during the last trimester failed to reduce lamb birth weights; rather those more-poorly fed had heavier lambs (Kenyon et al. 2011).

Prolactin

The plasma concentration of prolactin is a result of the effects of both pregnancy and photoperiod. Data from spring-lambing ewes indicate that the profile of prolactin concentration was consistent with the very few published New Zealand studies (Munro et al. 1980; Craven et al. 1994; Pearson et al. 1996, Peterson et al. 2019). Mating the ewes at different times, therefore, resulted in a peak in plasma prolactin concentration in the autumn-lambing ewes at an uncharacteristic time relative to the ambient photoperiod, in agreement with the results of Jenkinson et al. (1995) and Peterson et al. (2019). The timing of that peak may relate to the lower birth weight and greater wool growth in the autumn-lambing ewes.

Wool

In autumn-lambing ewes, clean fleece weights, midside WGR and MFD were significantly higher than those of spring-lambing ewes, whilst bulk and colour did not differ. Bulk was higher in single-bearing ewes (Table 2) than in twin-bearing ewes. Seasonal differences in pregnancy resulted in markedly different WGR at the same times of the year, being initially higher in the spring-lambing ewes (Fig. 2), presumably because of competition for nutrients between fleece and fetus in the autumn-lambing ewes. However, after the autumn lambing, WGR in those ewes was much higher than in the pregnant spring-lambing ewes. WGR did not decline during winter to the same extent and

Table 2 Effect of lambing season and litter size on wool-growth rate, fleece weight (FW), clean fleece weight (CFW), mean fibre diameter (MFD), bulk, and colour of the final fleece, lamb birth and weaning weights, and liveweight gain (LWG) to weaning (mean±SEM) in two groups of Romney × Border Leicester ewes mated to lamb in either autumn or spring.

	n	Wool growth rate[1] ($mg/cm^2/d$)	FW (kg)	CFW (kg)	MFD (μm)	Bulk (cm^3/g)	Colour (Y-Z)	Birth weight (kg)	Weaning weight (kg)	LWG (g/d)
Lambing season										
Autumn	25	374 ± 12^b	4.51±0.12^b	3.75±0.11^b	39.8±0.5^b	19.9±0.6	3.52±0.41	3.92±0.12^a	15.6±0.7^a	129±7^a
Spring	22	331 ± 14^a	4.05±0.13^a	3.26±0.12^a	38.2±0.5^a	20.4±0.6	3.62±0.46	5.27±0.13^b	25.3±0.7^b	229±7^b
Litter Size										
Single	33	361 ± 10	4.32±0.10	3.53±0.09	38.9±0.4	21.1±0.4^b	3.50±0.32	5.08±0.12^b	22.8±0.7^b	202±7^b
Twin	14	344 ± 16	4.33±0.15	3.48±0.14	39.1±0.6	19.2±0.7^a	3.64±0.52	4.10±0.13^a	18.0±0.8^a	155±8^a

[ab] Means within main effects having different superscripts are significantly different (P < 0.05) [cd] Means within main effects having different superscripts are significantly different (P<0.001)

[1] Total mid-side wool growth from days 0-345.

Figure 2 **(a)** Wool-growth rate (mg/cm²/d)) and **(b)** mean fibre diameter (μ) measured using midside wool samples taken every seven weeks from 25 autumn-lambing ewes (♦———♦) and 22 spring-lambing ewes (■------■) grazed at pasture and managed for normal commercial production. Vertical bars represent the standard error of the mean.

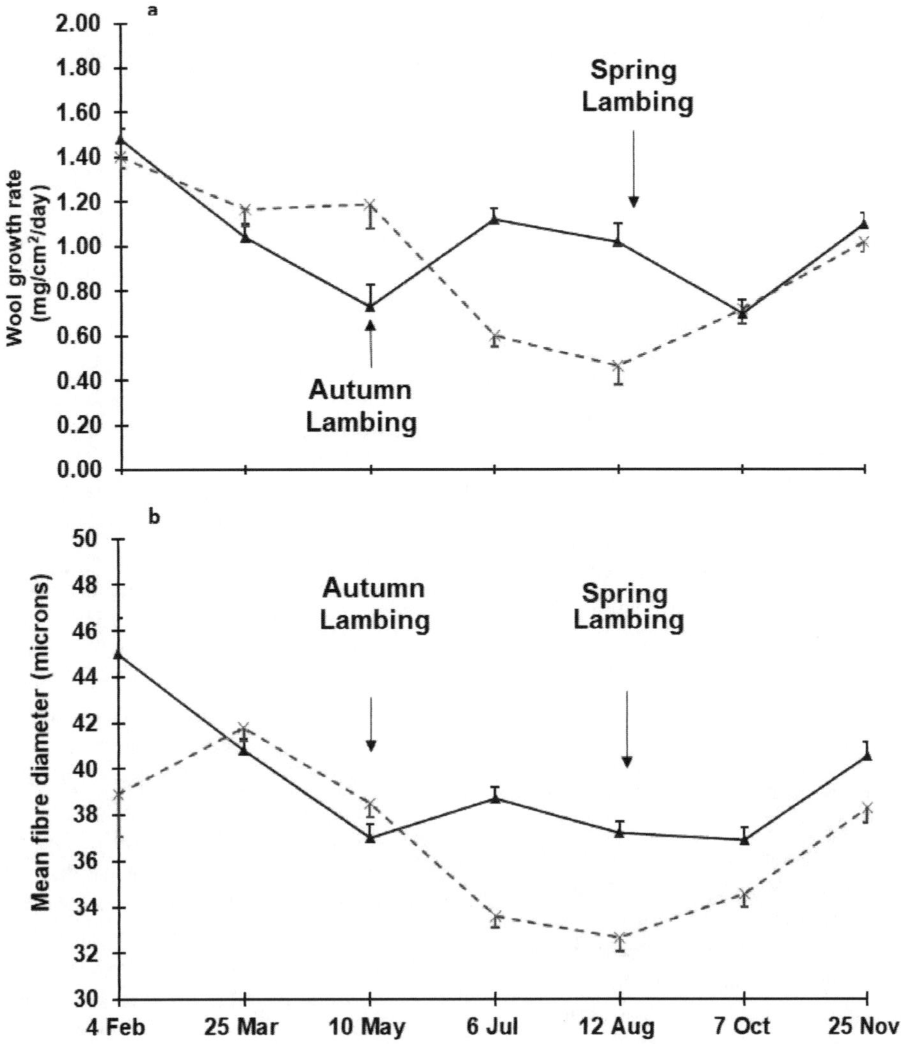

the growth curve was much flatter. Was this due simply to greater availability of nutrients for the fleece, or was WGR increased in the autumn-lambing ewes due to the concurrent high plasma concentration of prolactin?

Effect of pregnancy on wool growth: Pearson et al. (1999) detected an effect of pregnancy on individual fibre length growth rate, determined by autoradiography, as early as 21-35 days after joining, suggesting a hormonal influence on WGR at an early stage of pregnancy. In the second month of pregnancy, fleece production and MFD declined below that of non-pregnant ewes and continued to diverge until parturition. In the non-pregnant group, wool production increased from August, but did not increase in lambing ewes until after parturition, the rapid increase in growth coinciding with the disappearance of gestational hormones from the circulation and despite an increase in the metabolic requirement of lactation compared to that of the growing fetus (Rattray, 1986). In the results of Pearson et al. (1999), the inhibitory effect on wool growth was associated with pregnancy rather than with lactation. All three of their measures of wool growth (patch wool weight, fibre diameter and fibre length) indicated that wool growth increased markedly following parturition, which is consistent with the results of the present trial (Fig. 2), even though our two groups lambed in a different photoperiod (compared to each other). Thus, although they did not identify the hormones affecting wool growth, the data of Pearson et al. (1999) suggested that pregnancy is associated with an early hormonally mediated depression in follicle output superimposed on the decline associated with winter photoperiod, and the same is likely true in the present trial.

Effect of photoperiod and prolactin on wool growth: Prolactin concentrations (Fig.1b) in the spring-lambing ewes followed changes in photoperiod as expected, as did the seasonal pattern of wool growth (Fig. 2) in that group. However, the autumn-lambing ewes exhibited a peak in prolactin concentrations in May, presumably as a result of the changes in prolactin related to parturition and lactogenesis (Peterson et al. 1990; 1994). Given the established role of prolactin in controlling WGR (Pearson et al. 1999; Kendall et al. 2006), it is likely that the "unseasonal" peak in prolactin concentration is responsible for the concomitant increase in WGR. Follicle recrudescence during the spring has been linked to increasing photoperiod and plasma PRL concentration in a variety of species (Pearson et al. 1996) so it likely that the May increase in WGR is due to a signal from prolactin.

Birth weight

The present trial confirmed the results of previous trials in which autumn lambs were lighter than spring lambs (Reid et al. 1988; Peterson et al. 1990; Morris et al. 1993). Our hypothesis was that high summer prolactin concentration in the autumn-lambing ewes may have been responsible for the lighter birth weights in autumn. Figure 1b shows that, during pregnancy, autumn-lambing ewes were exposed to prolactin concentrations about five times higher than those in the spring-lambing ewes, lending support to our hypothesis. Furthermore, prolactin has been shown in lactating animals to optimise the action of IGF-I by repressing IGFBP-5, which is an inhibitor of IGF action (Wilde et al., 1999; Flint & Knight, 1997; Sorensen & Knight 2002). IGF-1 has been shown, in mice and rats, to increase fetal growth by removing maternal constraint (Gluckman et al. 1992). Nevertheless, this is just speculation and the mechanism is, as yet, unknown. Whilst there is no published evidence that prolactin *per se* increases fetal growth, in contrast, consumption of ergot alkaloids produced by endophytes resulted in decreased serum prolactin concentrations, vasoconstriction and reduced fetal growth (Britt et al. 2019). Further evidence against our hypothesis was the finding that elimination of the summer prolactin surge in autumn-lambing ewes (using bromocriptine) failed to prevent low autumn birth weights (Peterson et al. 2019).

Conclusions

It was not possible to identify the primary causes of wool growth (quantity and quality) differences between ewes lambing at different times of year (i.e., to identify causes of the low WGR response observed in autumn-lambing ewes, in particular to separate possible nutritional effects from photoperiod effects). However, the low WGR response was concomitant with a high plasma concentration of prolactin. Similarly, it was not possible to determine the cause of lower lamb birth weights in autumn, but we have shown that they follow high plasma prolactin concentrations during pregnancy.

Acknowledgements

Biological and iodination grade ovine PRL were donated by NIADDK and supplied through the National Hormone and Pituitary Program, University of Maryland School of Medicine. The rabbit anti-oPRL antiserum, was donated by Dr DFM. van de Wiel, "Schoonoord" Research Institute for Animal Husbandry, Zeist, The Netherlands. The Wool Industry Charitable trust provided financial support for the programme. Thanks are due to Animal Research Unit staff for care of the animals and to the late Phil Pearce for laboratory assays.

References

Bassett JM, Bomford J, Mott JC 1988. Photoperiod: An important regulator of plasma prolactin concentration in fetal lambs during late gestation. Quarterly Journal of Experimental Physiology 73: 241-244.

Bigham ML, Sumner RMW, Cox NR 1984a. The measurement and sources of variation of loose wool bulk in Romney, Coopworth, and Perendale breeds. New Zealand Journal of Experimental Agriculture 12: 23-28.

Bigham ML, Sumner RMW, Cox NR 1984b. Sources of variation and experimental design criteria for wool colour measurement applied to Romney, Coopworth, and Perendale flocks. New Journal of Experimental Agriculture 12: 231-236.

Britt JL, Greene MA, Bridges Jr WC, Klotz JL, Aiken GE, Andrae JG, Pratt SL, Long N, Schrink FN et al. 2019. Ergot alkaloid exposure during gestation alters. I. Maternal characteristics and placental development of pregnant ewes. Journal of Animal Science 97: 1874-1890.

Craven AJ, Parry AL, Wildermoth JE, Pearson AJ 1994. The effect of long-day photoperiod treatments on plasma prolactin and wool follicle activity in New Zealand Wiltshire sheep. Proceedings of the New Zealand Society of Animal Production 54: 135-138

Ebling FJP, Wood R, Suttie JM, Ade TE, Foster DL 1989. Prenatal photoperiod influences neonatal prolactin secretion in the sheep. Endocrinology 125: 84391.

Edmunds AR 1995. OFDA measurement of additional fibre parameters; some preliminary results. International Wool Textile Organisation Technical committee. Report Number 19, June 1995.

Flint DJ, Knight CH 1997. Interactions of prolactin and growth hormone (GH) in the regulation of mammary gland function and epithelial cell survival. Journal of Mammary Gland Biology and Neoplasia 2: 41-48.

Gargosky SE, Owens JA, Walton PE, Owens PC, Wallace JC, Ballard FJ 1991. Administration of insulin-like growth factor-I, but not growth hormone increases maternal weight gain in late pregnancy without affecting fetal or placental growth. Journal of Endocrinology 130: 395-400.

Gilmour AR 1990. REG – a generalised least squares program. Version 90.12 New South Wales Agriculture and Fisheries, Australia.

Gluckman PD, Morel PCH, Ambler GR, Breier BH, Blair HT, McCutcheon SN 1992. Elevating maternal insulin-like growth factor-1 in mice and rats alters the pattern of fetal growth by removing maternal constraint. Journal of Endocrinology 134: R1-R3.

Jenkinson CMC 1994. The effect of season on placental development and fetal growth in sheep. Dissertation for Postgraduate Diploma in Agricultural Science, Massey University.

Jenkinson CMC, Peterson SW, Mackenzie DDS, McCutcheon SN 1994. The effects of season on placental development and fetal growth in sheep. Proceedings of the New Zealand Society of Animal Production 54: 227-230.

Jenkinson CMC, Peterson SW, Mackenzie DDS, McCutcheon SN 1995. Seasonal effects on birth weight in sheep are associated with changes in placental development. New Zealand Journal of Agricultural Research 38: 337-345.

Jenkinson CMC, Min SH, Mackenzie DDS, McCutcheon SN, Breier BH, Gluckman PD 1999. Placental development and fetal growth in growth hormone-treated ewes. Growth Hormone & IGF Research 9: 11-17.

Kendall PE, Sumner RMW, Pearson AJ 2006. Environmental and physiological mechanisms underlying wool growth rhythms in coarse wool sheep. Proceedings of the New Zealand Society of Animal Production 66: 144-148 References 155-161.

Kenyon PR, Pain SJ, Hutton PG, Jenkinson CMC, Morris ST, Peterson SW, Blair HT 2011. Effects of twin-bearing ewe nutritional treatments on ewe and lamb performance to weaning. Animal Production Science 51: 406-415.

McCoard SA, Peterson SW, Jenkinson CMC, Campbell JT, McCutcheon SN 1996. Seasonal effects on fetal growth in sheep. Proceedings of the New Zealand Society of Animal Production 56: 297-300.

McCutcheon SN, Holmes CW, McDonald MF 1981. The starvation-exposure syndrome and neonatal lamb mortality. A review. Proceedings of the New Zealand Society of Animal Production 41: 209-217.

Munro CJ, McNatty KP, Renshaw L 1980. Circa-annual rhythms of prolactin secretion in ewes and the effect of pinealectomy. Journal of Endocrinology 89: 83-89.

Morris ST, Blair HT, Parker WJ, McCutcheon SN 1993. Evaluation of Border Leicester × Romney (BR), Poll Dorset × BR, and Suffolk × BR ewes for out-of-season lambing. New Zealand Journal of Agricultural Research 36: 349-362.

Morris ST, McCutcheon SN, Blair HT, Parker WJ 1994. Effect of lambing policy and ewe breed cross on wool growth patterns and wool quality. New Zealand Journal of Agricultural Research 37: 65-78

Pearson AJ, Parry AL, Ashby MG, Choy VJ, Wildermoth JE Craven AJ 1996. Inhibitory effect of increased photoperiod on wool follicle growth. Journal of Endocrinology 148: 157-166.

Pearson AJ, Kendall PE, Ashby MG, Wildermoth JE 1999. Fleece production patterns in Romney ewes: effects of photoperiod, pregnancy and lactation. Proceedings of the New Zealand Society of Animal Production 59: 30-33.

Peterson SW 1992. The role of prolactin in the control of ovine lactogenesis. PhD thesis, Massey University, Palmerston North, New Zealand.

Peterson SW, Wigati S, Jenkinson CMC 2019. Effects of maternal bromocriptine and melatonin treatments during early gestation on fetal lamb growth New Zealand Journal of Animal Science and Production 79: 74-79.

Peterson SW, Mackenzie DDS, McCutcheon SN 1990. Milk production and plasma prolactin levels in spring- and autumn-lambing ewes. Proceedings of the New Zealand Society of Animal Production 50: 483-485.

Peterson SW, Mackenzie DDS, McCutcheon SN, Lapwood KR 1994. The effect of peripartum administration of ovine prolactin on lactogenesis in autumn-lambing ewes. Proceedings of the New Zealand Society of Animal Production 54: 115-118.

Rattray PV 1986. Feed requirements for maintenance, gain and production. In: McCutcheon SN, McDonald MF, Wickham GA eds. Sheep production Vol. 2. Feeding, growth and health, NZIAS. pp 75-109.

Reid TD, Sumner RMW, Wilson VLD 1988. Performance parameters in an autumn-lambing ewe flock. Proceedings of the New Zealand Society of Animal Production 49: 255-260.

Sorensen A, Knight CH 2002. Endocrine profiles of cows undergoing extended lactation in relation to the control of lactation persistency. Domestic Animal Endocrinology 23: 111-123.

van Landeghem AAJ, van de Wiel DFM 1978. Radioimmunoassay for porcine prolactin: plasma levels during lactation, suckling and weaning and after TRH administration. Acta Endocrinologica 88: 653-667.

Wilde CJ, Knight CH, Flint DJ 1999. Control of milk secretion and apoptosis during mammary involution. Journal of Mammary Gland Biology and Neoplasia 4: 129-136.

A comparison of the ewe reproductive performance of Romney ewes bred to Romney or Wiltshire rams and the growth of their progeny to weaning

RA Corner-Thomas[a]*, AL Ridler[a], PR Kenyon[a], HT Blair[a] and ST Morris[a]

[a]*School of Agriculture and Environment, Massey University, Palmerston North, New Zealand.*

**Corresponding author.* Email: r.corner@massey.ac.nz

Abstract

This study reports the results of the first cross breeding of Romney ewes with Wiltshire rams as part of a grading-up transition. Romney mixed-age ewes were bred to either Romney (n=200) or Wiltshire rams (n=349) in March 2020. Ewe breeding pattern (cycle bred), reproductive performance (pregnancy and fecundity rates) and lambing performance (lambing rate, lamb birth and weaning live weight and lamb survival) were recorded. The reproductive performance of ewes bred to either Romney or Wiltshire rams showed few differences. There were no differences (P>0.05) in pregnancy, fecundity or lambing rates of ewes bred to either Romney or Wiltshire rams although minor differences (P<0.05) in the breeding pattern were observed. Lamb live weights at birth and in mid-lactation did not differ (P>0.05) among single, twin and triplet lambs sired by either Romney or Wiltshire rams, however at weaning Romney-sired triplet lambs were heavier than Wiltshire triplets (P<0.05). At weaning, a small percentage (1.6%) of Wiltshire lambs showed early signs of fleece shedding. Combined, these results suggest that in this flock there was little impact of introducing Wiltshire rams to a mixed-age Romney ewe flock on ewe reproductive performance and the growth of the F_1 progeny to weaning.

Keywords: Wiltshire; Romney; reproduction; growth; shedding

Introduction

The New Zealand sheep industry was founded on the production of wool, exports accounting for more than a third of New Zealand's export earnings (Easton 2010) until 1882 when the first shipment of refrigerated sheep meat allowed the export of meat to the United Kingdom (Cottle 2010). Since the late 1980s, however, the percentage of export earnings attributed to wool has decreased from approximately 13% to 3% in the mid 2000s (Easton 2010). In the 2018-19 survey, the Beef + Lamb Economic Service reported that wool contributed an average of 3% to the farm income of New Zealand sheep and beef farms.

In New Zealand, the majority of sheep are dual purpose producing both meat and wool (Beef + Lamb NZ 2020). Blair (2011) estimated that 59% of farms had Romney-based breeds which have been reported to produce fleeces with a fibre diameter of between 33 and 37 µm (Morris et al. 1996). Currently, 90% of the wool produced in New Zealand is classed as coarse or >30 µm (Beef + Lamb New Zealand 2021a; Simm 1998). Over the last five years the price paid for clean coarse wool has decreased from $6.14/kg in 2016 to $1.88/kg in January 2021 (JDJL Limited 2021). At this price farmers are not able to recover the costs of shearing and as a result, are investigating alternative options (Beef + Lamb New Zealand 2021b).

One option for farmers to reduce the impact of low wool prices on farm income is to introduce of a self-shedding breed such as the Wiltshire into the flock, to eliminate the need for shearing (Slee 1959). The Wiltshire breed in New Zealand originated from four Wiltshire Horn x Poll Dorset first cross ewes and a Wiltshire Horn ram imported from Australia in 1972, from which a stabilised polled Wiltshire breed was established (Morrison 2010). The heritability of fleece shedding in the New Zealand Wiltshire sheep has been reported to be high at 0.52 (Johnson et al. 2007). Due to this high heritability, a low-cost and medium-term strategy to transition to a Wiltshire flock is by using a grading-up approach (Slee 1959). Grading up involves continued use of Wiltshire rams across an existing ewe flock and the selection of replacement ewe lambs for greater shedding (Nicholas 2009). It has been predicted that this transition to a shedding flock would take three to four generations (Nicholas 2009; Slee 1959). Farrell et al. (2020) predicted that grading up from a Romney-based flock to a fully shedding Wiltshire flock could be achieved within 15 years, with shearing expenses eliminated after seven years. In addition, they predicted that once the flock had reached the desired final cross of 7/8Wiltshire the income from the sheep enterprise would be greater than that of the base Romney flock. The model, however, had limitations as data on shedding sheep in New Zealand were scarce, which may have reduced the accuracy of the model predictions. This study was designed, therefore, to determine the reproductive performance of Romney ewes bred to Wiltshire rams and the growth and shedding of the F_1 progeny of a flock in the first year of a grading-up transition.

Materials and methods

The study was conducted at Massey University's Riverside farm located 10km north of Masterton on the east coast of the North Island of New Zealand (40°50'31"S 175°37'04"E). In March 2020, Romney ewes (n=549) were bred to either Romney (n=200 ewes) or Wiltshire rams (n=349 ewes). The ram to ewe ratio used was in excess of the industry recommendation of 1:100 (Geenty, 2017) at 1:66 for the Romney rams and 1:87 for the Wiltshire rams. All rams were fitted with a mating harness and crayon to identify ewes that had been bred during the first or second oestrous cycles. Breeding began on the 25th of March (pregnancy day 1; P1) and rams remained with the ewes for two oestrous cycles (a total of 34 days) until the 28th of April (P35). The ram-harness crayon colour was changed at P18 and the presence of colour marks on the rump of the ewes was recorded at P35. After rams were removed, all ewes were managed as a single mob throughout pregnancy and until lambing. Pregnancy diagnosis was conducted 107 days after the start of breeding (P107). Pregnancy status (pregnant/non-pregnant) and the number of fetuses (0, 1, 2, or 3) was determined using transabdominal ultrasound conducted by a commercial operator. Ewes were weighed prior to breeding (P1), at ram removal (P35), pregnancy diagnosis (P107), in mid-lactation (Day 52 of lactation; L52) and at weaning (L108). Ewe body condition was scored on a scale from 1 to 5 as described by Jefferies (1961) at weighing events conducted on P1, P107, L52 and L108.

One week prior to the expected start of lambing, ewes were randomly allocated to one of 16 lambing paddocks at a density of 10 to 12 ewes per hectare. Each lambing paddock had a mix of pregnancy ranks (single, twin or triplet) and ewes bred to either Romney or Wiltshire rams. The lambing period began on the 11th of August (L1) with the final lamb born on L51. Lambs were ear tagged within 24 hours of birth and had their weight, birth rank and sex recorded. At L52, all lambs born prior to L47 were tail docked and ram lambs were ear marked and made short-scrotum using an elastrator and a rubber ring (Elastrator ring, Heiniger, New Zealand). The single ewe lamb born after L47 was weighed but not tail docked. Ewe and lamb live weights were recorded mid-lactation (L51) and at weaning (L108).

At weaning (L108; 26th November 2020) short-scrotum ram lambs were weighed and separated from ewe lambs and removed from the study. Of the 438 ewe lambs present at weaning, 374 (Wiltshire sired n=243 and Romney sired n=128) were retained as replacement animals. Fleece shedding scores were conducted on all the replacement ewe lambs sired by Wiltshire rams (n=243) and a random sample of Romney-sired ewe lambs (n=19). The shedding score (0 to 5: no shedding to full shedding) was conducted as per the method of Johnson et al. (2007) and O'Connell et al. (2012).

Statistical analysis

All analyses were conducted using SAS version 9.4 (SAS Institute, Cary, USA). Analyses of ewe reproductive performance (cycle marked (first, second, both cycles or not marked), fertility (pregnant / non pregnant), fecundity (number of fetuses), lambing (lambed / not lambed), ewe body condition score (BCS), and lamb survival to weaning (lived / died) were analysed using generalised models which included the fixed effect of sire breed. Models of variables with two outcomes (fertility, lambing and lamb survival) used a binomial distribution and logit transformation whereas models with ordinal outcomes (cycle, fecundity) used a Poisson distribution. Ewe BCS was analysed using a generalised model that allowed for repeated measures which included the fixed effects of day (P1, P107, L52 and L108) and sire breed (Romney or Wiltshire) and their two-way interaction. Means were back-transformed using a logit transformation and presented as the back-transformed mean with the 95% confidence interval in parentheses.

Analyses of ewe and lamb live weight were conducted using the mixed procedure allowing for repeated measures. The model for ewe live weight included the fixed effects of day and sire breed and their two-way interaction. The model for lamb live weight included the fixed effect of day of lactation (L0, L52 or L107), sire breed, birth rank (1, 2 or 3) and sex of the lamb (ewe or short scrotum) and the three-way interaction of day by sire breed by birth rank. Lamb growth rates from birth to weaning were analysed using a mixed model that included the fixed effect of sire breed and included lamb birth weight as a covariate. Results are presented as the mean and standard error of the mean.

The frequency of ewe lamb fleece shedding scores could not be statistically analysed since all ewe lambs in the subsample sired by Romney rams had a score of 0. A summary of the frequency of shedding scores have been reported.

Results

Ewe live weight and body condition score

Ewe live weight and BCS did not differ (P>0.05) among ewes bred to Romney or Wiltshire rams throughout the study (Table 1). During pregnancy, ewe live weights and BCS increased to the time of pregnancy diagnosis (P107)

Table 1 Ewe live weight (kg; mean ± SEM) and body condition score (BCS, mean with 95% confidence interval in parentheses)) of ewes bred to either Romney or Wiltshire rams at mating (pregnancy day 1; P1), ram removal (P35), pregnancy diagnosis (PD; P107), mid-lactation (lactation day 51; L52) and weaning (L108)

	Sire breed			
Weighing event	n	Romney	n	Wiltshire
Live weight (kg)				
Pre-breeding (P1)	184	67.7 ± 0.5 [ab]	328	66.6 ± 0.4 [a]
Ram out (P35)	184	67.2 ± 0.5 [a]	328	66.5 ± 0.4 [a]
PD (P107)	184	69.5 ± 0.5 [cd]	328	68.9 ± 0.4 [bc]
Mid-lactation (L52)	184	67.0 ± 0.5 [a]	331	66.7 ± 0.4 [a]
Weaning (L108)	169	70.2 ± 0.5 [d]	298	69.8 ± 0.4 [cd]
BCS				
Mating (P1)	184	2.65 (2.59 - 2.72) [b]	328	2.72 (2.57 - 2.87) [bc]
Ram out (P35)	184	2.90 (2.82 - 2.97) [d]	328	2.91 (2.85 - 2.98) [d]
PD (P107)	184	2.31 (2.23 - 2.39) [a]	331	2.37 (2.30 - 2.44) [a]
Mid-lactation (L52)	169	2.58 (2.48 - 2.68) [b]	298	2.66 (2.58 - 2.73) [bc]

[abcd] within each variable (live weight and BCS) means with different superscripts are significantly different P<0.05

Ewe reproductive performance

Table 2 Reproductive performance (percentage with 95% confidence interval in parentheses) of ewes bred to Romney or Wiltshire rams including cycle marked by the ram (first, second, returned to service or not marked;)), diagnosed with single, twin or triplet fetuses, fertility rate (ewes pregnant per ewe presented for breeding), fecundity rate (fetuses per 100 ewes presented for breeding), lambing rate (lambs born per ewe presented for breeding) and percentage of ewes lambed (had lambing record).

	Ram breed			
	n	Romney	n	Wiltshire
Cycle marked by ram				
First (%)	159	81.5 (75.5 - 86.4) [a]	302	88.8 (85 - 91.8) [b]
Second (%)	25	12.8 (8.8 - 18.3)	26	7.6 (5.3 - 11)
Returned to service (%)	0	0 (0 - 0)	14	4.1 (2.5 - 6.8)
Not marked (%)	11	5.6 (3.2 - 9.9)	12	3.5 (2.0 - 6.1)
Reproductive performance				
Single (%)	50	25.6 (20 - 32.2)	90	26.5 (22.1 - 31.4)
Twin (%)	141	72.3 (65.6 - 78.1)	239	70.3 (65.2 - 74.9)
Triplet (%)	4	2.1 (0.8 - 5.3)	11	3.2 (1.8 - 5.7)
Fertility rate (%) *	195	100	340	100
Fecundity rate (%)	195	176.4 (158.7 - 196.1)	340	176.8 (163.2 - 191.5)
Lambing rate (%)	194	176.3 (158.6 - 196.0)	331	175.2 (161.5 - 190.1)
Ewes lambed (%)	193	99.0 (96.0 - 99.7)	326	95.9 (93.2 - 97.6)

[abcd] means with different superscripts are significantly different P<0.05

* Statistical analysis was not conducted as all ewes were diagnosed as pregnant

and decreased after lambing, reaching their lowest value at mid-lactation, and then increased to the time of weaning.

A greater percentage (P>0.05) of ewes that were bred to Wiltshire rams were marked with the mating-harness crayon in the first 17-day oestrous cycle compared with ewes bred to Romney rams (Table 2). There was a tendency (P=0.052) for a lower percentage of the Wiltshire-bred ewes to be bred in the second cycle. The percentage of ewes that were marked by the ram in both cycles (returned to service) or that were not marked by the ram did not differ between ram breeds (P<0.05).

At pregnancy diagnosis, all ewes (100%) bred to either the Romney or Wiltshire rams were diagnosed as pregnant (Table 2). There were no differences in the percentage of ewes diagnosed as single-, twin- or triplet-bearing among ewes bred to either Romney or Wiltshire rams and, therefore, no difference in the fecundity rate (P>0.05). The percentage of ewes that had lambing records was similar between ewes bred to either Romney or Wiltshire rams (P>0.05).

Lamb live weight, growth and survival

Lamb live weights at birth and mid lactation were greater (P<0.05) for single-born lambs than for twin and triplets, which did not differ (P>0.05, Table 3). At weaning, Romney-sired single lambs were heavier (P<0.05) than either twin or triplet lambs, however, among Wiltshire-sired lambs, singles were heavier than twins which in turn were heavier than triplets (P<0.05).

At birth and mid-lactation, the live weights of single, twin and triplet lambs sired by either Romney or Wiltshire rams did not differ (P>0.05, Table 3). At weaning, however, triplets sired by Romney rams were heavier (P<0.05) than those sired by Wiltshire rams.

For both Romney- and Wiltshire-sired lambs, liveweight gains from birth to weaning were greater (P<0.05) for singles than for twin or triplets, which did not differ (P>0.05, Table 32). The liveweight gain of single lambs was greater for Wiltshire-sired lambs than for Romney (P<0.05). Twin and triplet lambs had similar live weight gains regardless of the breed of their sire (P>0.05).

Lamb survival rates to weaning did not differ (P>0.05) among Romney 84.2% (79.9–87.7%) and Wiltshire sired lambs 87.2% (84.1–89.7%). The survival of triplet lambs (72.5% (52.9–86.0%)) was lower (P<0.05) than that of single and twin lambs which did not differ (81.7% (74.1–87.4%) *vs.* 87.6 (84.9–89.8%); P>0.05). Ewe lambs tended (P=0.08) to have greater survival rates than did ram lambs (87.7 (84.2–90.5%) *vs.* 83.5 (79.8–86.7%), respectively)).

Lamb shedding scores

At weaning, four Wiltshire-sired ewe lambs (1.6%) of the 243 scored showed signs of wool shedding particularly on the neck. All four lambs were scored 1 on the scale of 0

Table 3 Live weight (kg; mean ± SEM) of single, twin and triplet lambs sired by either Romney or Wiltshire rams at birth (L0), mid-lactation (L52) and weaning (L107) and liveweight gain (g/day) from birth to weaning

		Live weight (kg)						Liveweight gain (g/day)
Sire breed	Birth rank	n	L0	n	L52	n	L107	L0 to 107
Romney	Single	49	6.2 ± 0.4^b	39	18.0 ± 0.4^d	38	31.9 ± 0.5^g	277.4 ± 6.5^b
	Twin	287	5.2 ± 0.2^a	251	14.2 ± 0.2^c	246	28.0 ± 0.2^f	251.4 ± 2.5^a
	Triplet	12	4.3 ± 0.8^a	10	15.1 ± 0.9^c	10	28.6 ± 0.9^f	251.8 ± 12.5^a
Wiltshire	Single	92	6.1 ± 0.3^b	80	17.4 ± 0.3^d	80	32.4 ± 0.3^g	296.6 ± 4.5^c
	Twin	444	5.2 ± 0.1^a	397	14.1 ± 0.1^c	395	28.4 ± 0.1^f	255.5 ± 2.0^a
	Triplet	30	4.3 ± 0.5^a	19	13.4 ± 0.6^c	18	26.6 ± 0.7^e	241.6 ± 9.3^a

[abcdef] means with different superscripts are significantly different P<0.05

to 5. The four ewe lambs that were scored 1 were twin-born and were 91 (n=1), 93 (n=1) and 100 days of age (n=2). The age range of all the progeny born in the study ranged from 57 to 107 days. All of the subsample of Romney-sired ewe lambs were scored 0.

Discussion

The reproductive performance of Romney ewes bred to Wiltshire rams showed few differences to those bred to Romney rams. The breeding pattern showed that, although there were more ewes marked by Wiltshire rams in the first oestrous cycle compared to Romney rams, there was no difference in pregnancy and fecundity rates at pregnancy diagnosis. The pregnancy scanning percentage of the Romney flock at Riverside farm has been reported to range between 170 and 190% (Kenyon et al. 2011; Pettigrew et al. 2020; Ramírez-Restrepo et al. 2005). Given the high fecundity rate reported for ewes at Riverside farm, the low ram-to-ewe ratio and the use of teams of rams it was not expected that there would be any difference in the fecundity of Romney ewes bred to Wiltshire or Romney rams.

The live weights of lambs in the current study did not differ among lambs sired by Romney or Wiltshire rams except at weaning (L108) when triplet-born lambs sired by a Wiltshire ram were lighter than their counterparts sired by a Romney ram. Few data have been published on the live weight of Wiltshire lambs. Sumner et al. (2012) reported that Wiltshire lambs had a mean birth weight of 4.3 kg and at weaning weight of 25.1 kg, which was lighter than in the current study. Scobie et al. (2016), however, reported similar lamb weaning weights for lambs born to a line of Wiltshire sheep selected for increased fleece weight, but lower live weights for those born to the decreased-fleece-weight line. Lamb liveweight gains from birth to weaning showed that, although there were no differences in the weaning weight of single lambs sired by Wiltshire or Romney rams, there was a 19 g/day greater growth rate for the progeny of the Wiltshire rams.

Lamb survival in the current study was similar for lambs sired by Wiltshire and Romney rams. Survival rates were greater than those reported by Sumner et al (2012) but were within the range reported for studies using Romney ewes at Riverside farm (Kenyon et al. 2011; Pettigrew et al. 2020). As would be expected, the survival of triplet lambs was lower than that of single and twin lambs, however, in the current study the survival of single and twin lambs was similar. The survival rates in the current study were lower for single lambs than previously reported but similar for twin and triplet lambs (Morris & Kenyon 2004; Pettigrew et al. 2020).

At weaning in late-November, very few lambs sired by Wiltshire rams showed signs of fleece shedding. This is perhaps not surprising as O'Connell et al. (2012) concluded that shedding was best observed in Wiltshire lambs during January compared with December. Shedding is driven primarily by day length which influences prolactin concentrations (O'Connell et al. 2012; Pearson et al. 1996). O'Connell et al. (2012) also observed that among Wiltshire lambs, shedding scores were greater for singles than for twins which in turn were greater than for triplets. They also reported that older lambs had greater shedding scores than did younger lambs. Slee (1959) reported that adult F_1 Wiltshire x Blackface animals showed a range of shedding scores from 0 (no shedding) to 5 (90 to 100% of the body area shed), therefore, there is potential for the progeny in the current study to show fleece shedding as they age. Further observations are required as the lambs age before firm conclusions can be drawn.

Conclusion

The results of this study suggest that for the flock investigated, there was little effect of introducing Wiltshire rams on the reproductive performance of Romney ewes and their progeny. Further studies are required to determine the variability among flocks before firm conclusions can be made. Lambs showed few signs of wool shedding, however, given that the observation was conducted a month prior to the longest day there had been little opportunity for the influence of the longer day lengths to have an effect.

Acknowledgements

This research was funded by the L A Alexander Trust, Sydney Campbell Trust and the Massey Foundation.

References

Beef + Lamb New Zealand 2021a. Wool exports by month 2019 to 2020. Retrieved 20 April 2021, from https://beeflambnz.com/data-tools/wool

Beef + Lamb New Zealand 2021b. Benchmark your farm. Retrieved 20 April 2021, from https://beeflambnz.com/data-tools/benchmark-your-farm

Beef + Lamb NZ 2020. Compendium of New Zealand Farm Facts. Wellington, New Zealand, Beef + Lamb New Zealand. 44th Ed. 28 p.

Blair H 2011. Ram breeding in New Zealand two decades after the introduction of exotic sheep breeds. Proceedings of the Association for the Advancement of Animal Breeding and Genetics 19: 407-410.

Cottle DJ 2010. International Sheep and Wool Handbook. Nottingham University Press. Nottingham, United Kingdom.

Easton B 2010. Wool, meat and butter exports, 1853–2000. Retrieved 20 April 2021 http://www.TeAra.govt.nz/en/graph/24323/wool-meat-and-butter-exports-1853-2000

Farrell LJ, Morris ST, Kenyon PR, Tozer PR 2020. Modelling a transition from purebred Romney to fully shedding Wiltshire–Romney crossbred. Animals 10: 2066.

Geenty K 2017. Making every mating count. Wellington, New Zealand, Beef + Lamb New Zealand. 94 p.

JDJL Limited 2021. Wool indicator prices. Retrieved 21/01/2021 2021, from http://www.interest.co.nz/Charts/Rural/Lamb%20Y

Jefferies BC 1961. Body condition scoring and its use in management. Tasmanian Journal of Agriculture 32: 19-21.

Johnson P, O'Connell D, Dodds K, Sumner R, Mcewan J, Pearson A 2007. Wool shedding as a trait for genetic improvement using marker assisted selection? Proceedings of the Association for the Advancement of Animal Breeding and Genetics. PP 541-544.

Kenyon PR, Maloney SK, Blache D 2014. Review of sheep body condition in relation to production characteristics. New Zealand Journal of Agricultural Research 57: 38-64.

Kenyon PR, Van der Linden DS, West DM, Morris ST 2011. The effect of breeding hoggets on lifetime performance. New Zealand Journal of Agricultural Research 54: 321-330.

Morris CA, Johnson DL, Sumner RMW, Hight GK, Dobbie JL, Jones KR, Wrigglesworth AL, Hickey SM 1996. Single trait selection for yearling fleece weight or liveweight in romney sheep – correlated responses in liveweight, fleece traits, and ewe reproduction. New Zealand Journal of Agricultural Research 39: 95-106.

Morris ST, Kenyon PR 2004. The effect of litter size and sward height on ewe and lamb performance. New Zealand Journal of Agricultural Research 47: 275-286.

Morrison J 2010 August 2010. Treasure diversity. Rare Breeds NewZ: 7-10.

Nicholas FW 2009. Introduction to Veterinary Genetics, John Wiley & Sons.

O'Connell D, Scobie D, Hickey S, Sumner R, Pearson A 2012. Selection for yearling fleece weight and its effect on fleece shedding in New Zealand wiltshire sheep. Animal Production Science 52: 456-462.

Pearson A, Parry A, Ashby M, Choy V, Wildermoth J, Craven A 1996. Inhibitory effect of increased photoperiod on wool follicle growth. Journal of Endocrinology 148: 157-166.

Pettigrew EJ, Hickson RE, Blair HT, Griffiths KJ, Ridler AL, Morris ST, Kenyon PR 2020. Differences in lamb production between ewe lambs and mature ewes. New Zealand Journal of Agricultural Research: 1-14.

Ramírez-Restrepo CA, Barry TN, Pomroy WE, López-Villalobos N, Mcnabb WC, Kemp PD 2005. Use of lotus corniculatus containing condensed tannins to increase summer lamb growth under commercial dryland farming conditions with minimal anthelmintic drench input. Animal Feed Science and Technology 122: 197-217.

Scobie D, O'Connell D, Noble A, Greer AW 2016. The influence of previous reproduction on subsequent fertility in multiparous ewes. Proceedings of the New Zealand Society of Animal Production 76: 155-158.

Simm G 1998. Genetic improvement of cattle and sheep, Farming press, Miller freeman UK Ltd, Ipswich, United Kingdom. 433 pp.

Slee J 1959. Fleece shedding, staple length and fleece weight in experimental Wiltshire horn-Scottish blackface sheep crosses. The Journal of Agricultural Science 53: 209-223.

Sumner RMW, Scobie DR, O'Connell D, Henderson HV 2012. Relative performance of Wiltshire and Perendale sheep. Proceedings of the New Zealand Society of Animal Production. 72: 28-34.

Impact of winter near-maintenance or non-restricted feeding on growth and intramuscular fat content of yearling rams

PL Johnson*, K Knowler, KM McRae, WE Bain, G Pile, CR Craigie and SJ Rowe

Invermay Agricultural Centre, AgResearch Ltd, Private Bag 50034, Mosgiel, New Zealand

**Corresponding author:* Email: tricia.johnson@agresearch.co.nz

Abstract

There is increased interest in genetic improvement in intramuscular fat in lamb due to its positive association with meat quality. Live-animal measures such as ultrasound or computed tomography scanning can enable genetic selection. The optimum time at which such measurements should take place is unknown. This study investigated levels of intramuscular fat in year-old rams. During the winter of 2020, ram lambs were fed on either a just-above-maintenance pasture diet (JAM) (n=119) with a target weight gain of 50 g/day, or non-restricted feeding (GAIN) (n=120) with a target weight gain of 180-200 g/day, with all rams slaughtered after 10 weeks of differential feeding. Both groups of rams were fed a ryegrass-dominant pasture. The target for the GAIN group was achieved with an average growth rate of 182±42 g/day (range 76 to 293 g/day), however, for JAM group the average growth rate was 24±64 g/day (range -212 to 126 g/day), with a third of the animals losing weight. The JAM group were split according to whether they had gained (JAM – Gain) or lost weight (JAM – Loss) for the analysis. There were differences ($P<0.001$) at the time of slaughter in early spring in carcass weight between groups (23.7±0.35, 19.5±0.40 and 17.0±0.47 kg for GAIN, JAM – Gain and JAM – Loss respectively). The levels of intramuscular fat were low across all groups compared to levels normally observed in 6-8-month old ram lambs and with further differences ($P<0.001$) between the three groups (2.1±0.08, 1.9±0.09 and 1.4±0.10 kg for GAIN, JAM – Gain and JAM – Loss respectively). More work is required to understand management systems that allow improved expression of the trait. Additionally, careful consideration needs to be given to winter grazing management of all young stock, as there is considerable variability in individual animals handling of winter maintenance feeding.

Keywords: lamb; maintenance feeding; intramuscular fat

Introduction

The trait of intramuscular fat (IMF) in lambs has been associated with consumer liking of sheepmeat, with 3% IMF identified as optimum for New Zealand lamb (Realini et al. 2021). Intramuscular fat has been identified as a trait with high heritability in New Zealand lamb (Johnson et al. 2018) and there is interest in incorporating the trait into genetic improvement programmes. The gold standard for measuring IMF is to directly measure its levels in muscle post-slaughter. However, this means that the animal is no longer alive to be considered for selection, or that relatives must be generated and measured (e.g., a progeny test). Therefore, establishing live-animal predictor measurements for IMF, including ultrasound and computed tomography scanning, would be an advantage. Given IMF has slow accretion to the point that it can be visually observed, the age at which the scanning should take place requires investigation, including what the levels are in yearling rams and whether it is a more appropriate time to measure IMF to get the full expression of the trait. Yearling rams present the last time at which data collection is possible on cohorts of rams, as ram sales take place at approximately 15 months of age.

The feed requirements of livestock for different growth paths and physiological states have been calculated and modified over time as new knowledge becomes available (Nicol & Brookes 2007). For a given physiological state, the basic equations contain information on the live weight and growth rate of the animals. For maintenance feeding, energy requirements are those that satisfy basal metabolism, grazing and walking requirements. (Nicol & Brookes 2007). However, in all such models, an "average" animal is considered and there is evidence that individual animals need different amounts of feed to achieve physiological outcomes, which has been termed as variation in feed efficiency and demonstrated specifically in sheep (Johnson et al. 2016).

The objective of the study was to investigate the variability in IMF in yearling rams that had been placed on different feeding regimes in the ten weeks before slaughter. The feeding regimes representing typical winter scenarios of feeding just-above-maintenance (target 50 g/day) and achieving liveweight gain (target 180-200 g/day). This methodology also enabling the impact of the different feeding regimes on growth rates, and other carcass and meat quality traits to be investigated.

Materials and methods

Animals and pre-slaughter management

Permission for this trial was granted by the AgResearch Ethics Committee (Ethics Number 15096).

The animals used in this study were 10-month-old ram lambs born in 2019, sourced from the AgResearch Progeny Test flock (Brito et al. 2017). Given that the animals were from a progeny-test resource, lifetime information was available on individual animals, including birth and rearing rank, age of dam, and dam and sire.

Prior to July 2020, the 240 animals available for the study had been run as one mob. In early July 2020, the animals were randomly allocated to two grazing treatment groups. The animals were grazed at the AgResearch Invermay Farm near Mosgiel, New Zealand. The pasture was predominantly ryegrass with starting pasture covers of approximately 2500 kgDM/ha. The first grazing treatment was set for the lambs to be growing at just above maintenance (JAM), gaining only 50 g/day, which was achieved through controlled grazing using recommendations based on the Beef + Lamb New Zealand Feed Smart App (https://beeflambnz.com/news-views/feedsmart-app-making-feed-management-easier). The program predicts the animal intake requirement (kgDM/head/day) according to the information supplied by the user on live weight and target growth rate, and provides guidance on how long animals should remain in paddocks to achieve that intake depending on pasture covers. The second grazing mob were targeted to gain live weight at 180-200 g/day (GAIN). This was achieved by selecting paddocks with high-quality leafy pasture, with the animals regularly shifted to new paddocks. Whilst higher growth rates can be achieved at other times of the year, the target of 180-200 g/day was representative of high growth for winter feeding on pasture. The animals were weighed every three weeks throughout the trial period, to ensure target levels of weight gain were being achieved at the group level.

Slaughter and post-slaughter measurements

In mid-September, a pre-slaughter live weight was recorded, and the animals were trucked and processed at the Alliance Group Ltd meat processing plant. Post-slaughter, hot carcass weight was recorded. Dressing-out percentage (DO was calculated as carcass weight/pre-slaughter live weight*100). The carcasses were scanned through the VIAscan® system, which is a two-dimensional imaging system that estimates the lean content of the carcass and provides a proxy for fat depth (VSGR) (Hopkins et al. 2004). The circumference of the buttocks was measured as described by Brito et al. (2017).

The day following slaughter, a single boneless loin (*M. longissimus lumborum*) was collected from each carcass. The boneless loin was cut in three to provide samples for either IMF (a 2-cm slice), shear force (SHF, as a measure of tenderness), or for the collection of colour, pH and marbling score data. Analysis of IMF was carried out using a direct transmethylation approach on freeze-dried and ground muscle powder, as described by Agnew et al. (2019), with the resulting data back-transformed to provide estimates of the levels of individual fatty acids. Within this paper, only summary traits of the percent saturated fatty acids, mono-unsaturated fatty acids and poly-unsaturated fatty acids together with total IMF are reported. All other meat-quality traits were measured as described by Brito et al. (2017). The loin slices were placed on small plastic trays and wrapped using semi-permeable cling film and stored at 4°C with colour (L24, A24 and B24) measured 24 hours after blooming using a Minolta Chromometer (Konica Minolta Sensing, Inc., Osaka Japan). Marbling was visually scored on the loin slices (1-5 scale) 96 hours after the loins had been sliced. The loin sample for shear force (SHF) was frozen 24 hours after slaughter, and subsequently measured using the MIRINZ protocol.

Statistical analysis

Preliminary analysis of the data revealed that there was significant variation in the growth rates of the rams within the JAM group, and although on average they had gained 24±64 g/day, the range was -212 to 126 g/day. Thus, the animals were split into three groups for analysis: those that were in the JAM group and lost weight (JAM – Loss), those that were in the JAM group and gained weight (JAM – Gain) and those that were in the GAIN group (no animals

Figure 1 Boxplot of the distribution in liveweight gain for ram lambs that were differentially fed through late winter either to achieve liveweight gain (GAIN) or to be fed Just Above Maintenance (JAM), with the JAM group futher split into those that lost weight (JAM - Loss) or gained weight (JAM - Gain).

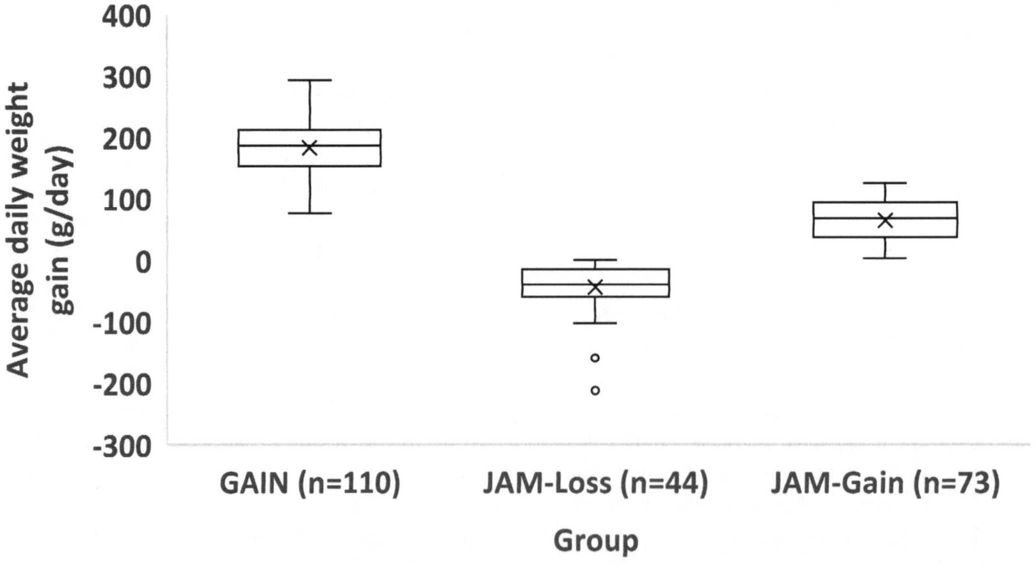

in the GAIN treatment lost weight, with the lowest weight gain being 76 g/day). The distribution of growth rates is shown in Fig. 1.

Analysis of the data was carried out using the mixed-model procedure in SAS (SAS, 2004). In all models, age of dam, birth-rearing rank and treatment group were fitted as fixed effects, while sire was fitted as a random effect. For growth rate, final live weight and carcass weight, starting live weight was fitted as a covariate. For the remaining carcass traits (DO, VSGR and butt circumference) and selected meat quality traits (SHF and IMF traits), two models were fitted: one including carcass weight as a covariate and one without carcass weight as a covariate. For SHF, pH and pH^2 were also fitted as covariates to account for the influence of the quadratic relationship between tenderness and pH (Brito et al. 2017). For the colour traits, pH was fitted as a covariate. For the fatty acid composition traits, total IMF was also fitted as a covariate, as the composition of IMF is known to vary at different levels of IMF (Wood et al. 2008).

Results and discussion

The original objective of the study was to investigate the variability in IMF in yearling rams that had been placed on different feeding regimes in the ten weeks before slaughter. The two different feeding regimes included unrestricted access to pasture allowing for high growth rates on grass winter feeding only (182±42 g/day) or feeding for just above maintenance (24±64 g/day).

The results of the analysis for all liveweight, carcass and meat-quality traits are in Table 1. Overall, the levels of IMF were very low, as assessed by both visual marbling score and measured IMF, even for the GAIN group. The average IMF for New Zealand lamb processed at approximately six months of age has been shown to be 2.6% across various progeny tests (Craigie et al. 2017), whereas it was only 2.1% in the GAIN group and as low as 0.8% in the JAM – Loss group in this current study. It was anticipated that, given that the animals in this study were six months older than the animals of Craigie et al. (2017), they should have had higher levels of intramuscular fat.

For all models considering IMF and its composition, treatment group was significant with or without carcass weight in the model, indicating that the observed differences were over and above differences in carcass weight alone. Animals that were in the JAM-Loss group had very low levels of both marbling and IMF. The GAIN group had higher levels of marbling and IMF relative to the JAM-Gain group, although there was no significant difference in marbling score between the JAM-Gain and GAIN group when carcass weight was fitted in the model. The differences between the groups in IMF composition were consistent between models for saturated fatty acids, with a difference across groups and JAM-Loss animals having the lowest levels of saturated fatty acids. Results were also consistent for polyunsaturated fat acids between models, with the JAM-Loss group having higher ($P<0.001$)

Table 1 Least squares means for liveweight, carcass and meat-quality traits for yearling rams that had been differentially fed for ten weeks prior to slaughter to achieve liveweight gain (GAIN) or just above maintenance (JAM), with the JAM group split according to weight lost (JAM- Loss) or gained (JAM-Gain). Two different models were fitted for the carcass and select meat quality traits, one excluding and one including carcass weight as a covariate.

	Carcass weight not fitted in the model[1]				Carcass weight fitted in the model[1]				
	JAM - Loss	JAM - Gain	GAIN	Group P value	JAM - Loss	JAM - Gain	GAIN	Carcass P value	Group P value
Starting live weight (kg)	50.8±0.97	50.0±0.83	50.2±0.75	0.6725					
Final live weight (kg)[2]	45.2±0.73[a]	50.4±0.84[b]	59.2±0.49[c]	< 0.001					
Average Daily Gain (g/day)[2]	-50.5±1.03[a]	24±1.01[b]	151±0.70[c]	< 0.001					
Carcass weight (kg)[2]	16.0±0.38[a]	18.1±0.43[b]	22.0±0.26[c]	< 0.001					
Dressing out (%)	35.1±0.38[a]	35.8±0.32[b]	37.6±0.29[c]	< 0.001					
VIAscan GR (mm)	1.3±0.41[a]	1.7±0.34[b]	3.7±0.32[c]	< 0.001	2.8±0.36	2.1±0.27	2.1±0.30	< 0.001	0.0933
VIAscan Total (%)	59.5±0.49	59.8±0.41	59.3±38	0.3810	59.6±0.54	59.8±0.41	59.1±0.45	0.6461	0.3049
Butt circumference (cm)	63.2±0.65[a]	65.9±0.74[b]	69.1±0.4[c]	< 0.001	66.7±0.31	67.3±0.33	66.7±0.22	< 0.001	0.1296
pH	5.86±0.03[b]	5.75±0.03[a]	5.75±0.03[a]	0.0015					
Colour L*	41.2±0.32	41.1±0.27	40.8±0.25	0.3417					
Colour a*	14.7±0.25[b]	14.6±0.21[b]	13.6±0.191	< 0.001					
Colour b*	6.5±0.21[a]	6.3±0.17[a]	7.0±0.15[b]	< 0.001					
Shear force (kgF)	8.1±0.47b	8.5±0.39[b]	7.69±0.35[a]	0.0427	8.0±0.51	8.5±0.39	7.8±0.42	0.2494	0.1755
Marble Score	1.4±0.10[a]	1.9±0.09[b]	2.1±0.08[c]	< 0.001	1.5±0.11[a]	1.9±0.09[b]	2.0±0.09[b]	< 0.001	0.0002
Intramuscular fat (%)	0.8±0.12[a]	1.3±0.10[b]	2.1±0.09[c]	< 0.001	1.1±0.12[a]	1.4±0.10[b]	1.8±0.10[c]	< 0.001	< 0.001
IMF - Saturated fat (%)	40.7±0.5[a]	42.7±0.39[b]	43.6±0.35[c]	< 0.001	40.6±0.53[a]	42.7±0.39[b]	43.7±0.41[c]	0.0084	< 0.001
IMF - Monounsaturated fat (%)	32.7±0.55[a]	35.4±0.46[b]	35.1±0.44[b]	< 0.001	33.3±0.55[a]	35.5±0.44[b]	34.3±0.45[a]	< 0.001	< 0.001
IMF - Polyunsaturaated fat (%)	18.5±0.70[b]	14.7±0.55[a]	14.3±0.51[a]	< 0.001	18.0±0.7[b]	14.7±0.54[a]	15.0±0.57[a]	< 0.001	< 0.001

[1]Values with different superscripts were significantly different from each other ($P<0.05$)
[2]Starting live weight fitted as covariate

levels of PUFA compared to the other two groups (which were not significantly different from each other). For monounsaturated fatty acids, the results differed between the models, but in both models, values were lowest for the JAM-Loss group. These results support the review of Wood et al. (2008), whereby at low levels of IMF most of the fatty acids are phospholipids associated with muscle structure (mostly PUFA), whereas at higher levels of IMF the predominant fatty acids are those associated with triacylglycerides (mostly SFA).

The DO was very low (35-37%) compared to values normally observed in six-to-eight-month-old ram lambs (42.5% Johnson et al. (2008)). This significantly lower DO indicates that a higher proportion of the live weight of the animals was made up of non-carcass components, including the internal organs. Additionally, the levels of subcutaneous fat were very low, so low traditional GR measurements (Kirton 1989) could not be made. The levels detected using the VIAscan system were, on average, less than half of those observed in non-castrated male lambs, using the same system, that were processed at seven months of age (Johnson et al. 2017). Variation in carcass fatness is significantly related to variation in DO, with DO increasing with increasing carcass fatness at a set live weight (Litherland et al. 2010). Whether the very lean carcasses are the sole reason for the low DO cannot be inferred in this data set but will have been a contributing factor.

For the carcass traits (DO, VSGR, buttocks circumference) and SHF, there were differences ($P<0.001$ for carcass traits and $P<0.05$ for SHF) between the groups when carcass weight was not fitted as a covariate in the analysis. The differences, however, were no longer significant when carcass weight was fitted as a covariate and, as such, the differences were the result of differences in carcass weight. For pH, the pH of the JAM-Loss animals was higher ($P=0.0015$) than that of the other two groups (which were not significantly different from each other). This is biologically plausible, as animals losing weight prior to slaughter likely have reduced energy and glycogen stores to drive post-mortem decline in pH, although this phenomenon was not observed by Devine et al. (1983) who investigated variation in pH in animals that had undergone weight loss prior to slaughter.

The results highlight that care needs to be taken when feeding animals to maintenance levels. Whilst the animals in the JAM group were gaining weight on average, there was approximately one-third of them that lost weight. Whether the weight loss was due to inefficient animals (Johnson et al. 2016) not being offered enough feed to fulfil their maintenance requirements, or other causes, cannot be determined from this study. Whichever the case, it does demonstrate that there is considerable individual variability in how animals handle winter maintenance feeding, and more feed will ultimately be required for some individuals to regain the weight and body condition lost if they are to be retained. These results provide evidence that individual animal monitoring through winter is warranted to identify such animals and to manage them accordingly.

Conclusion

The results of this study indicate that the levels of IMF observed in yearling rams is low after winter feeding, and influenced by pre-slaughter feeding regimes. If live animal measures of IMF are to be optimised, more research is needed to understand the age and management systems required to ensure that the trait is being expressed. Additionally, there is significant variation in how individual animals perform under winter maintenance feed regimes, specifically that some animals are compromised, and that monitoring is required to identify such animals to prevent the animals from losing weight.

Acknowledgements

This work was supported by the MBIE Endeavour Research Programme 'Capturing the Value of New Zealand Red Meat' (Contract Reference: C10x1602) and AgResearch SSIF funding. The authors gratefully acknowledge the team at the Alliance Group Ltd Lorneville Plant for their assistance processing the animals, especially Johanna Phillips.

References

Agnew MP, Craigie CR, Weralupitiya G, Reis MM, Johnson PL, Reis MG 2019. Comprehensive evaluation of parameters affecting one-step method for quantitative analysis of fatty acids in meat. Metabolites 9: 189.

Brito LF, McEwan JC, Miller SP, Bain WE, Lee M, Dodds KG, Newman S-A, Pickering NK, Schenkel FS, Clarke, SC 2017. Genetic parameters for various growth, carcass and meat quality traits in a New Zealand sheep population. Small Ruminant Research 154: 81-91.

Craigie CR, Agnew MP, Stuart A, Shorten P, Reis M, Taukiri K, Johnson PL 2017. Intramuscular fat content of New Zealand lamb M. longissimus lumborum. Proceedings of the New Zealand Society of Animal Production 77: 189-193.

Devine CE, Chrystall BB, Davey CL, Kirton AH 1983. Effects of nutrition in lambs and subsequent postmortem biochemical changes in muscle, New Zealand Journal of Agricultural Research 26: 53-57.

Hopkins DL, Safari E, Thompson JM, Smith CR 2004. Video image analysis in the Australian meat industry - precision and accuracy of predicting lean meat yield in lamb carcasses. Meat Science 67: 269-274.

Johnson PL, Kenyon PR, Burnham DL, West DM 2008. To wether - short scrotum or leave lambs as entire rams? Revisiting an old question using new genetics. Proceedings of the New Zealand Society of Animal Production 67: 44-47.

Johnson PL, Miller SP, Knowler KJ 2016. Preliminary investigations into the trait of residual energy intake in sheep. Proceedings of the New Zealand Society of Animal Production 75: 34-37.

Johnson PL, Stuart AD, Mackie A 2017. BRIEF COMMUNICATION: High ultimate pH in short-scrotum male lambs processed during the breeding season. Proceedings of the New Zealand Society of Animal Production 77: 194-196.

Johnson PL, Agnew MP, Shorten PR, Reis MM, Cullen N, Craigie CR 2018. Preliminary genetic parameter estimates for intramuscular fat percentage in New Zealand lamb. Proceeding of the World Congress on Genetics Applied to Livestock Production 11: 611.

Kirton AH 1989. Principles of classification and grading. In: Purchas, RW Butler-Hogg BW, Davies AS eds Meat Production and Processing. New Zealand Society of Animal Production Occasional Publication Number 11. pp 143-157.

Litherland AJ, Dynes RA, Moss RA 2010. Factors affecting dressing-out percentage of lambs. Proceedings of the New Zealand Society of Animal Production 70: 121-126.

Nicol AM, Brookes IM 2007 The metabolisable energy requirements of grazing livestock In: Rattray PV, Nicol AM, Brookes IM eds Pasture and supplements for grazing dairy cattle New Zealand Society of Animal Production, Hamilton New Zealand Occasional Publication 14. pp 151-172.

Realini CE, Pavan E, Johnson PL, Font-I-Furnols M, Jacob N, Agnew M, Craigie CR, Moon CD 2021. Consumer liking of M. longissimus lumborum from New Zealand pasture-finished lamb is influenced by intramuscular fat. Meat Science 173: 108380.

SAS. (2004). SAS/STAT 9.1 User's Guide. SAS Institute Inc, Cary, North Carolina, USA.

Wood JD, Enser M, Fisher AV, Nute GR, Sheard PR, Richardson RI, Hughes SI, Whittington SM 2008. Fat deposition, fatty acid composition and meat quality: A review. Meat Science 78: 343-358.

Effects of alternative forages on nitrate leaching under intensive sheep grazing

S Maheswaran*, LM Cranston, JP Millner, DJ Horne, JA Hanly, PR Kenyon and PD Kemp

School of Agriculture & Environment, Massey University, Palmerston North, New Zealand

**Corresponding author.* Email: *S.Navarathnam@massey.ac.nz; L.Cranston@massey.ac.nz*

Abstract

There is increasing interest in the use of alternative forages to improve animal performance and reduce nitrogen (N) leaching on farms. A system-scale field trial, which included 20 hydrologically isolated drainage plots (800 m^2/plot) was carried out at Massey University's 'Keeble' farm to investigate the effects of forage type on ewe and lamb performance and the quantity of N lost in drainage. Four forage types (treatments) were studied: perennial ryegrass/white clover (RGWC), plantain/WC (PWC), Italian ryegrass/WC (IRWC) and swedes (Brassica). The forage types were included as ~24% area of a self-contained farmlet, with the remaining area (~76%) being RGWC. Ewes were weighed at mating, pre-lambing and weaning, and lambs were weighed at weaning. Drainage (mm) from each plot was measured using a tipping-bucket flow meter and sub-samples of water were analysed for nitrate-N, ammonium-N and total-N losses. The forage types had no effect (P>0.05) on ewe or lamb live weight at any occasion. The average nitrate-N losses for RGWC (0.77 kg N ha^{-1}), PWC (0.45 kg N ha^{-1}) and IRWC (0.56 kg N ha^{-1}) treatments did not differ (P>0.05) and was lower than the brassica treatment (6.36 kg N ha^{-1}).

Keywords: drainage water; forage species; live weight; nitrogen loss; sheep grazing

Introduction

Globally, there is increasing concern about the contamination of waterways with nutrients, in particular, the nitrogen (N) leached from agricultural land (Houlbrooke et al. 2004). Nitrate (NO_3^-) is the most common form of N contaminant in drainage water (Haynes & Williams 1993). Urine N deposited by sheep on the pasture is the main source of NO_3^- leaching in the sheep grazing pasture system (Williams & Haynes 1990). Surplus N (not required for plant growth) in urine patches is susceptible to leaching when drainage occurs (Haynes & Williams 1993). Nitrate leaching is widely believed to be significantly lower from sheep farming systems compared to dairy cow systems (Williams & Haynes 1994). However, in order to remain economically viable, sheep production systems in New Zealand have intensified over time with greater use of N fertiliser and higher stocking rates (Morris & Kenyon 2014). Therefore, it is possible that N leaching losses are greater than historical figures would indicate and should be reassessed under modern intensive practices.

There is increasing interest in the use of alternative forages in pastoral agriculture in New Zealand. Some of these forages are used to reduce N leaching, i.e., improve environmental outcomes (e.g., plantain; *Plantago lanceolata* L. and Italian ryegrass; *Lolium multiflorum* Lam), while others are employed to increase animal performance (e.g., winter crops). There is evidence that the use of plantain on dairy farms can help to mitigate N losses as a result of a decrease in the concentration and amount of urinary N excreted, due to an increase in urine volume and decrease in ammonia production in the rumen (Navarrete et al. 2016). It is known that rapidly growing pasture species, such as Italian ryegrass, display greater N uptake during winter and, therefore, reduce N leaching compared to a perennial ryegrass (*Lolium perenne* L)/ white clover (*Trifolium repens*) sward when grazed by dairy cows (Woods et al. 2016). In contrast, winter crops such as brassicas (e.g., swedes and turnips) produce large quantities of forage (e.g., 8-15 t DM ha^{-1}) and are grazed with high stocking rates during winter when grass growth rate is low (Lucci et al. 2013). Following grazing, the land is left bare until conditions allow for re-sowing in spring. This process of grazing winter brassica species may return large amounts of N to the soil during winter, and subsequently may result in very high rates of N leaching (Monaghan et al. 2013).

Currently, there are no studies of leaching under modern, intensive sheep production systems, particularly when alternative forages are grazed. Further, the limited previous studies carried out estimated N leaching indirectly through animal data (e.g., based on grazing study data) or the use of lysimeters or suction cups. The present study utilises hydrologically isolated plots, which allow for the complete capture of all leached nutrients and drainage water. The aim of this study was to quantify and compare N leaching under sheep grazing on four different forage types (perennial ryegrass (RGWC), plantain/white clover (PWC), Italian ryegrass/white clover (IRWC) and brassica), and to compare the sheep performance in farm systems that incorporated alternative forages. These forages were selected for their ability to reduce N leaching and/or improve animal performance.

Materials and methods

This study was carried out at Massey University's Keeble farm, 5 km southeast of Palmerston North, Manawatu, New Zealand (40°24'02.0"S 175°35'52.8"E) with the approval of the Massey University Animal Ethics Committee (MUAEC 19/20). The site is located in a flat to easy rolling landscape (c. 3% slope) on the Tokomaru silt loam soil, a Fragic Perch-gley Pallic Soil (Hewitt 1998). This paper presents the data collected during 2020.

Experimental design and treatments

The design included four farmlets (each approximately 3.32 ha). Three of the farmlets had 0.81 ha (24% of the area) sown in one of three alternative forages, i.e., IRWC or PWC or brassica (swedes; *Brassica napus* ssp. *Napobrassica*). The remaining area of each farmlet was in RGWC. The fourth farmlet was entirely RGWC (3.31 ha). Each farmlet was stocked at 14 ewes/ha for the entirety of the experiment (46 ewes per farmlet) and was reflective of an intensive sheep-farming operation for this farm class and area (Beef + Lamb New Zealand Economic Service 2021).

The research site included an area with 20 drainage plots (five plots/treatment). Each plot was 40 m x 20 m and contained a hydrologically isolated mole pipe drainage system. The drainage plots were sown with the pasture species reflective of their treatment. These pastures were sown in the autumn of 2019 (21 March 2019) with a roller drill and chain harrows. The plantain/WC and IRWC plots were over-sown in April 2020. The brassica was sown by direct drilling in the spring of 2019. Nitrogen fertiliser, in the form of urea (46% N), was applied to the brassica plots at the rate of 30 kg N ha^{-1}in March. Nitrophoska (12% N) and urea were applied to the other three treatment plots at the rate of 30 kg N ha^{-1} in April and October, respectively. The plots were grazed according to the cumulation of forage and best grazing management practices for these species.

Ewe management and measurements

A total of 184 pregnant mixed-age Romney ewes were selected from Massey University's "Keeble" flock at the start of the study (27 June 2019). The ewes were stratified by live weight (LW) and number of fetuses carried (single/twin/triplet) and allocated to the four forage treatments (n=46 at start) to ensure that treatments did not differ in LW or number of fetuses carried. Ewes that died or were culled were replaced in February 2020, with ewes of a similar LW. Romney rams were introduced on 9 April 2020 for a period of 34 days (two rams per farmlet of 46 ewes). The rams were rotated between mobs on Day 17, to avoid the impact of ram performance on pregnancy rate. All ewes were pregnancy diagnosed using trans-abdominal ultrasound on 1 July 2020 and diagnosed as non-pregnant, single, twin- or triplet-bearing. Three ewes diagnosed as non-pregnant were replaced by pregnant ewes of similar stage of gestation.

Each mob of ewes was rotationally grazed within their farmlet. The duration of grazing and the interval between paddocks was adjusted based on pasture growth rate and cover, which were assessed visually. When the ewes grazed the drainage plots, each mob was separated into five mobs to enable all five replicates to be grazed at the same time (approximately nine ewes/replicate). All replicates within a forage treatment were always grazed for the same number of days and ewes were moved when the target post-grazing residual was reached in any replicate. Where appropriate, treatments on the drainage plots were grazed at once to minimise the effect of differing weather conditions (see Fig. 1). However, the brassica crop was grazed continuously during winter, after the ewes were gradually adjusted to the diet using on-off grazing with increasing grazing durations of brassica each day (2, 4, 6, 8, hours on days one through four) before continuous grazing. Due to mild winter conditions in 2020, the brassica yield was considerably higher than had been estimated, and ewes were unable to graze all the bulb material before set-stocking, with an estimated 5,461 kg DM ha^{-1} remaining.

Approximately one week before the planned start of lambing (26 August 2020), ewes were set stocked on the RGWC area of each farmlet at a stocking rate of 18

Figure 1 Number of days sheep grazed each of the forage treatments (RGWC: perennial ryegrass/white clover; IRWC: Italian ryegrass/ white clover; PWC: plantain/ white clover; brassica) on the hydrologically isolated drainage plots during 2020.

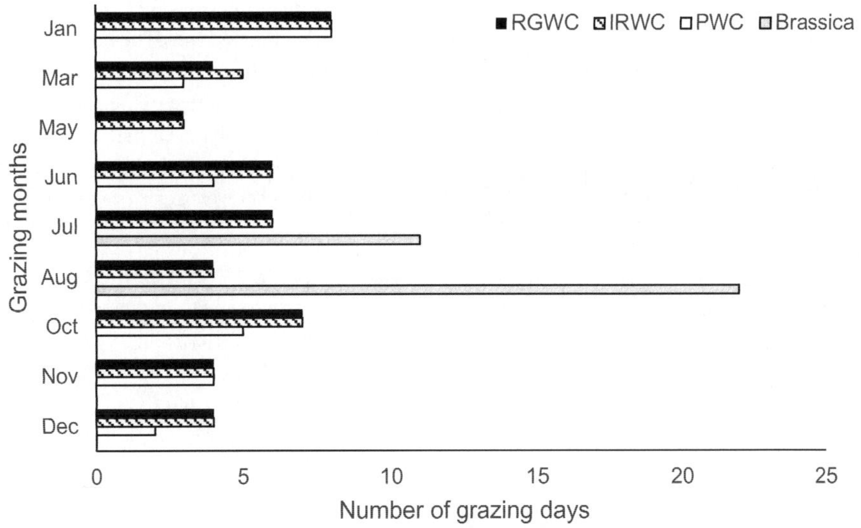

ha[1], until the end of lambing (15 October 2020). Ewes and lambs within each treatment were then rotationally grazed as one mob until weaning (12 December 2020). Ewes were weighed at mating, pre-lambing, and weaning. Lambs were also weighed at weaning.

Drainage water volume measurements and water analysis

Rainfall data were sourced from the National Institute of Water and Atmospheric Research (NIWA) site which was located approximately 2 km from the research area. Drainage water from each isolated plot was channelled through drainage pipes into individual tipping-bucket flow meters located in sampling pits nearby. The flow rate of drainage water was measured with an individual tipping bucket (~5 L) for each plot. Each tipping bucket was calibrated dynamically to account for large tip volumes at high flow rates (Humphrey et al. 1997). All tipping buckets were instrumented with data loggers to provide continuous flow rate measurements. Water from the bucket was tipped over a pipe with small hole on it which passed the sub-sample (~0.5 ml) to a larger container for storage until the completion of the drainage event. After the completion of each drainage event, water samples (~100 ml) were collected manually from the storage container for water quality analysis.

Approximately 50 ml of the subsample was filtered through a 0.45 µm filter. Filtered samples were analysed for NO_3^--N and ammonium-N (NH_4^+-N), using a Technicon Auto Analyser (Blakemore et al. 1987). The unfiltered samples were analysed for total N (TN) content using the persulphate digestion method of Hosomi and Sudo (1986). The amount of NO_3^--N, NH_4^+-N and TN losses (kg ha^{-1}) were calculated as the product of the measured drainage quantity and their concentrations.

Statistical analysis

All data were analysed in SAS 9.4 (Statistical Analysis System, version 9.4; SAS Institute Inc., Cary, NC, USA). Live weight of ewes was subjected to an analysis of variance using the MIXED procedure. The model included the fixed effects of measurement time, forage treatment and the interaction between measurement time and forage treatment. Lamb LW, cumulative drainage (mm), NO_3^--N concentrations and cumulative losses of TN, NO_3^--N and NH_4^+-N were analysed using proc GLM with forage treatment as fixed effect and means were separated using LSD procedure. Drainage (mm) from one of the PWC plots was five times greater than the average of the other plots and was deemed to be an outlier caused by a leak in a water pipe. Therefore, in the calculation of cumulative drainage and N losses, the data collected from this particular plot were omitted.

Results

Rainfall and drainage

In 2020, the spring was very wet compared to winter, with a particularly high monthly rainfall (136.7 mm) being recorded in September (Fig. 2a). The drainage season in 2020 began in late April and ended in December. The average annual quantity of water that drained across all plots was 83 mm. Throughout the season, there were 13 drainage events from the mole-pipe drainage system, ranging between 0.13 to 19.0 mm (average across all plots). Large drainage events were measured in September and December (Fig. 2a), and the majority of drainage occurred in December (average of 33 mm).

The drainage (mm) of all the treatments remained consistent during autumn (April to June) (average of 1 mm) and winter (July to September) (average of 28 mm), however, during spring (October to December), differences in drainage (mm) was observed between treatments. In the last three months of the year, high (76.2 mm; $P<0.05$) and low (32.9 mm $P<0.05$) drainage (mm) were observed from the brassica and PWC treatments, respectively. Overall, the annual cumulative drainage of the brassica (103 mm) was 7 to 20% greater than that of the other treatments (Fig. 2b).

Nitrate-N concentrations

The average NO_3^--N concentrations of all treatments ranged from 2.7 mg L^{-1} (PWC) to 6.5 mg L^{-1} (brassica) (Fig. 2b). The concentration of NO_3^--N (mgL^{-1}) in drainage water fluctuated between treatments over the year. For RGWC, IRWC and PWC treatments, relatively high NO_3^--N concentrations were observed at the commencement of drainage events in April (RGWC:15.3 mg L^{-1}; IRWC:14.7 mg L^{-1}; PWC 10.5 mg L^{-1}) and, subsequently, when drainage commenced again in June (RGWC:7.3 mg L^{-1}; IRWC:17.3 mg L^{-1}; PWC 6.6 mg L^{-1}). Whereas for the brassica, the highest NO_3^--N concentration of 16.5 mg L^{-1} occurred at the end of the drainage season in December. Overall, compared with the other treatments, the brassica treatment had higher ($P<0.05$) NO_3^--N concentrations in the last six drainage events.

Nitrate-N loss and cumulative NO_3^--N losses

Nitrate-N leached from all treatments remained less than 0.3 kg N ha^{-1} until late July (Fig. 2c). Then, with subsequent drainage events, the losses from the brassica treatment increased ($P<0.05$), while the losses from other treatments remained low. The cumulative leaching load of NO_3^--N kg ha^{-1} from the brassica plots (6.36 kg N ha^{-1}) was higher ($P<0.05$) than that of the other three treatments, with large differences observed particularly in the latter part of the drainage season (Fig. 2c). The annual NO_3^--N leaching loads from IRWC (0.56 kg N ha^{-1}) and PWC (0.45 kg N ha^{-1}) were very low and did not differ ($P>0.05$).

Cumulative NH_4^+-N and total N losses

Ammonium made a minor contribution to the total N leached, and across all treatments represented 7% of inorganic N. Total N losses for RGWC (2.6 kg N ha^{-1}), PWC (1.5 kg N ha^{-1}) and IRWC (2.1 kg N ha^{-1}) treatments were similar ($P>0.05$), but 73 to 84% lower ($P<0.05$) than that of the brassica treatment (16.8 kg N ha^{-1}).

Figure 2 Rainfall (mm) (a) in 2020 (NIWA/Ag Research, Palmerston North meteorological station, 2 km from study site), mean NO_3^--N concentrations (mg L^{-1}) and cumulative drainage (mm) (b), and mean NO3$^-$-N losses (g ha^{-1}) and cumulative NO_3^--N losses (kg ha^{-1}) (c) from perennial ryegrass/white clover, Italian ryegrass/white clover, plantain/white clover and brassica in 2020.

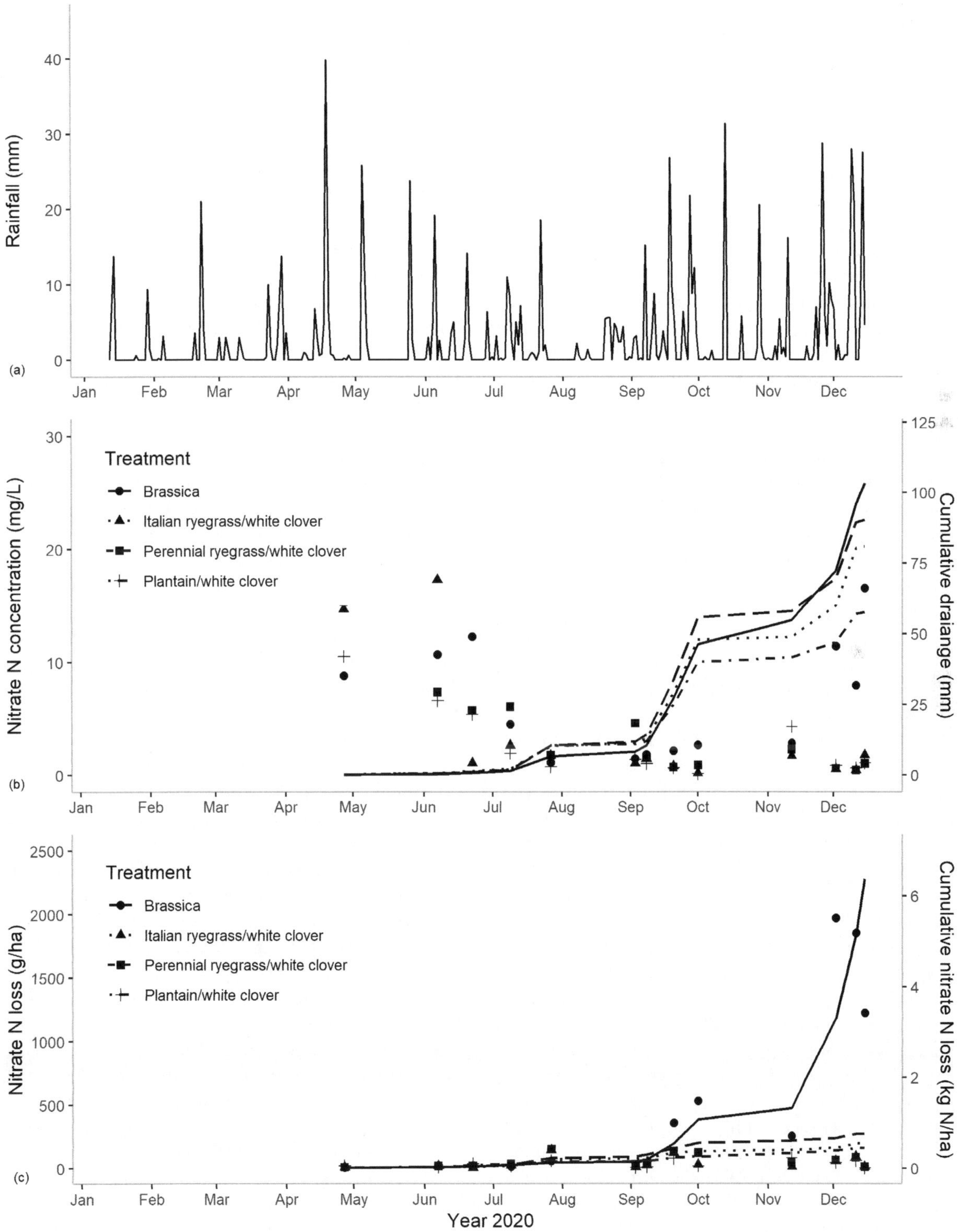

Table 1 Effect of forage treatment (RGWC: perennial ryegrass/white clover; IRWC: Italian ryegrass/ white clover; PWC: plantain/ white clover; brassica) on live weight (LW; kg) of ewes at mating, pre-lambing and weaning, and lambs at weaning.

Treatment	No. of ewes	Ewes LW (kg)			SEM	No. of lambs	Lambs LW (kg)	SEM
		Mating	Pre-lambing	Weaning			Weaning	
RGWC	40	72.3	95.1	75.6	1.40	62	27.3	0.49
IRWC	39	70.5	94.8	76.2	1.44	56	26.6	0.49
PWC	42	74.8	94.2	73.0	1.40	69	25.4	0.45
Brassica	43	69.7	89.3	74.6	1.42	62	26.6	0.46

Sheep performance and number of grazing days on alternative forages

Forage treatments had no effect (P>0.05) on ewe or lamb LW on any occasion (Table 1). The RGWC and IRWC plots were grazed nine times throughout the year (Fig. 1) for a total of 46 and 47 days, respectively. The PWC plots were grazed six times and the number of grazing days was relatively low (26 days). The brassica plots were grazed only once continuously during winter (33 days).

Discussion

There was no effect of forage treatment on ewe or lamb LW. Therefore, the impacts of N leaching under different forage types can be directly compared without confounding effects of different animal performance. The lack of an effect of treatment on LW is perhaps unsurprising given that each forage treatment made up 24% of each farmlet system. The brassica treatment had no observed benefit to animal performance, although the crop was not utilised fully. The excess bulb could have been used to stock more sheep or to graze for a longer period, and potentially improve animal performance.

The winter of 2020 was relatively warm with less rainfall, compared with the last five years (NIWA 2021). The length of drainage season was longer than reported in other studies conducted near the current research site (Magesan et al. 1996; Christensen et al. 2019). The drainage (mm) at the end of the season (October to December) was also higher than reported in the previous studies. The higher drainage (mm) measured for the brassica treatment plots was likely due to the lower evapotranspiration rates associated with bare soil during spring (Hanly et al. 2017).

The average NO_3^--N concentrations of drainage water collected from RGWC, IRWC and PWC treatments started high (above 5 mg L^{-1}) for early-drainage events and decreasing with successive drainage events over the winter months (Fig. 2b), as observed by Heng et al. (1991) and Magesan et al. (1996). It is likely that the NO_3^--N that had accumulated in the soil under urine patches, prior to the commencement of drainage events in April and between the drainage events in April and June, contributed to early NO_3^--N leaching for the RGWC, IRWC and PWC treatments (Fig. 2b). This demonstrates that grazing RGWC, IRWC and PWC treatments during late summer and autumn period is an important source of the NO_3^--N that accumulates in soil which, in turn, contributes to NO_3^--N leaching during the early stages of the drainage season. Due to the very low leaching losses observed in the present study, the N accumulation under sheep grazing pastures appears to be much lower than the N accumulation under cattle grazing pastures (Rodriguez et al. 2020). The brassica treatment, however, had higher concentrations of NO_3^--N leached during the initial stage of the drainage season which may have been due to the mineralised N that accumulated as a result of renewing the forage (spraying herbicide and direct drilling) (Fraser et al. 2013).

Grazing of RGWC, IRWC and PWC treatments during spring or late winter did not increase the NO_3^--N concentrations (Fig. 2b). Although drainage continued through into December, NO_3^--N concentrations remained <1.8 mg L^{-1}. This trend of drainage water NO_3^--N concentrations declining over the winter months has been previously reported in dairy cattle studies on the same soil type (Hanly et al. 2017; Christensen et al. 2019). Those studies suggested that lower NO_3^--N concentrations in spring and early summer were likely caused by a combination of factors, such as active pasture growth thereby increased NO_3^--N uptake and insufficient drainage volumes to move significant quantities of NO_3^--N into mole channels before the completion of the drainage season (Christensen et al. 2019).

The brassica treatment had greater NO_3^--N concentrations than other treatments at the end of the drainage season in December. The brassica plots were intensively grazed during winter after which the land was left bare for 17 weeks until re-sowing in late December. The lack of plant uptake likely resulted in higher amounts of NO_3^--N susceptible to N leaching under bare soil in subsequent drainage events. A dairy cow study has also shown that N leaching losses from forage-crop grazing during winter could be more than twice that under grazed RGWC (Monaghan et al. 2013). It is important to note that, in the current study, ewes began grazing the brassica plots in the latter part of the winter and had not completely eaten all the bulbs before set stocking for lambing. As a result, there is a risk of greater total NO_3^--N leaching if the ewes had been able to graze for a longer duration and harvest all the bulb material. It might also be argued that the drainage season in 2020 was protracted, therefore, the losses from the brassica were potentially greater than might have occurred in a more typical year. A daily soil-water balance model analysed using 30 years of data, indicated that only three out of 30 years had greater quantities of drainage after mid-October (Hanly et al. 2017). In addition, resowing of the subsequent forage crop in 2020 was delayed until 27 December 2020 due to prolonged wet soil conditions. To determine the losses that occur when brassica plots are resown more typically in mid spring multiple years of research is required. The amounts of NO_3^--N leached per ha were much lower than those previously reported (8 to 50 kg N ha^{-1}) for sheep grazing RGWC on the same soil type (Heng et al. 1991; Magesan et al. 1996; White et al. 1998). Differences in the stocking rate and grazing time may explain the differences among the sheep studies. For example, White et al. (1998) stocked 40-50 sheep in an area of 0.125 ha for a week in May 1990 and 21 sheep for 5 days in July 1991 and reported losses of 35 and 43 kg N ha^{-1}, respectively, while in the current study, stocking rate was kept constant (46 sheep in an area of 0.4 ha), and the number of grazing days was adjusted according to the amount of forage available. Lower NO_3^--N loss measured in the present study compared to the previous studies could also be due to lower drainage (mm) recorded in 2020 than those reported in the previous studies. The values reported by Heng et al. (1991), Magesan et al. (1996) and White et al. (1998) appear to be very high (~40 kg N ha^{-1}) and more recent measures of leaching under dairy cows (Hanly et al. 2017; Christensen et al. 2019) are more comparable to the present study (2 to 10 kg N ha^{-1}).

The N leaching values observed in the IRWC and PWC treatments were lower than that of RGWC treatment but did not differ significantly. Multiple years of studies with contrasting rainfall patterns are required to determine if differences might occur among forage treatments.

Conclusion

This study suggests N leaching under intensive sheep grazing of pasture-based forages (RGWC, IRWC, PWC) was very low on drained pallic soil. The grazing of a winter crop (brassica) is likely to result in a marked increase in N leaching in years where resowing of subsequent forage is delayed until the end of spring. Further, none of the alternative forages improved animal performance above the levels achieved on RGWC pasture.

Acknowledgments

The authors wish to thank D. Burnham, M. Osborne, B. Toes and Keeble farm staff for their technical assistance. This research was funded by L.A. Alexander Trust, C. Alma Baker Trust and Massey University.

References

Beef + Lamb New Zealand Economic Service 2021. On farm data & industry production Beef + Lamb New Zealand, New Zealand. Accessed: 8/3/2021. https://beeflambnz.com/data-tools/farm-classes

Blakemore L, Searle P, Daly B 1987. Methods for chemical analysis of soils. New Zealand Society of Soil Science, Lower Hutt, New Zealand. New Zealand soil Bureau scientific report: 80. 103 p.

Christensen CL, Hedley MJ, Hanly JA, Horne DJ 2019. Duration-controlled grazing of dairy cows. 2: Nitrogen losses in sub-surface drainage water and surface runoff. New Zealand Journal of Agricultural Research 62: 48-68.

Fraser P, Curtin D, Harrison-Kirk T, Meenken E, Beare M, Tabley F, Gillespie R, Francis G 2013. Winter nitrate leaching under different tillage and winter cover crop management practices. Soil Science Society of America Journal 77: 1391-1401.

Hanly JA, Hedley MJ, Horne DJ 2017. Effects of summer turnip forage cropping and pasture renewal on nitrogen and phosphorus losses in dairy farm drainage waters: A three-year field study. Agricultural Water Management 181: 10-17.

Haynes R, Williams P 1993. Nutrient cycling and soil fertility in the grazed pasture ecosystem. Advances in Agronomy 49: 119-199.

Heng L, White R, Bolan N, Scotter D 1991. Leaching losses of major nutrients from a mole-drained soil under pasture. New Zealand Journal of Agricultural Research 34: 325-334.

Hewitt A 1998. New Zealand Soil Classification. 2nd edition. Lincoln, New Zealand, Manaaki Whenua-Landcare Research New Zealand Ltd Press.

Hosomi M, Sudo R 1986. Simultaneous determination of total nitrogen and total phosphorus in freshwater samples using persulfate digestion. International Journal of Environmental Studies 27: 267-275.

Houlbrooke D, Horne D, Hedley M, Hanly J, Snow V 2004. A review of literature on the land treatment of farm-dairy effluent in New Zealand and its impact on water quality. New Zealand Journal of Agricultural Research 47: 499-511.

Humphrey M, Istok J, Lee J, Hevesi J, Flint A 1997. A new method for automated dynamic calibration of tipping-bucket rain gauges. Journal of Atmospheric and Oceanic Technology 14: 1513-1519.

Lucci G, Shepherd M, Carlson B, Lane M, Berry I 2013. Establishment of winter forage crops on pumice soils–experimental results and farmers' perspectives. Accurate and efficient use of nutrients on farms. Fertilizer and Lime Research Centre, Massey University, Palmerston North, New Zealand. Occasional Report, (26). 7 p.

Magesan G, White R, Scotter D 1996. Nitrate leaching from a drained, sheep-grazed pasture. I. Experimental results and environmental implications. Soil Research 34: 55-67.

Monaghan R, Smith L, De Klein C 2013. The effectiveness of the nitrification inhibitor dicyandiamide (DCD) in reducing nitrate leaching and nitrous oxide emissions from a grazed winter forage crop in southern New Zealand. Agriculture, Ecosystems & Environment. 175: 29-38.

Morris S, Kenyon P 2014. Intensive sheep and beef production from pasture - A New Zealand perspective of concerns, opportunities and challenges. Meat Science 98: 330-335.

Navarrete S, Kemp PD, Pain SJ, Back PJ, technology 2016. Bioactive compounds, aucubin and acteoside, in plantain (*Plantago lanceolata* L.) and their effect on in vitro rumen fermentation. Animal Feed Science and Technology 222: 158-167.

NIWA 2021. Annual Climate Summary 2020. NIWA (National Institute of Water and Atmospheric Research) National Climate Centre, New Zealand. Accessed: 8/4/2021. https://niwa.co.nz/climate/summaries/annual-climate-summary-2020

Rodriguez MJ, Kemp P, Navarrete S, Hanly J, Horne D, Bishop P 2020. NItrogen losses from plantain: What can we say?. Nutrient management in farmed landscapes. Fertilizer and Lime Research Centre, Massey University, Palmerston North, New Zealand. Occasional Report, (33). 8 p.

White R, Heng L, Magesan G 1998. Nitrate leaching from a drained, sheep-grazed pasture. II. Modelling nitrate leaching losses. Soil Research 36: 963-978.

Williams P, Haynes R 1990. Influence of improved pastures and grazing animals on nutrient cycling within New Zealand soils. New Zealand Journal of Ecology 14: 49-57.

Williams P, Haynes R 1994. Comparison of initial wetting pattern, nutrient concentrations in soil solution and the fate of 15 N-labelled urine in sheep and cattle urine patch areas of pasture soil. Plant and Soil 162: 49-59.

Woods R, Cameron K, Edwards G, Di H, Clough T 2016. Effects of forage type and gibberellic acid on nitrate leaching losses. Soil Use and Management 32: 565-572.

Genetic parameters for urination traits and their relationships with production traits of dairy cattle grazing temperate ryegrass and white-clover pastures

RC Handcock[a]*, LA Box[b], C Glassey[c] and DJ Garrick[a]

[a]*A.L. Rae Centre for Genetics and Breeding, Massey University, Hamilton 3214, New Zealand;* [b]*AgResearch, Ruakura Research Centre, Hamilton 3240, New Zealand;* [c]*DairyNZ Ltd, Private Bag 3221, Hamilton 3240, New Zealand*

**Corresponding author.* Email: R.C.Handcock@massey.ac.nz

Abstract

The objective of this study was to quantify genetic parameters of daily urination traits; urinary nitrogen (UN), urination volume (Uvol), urination number (Unum), average UN concentration (UNconc) and volume per urination event (VolEvent) including their genetic correlations with milk urea nitrogen concentration (MUN) and other production traits in New Zealand dairy cattle. Heritability estimates for urination traits measured on four consecutive days ranged from 0.17 to 0.35. Repeatability estimates ranged from 0.31 to 0.58, indicating that reliable prediction of urination measures would benefit from repeated measurement over a number of days. The heritability estimate for MUN was 0.29, similar to reports in the literature, whilst the estimate of repeatability (0.73) was greater than previously published values. Estimates of the genetic correlations between MUN and urination traits ranged from -0.39 to 0.41, however, due to the small number of animals in the current study, the confidence intervals were wide. The low genetic correlations may result from variation in voluntary feed intake from cow to cow that is a feature of pasture-based grazing systems. The results suggest that urination traits may be changed through selection. The low estimates of genetic correlations between MUN and urination traits suggest that MUN may not be a useful estimator of individual UN excretion in pasture-fed dairy cattle.

Keywords: urine traits; milk urea

Introduction

Temperate pasture-based (perennial ryegrass/white clover) dairy cattle typically consume forages that contain a high concentration of protein relative to energy, in excess of their daily requirements (Kolver & Muller 1998). Unless supplemented with a high-energy low-protein feed, these cows will excrete surplus nitrogen (N) principally in their urine (Selbie et al. 2015). One method to reduce N losses at a farm level would be a strategy that reduces the total daily quantity of urinary nitrogen (UN) excreted. An alternative N-mitigation strategy would be to identify animals that have more frequent, smaller urinations, as this would lead to a more uniform and dilute spread of UN deposited on pasture at a lower UN application rate (Kennett et al. 2020).

The direct measurement of UN, urine volume and number of urinations from grazing dairy cattle is difficult and expensive. It has been suggested that milk urea nitrogen concentration (MUN) estimated from routine herd tests could be used as an estimator of UN excretion per cow (Beatson et al. 2019).

The objective of this study was to quantify genetic parameters of urination traits and their genetic correlation with MUN and other production traits in New Zealand dairy cattle grazing temperate perennial ryegrass/white clover pastures.

Materials and methods

This experiment was conducted over two years and comprised various experimental runs in mid-late lactation (summer and autumn). Measurements of daily urination traits spanned Monday to Friday and comprised of groups of 30 cows split evenly into two cohorts of 15 cows per group.

Year One was conducted with the approval of the Ruakura Ethics Committee (#14634) on three dairy farms; Massey University's Dairy 4, situated near Palmerston North (40.2333° S, 175.3634° E), DairyNZ's Scott farm, situated near Hamilton (37.772° S 175.378° E) and AgResearch's Tokanui Dairy Research farm, situated near Te Awamutu, Waikato (38.0719° S 175.3273° E). Six experimental runs were completed between January and March 2019. Runs 1, 3 and 5 were completed at Dairy 4, Runs 4 and 6 were completed at Tokanui and Run 2 was completed at Scott farm. One cow was measured in two runs so records were on a total of 179 Holstein-Friesian-Jersey crossbred cows milked twice-a-day.

Year Two was conducted at Ashley Dene, situated near Burnham, Canterbury, New Zealand (43.6468° S 172.34679° E) between January and August 2020 with the approval of the Lincoln University Animal Ethics Committee (#2019-46). Six experimental runs were completed on a total of 180 Holstein-Friesian-Jersey crossbred cows milked twice a day.

In Year Two, animals were selected based on divergence in their parent average MUN breeding value (MUNBV; mg/dL), supplied by CRV (Beatson et al. 2019). The parent average MUNBV was estimated without including the cows own MUN phenotypes. The parent average is the most accurate estimate of genetic merit on the individual that is not confounded by environmental effects on the cow that may otherwise influence the repeatability of subsequent measurements.

Urination measures

Urine sensors (Betteridge et al. 2013; Shepherd et al. 2017a) of the same kind (Mk II, developed by AgResearch) were attached to the cows on a Monday after the morning milking (between 8 am and 10 am) and removed on Friday before the morning milking (between 5 am and 7 am). For data analysis, day was defined as the 24 hours from 9 am one day to 8:59 am the following day.

The urine sensors measure refractive index (RI), pressure, duration and time of each urination event. These data are used to estimate the urinary N concentration and volume of each urination event. The urinary nitrogen (UN) yield of each urination event (g N/event) was calculated as UN concentration (g/dL) x 10 x volume of the event (L). Mid-stream urine samples were collected at the start and end of each experimental run to calibrate the RI to urinary N concentration, as described by Shepherd et al. (2017b)

Over the four days, there were occasional periods when the urine sensor was not operational for the measurement of N concentration or was not functioning at all (due to being dislodged or blocked by faeces). For each cow-day, the number of urination events (Unum), cumulative urinary volume (Uvol) and cumulative UN was known for the total elapsed time that the urine sensor was functioning. These cumulative measures were divided by their respective total elapsed time of valid observations to give a per-minute value. The per-minute value was multiplied by the number of minutes in the day to get a cumulative measure representing daily Unum, Uvol and UN. For each cow-day the average volume per urination event (VolEvent) was calculated by dividing Uvol by Unum. Likewise, the average UN concentration (UNconc) was calculated by dividing UN by Uvol. Cow-days when the urine sensor was functioning for less than 50% of the day were excluded from the analysis (n=85 cow-days for Year One and n=187 cow-days for Year Two).

Milk production

In Year One, individual cow milk yields (MY) were recorded during each milking using individual test buckets and collecting the weight of milk in each bucket at Tokanui, using DeLaval milkmeters (FloMaster Pro) and its associated AlproTM herd and parlour management software (DeLaval International AB. Tumba, Sweden) for Scott farm and for Dairy 4. Milk samples were also collected from each milking starting Monday afternoon and ceasing Friday morning. Milk samples were transported daily to DairyNZ Hamilton for proportional milk composition (milk fat percentage, milk protein percentage and milk urea (MU) concentration; mg/dL) using the FOSS MilkoScan FT1 machines.

In Year Two, individual cow MY were automatically recorded each milking using the inbuilt Afimilk system with Afikim Afiflo 2000 milk meters. Milk samples were collected using the in-line fat sampler which is included with the Afikim Afiflo 2000 milk meters and then sub-sampled before being transported daily to MilkTestNZ Hamilton for proportional milk composition on the FOSS MilkoScan FT and 7 RM machines that use mid-infrared technique to measure milk components. The machines in both years were operated according to the ISO 9622:2013 guidelines (https://www.iso.org/standard/56874.html).

In both years MU concentration (mg/dL) was converted to MUN concentration (mg/dL) by multiplying by 0.466. Milk urea nitrogen yield (MUNY; grams) at each milking was calculated as MUN concentration (mg/dL) divided by 100 and multiplied by MY (L). Milk fat yield (FY), and milk protein yield (PY) at each milking were calculated by multiplying the milk fat percentage and MY or milk protein percentage and MY, respectively. The daily MY, MUNY, FY and PY were calculated as the sum of the afternoon and following morning yields to correspond to the 9 am to 8:59 am defined day for the urine sensors.

Two cows that were having extended lactations (>500 days) atypical of New Zealand pasture-based dairy cattle were removed from the dataset. Only cows that had data on urination traits had their milk traits included.

Genotypes

Cattle were genotyped by Weatherbys on an Illumina 50,000 SNP bovine panel. The small proportion of SNPs that were missing on any animal were imputed using the methodology of FImpute (Sargolzaei et al. 2014). Only mapped SNPs assigned to chromosomes 1 through 29 were included in the analysis (n=42,231).

Statistical analysis

Bayesian univariate repeatability models were run using the Julia for Whole-genome Analyses Software (JWAS) package (Cheng et al. 2018) run in a Julia computing environment (julialang.org). Inference was based on Markov chain Monte Carlo (MCMC) chains of 90,000 samples, retaining every 10th sample, after a burn-in of 10,000 samples which had been discarded.

The repeatability model equation was:

$$\boldsymbol{y} = \mathrm{RGD} + \mathrm{MUNBV} + \mathrm{age} \times \mathrm{year} + \mathrm{DIM} + \mathrm{pJ} + \mathrm{het} + \mathrm{AnimPerm} + \mathrm{SNPs} + \boldsymbol{e}$$

where $\boldsymbol{y}$ is the daily measurement on the trait of interest; RGD was the fixed class effect of run-group-day in which the animal was grazing; MUNBV was the fixed class effect of MUN BV group (Low, High or Year One); age × year was the interaction of the fixed class effects of age of the cow in years from birth to most recent parturition within year of the study; DIM was the fixed covariate of days in milk on day 1 of RGD; pJ was the Jersey breed proportion; het was the specific heterosis coefficient between Holstein-Friesian and Jersey (Dickerson 1973); AnimPerm is the random permanent effect of animal assumed to be independently and identically normally distributed with variance σ_c^2; SNPs are additive covariates for all of the 42,231 autosomal loci with effects independently and identically normally distributed with variance σ_α^2; and $\boldsymbol{e}$ is the residual effects independently and identically normally distributed with variance σ_e^2.

Co(variance) components were estimated by fitting the model equation pairwise using bivariate repeatability animal models.

The 95% credibility intervals were calculated by taking the 97.5^{th} percentile of the MCMC samples as the upper bound and the 2.5^{th} percentile as the lower bound.

Results

Descriptive statistics for the urination and milk traits are presented in Table 1. On average, after accounting for missing values, cows had approximately three days of urination traits included in the final dataset. The average daily UN excretion was 203.8 g with an average frequency of 12.3 urinations per day. On average, MUN concentration was 11.18 mg/dL.

The estimates of heritability for the five urination traits were moderate and ranged from 0.17 to 0.35 (Table 2). Repeatability estimates were greater than 0.3 for all traits studied, with milk traits tending to have higher repeatabilities than urination traits (Table 2). There were no discernible differences between the MUN BV groups (High minus Low) for any of the urination or milk traits, based on the 95% CI bordering zero (Table 2).

Genetic and phenotypic correlations among the five urination traits are in Table 3. Phenotypic correlations among three of those, namely UN, Uvol and Unum were moderately high and positive, suggesting that cows that excreted a high daily UN load would be doing so with more urinations and with a greater total volume per day. Although phenotypic correlations between UN and other urination traits were positive, they were negative between UNconc and other urination traits. The genetic correlations of UN with UNconc, Uvol and Unum and between Uvol and Unum were moderately high and positive (Table 3). The genetic correlation between UNconc and Uvol was moderately high and negative (Table 3).

The posterior mean of the genetic correlation between UN and MUN was 0.09, between UNconc and MUN was 0.41, whereas between MUN and the other three urination traits the genetic correlations were negative (-0.27 to -0.39; Table 4). Genetic correlations of MUNY with UN (0.24) and MUNY with UNconc (0.35) were low and positive (Table 4). Phenotypic correlations between MUN and urination traits were near zero, between -0.09 and 0.11 (Table 4). Likewise, phenotypic correlations between MUNY and urination traits were low and near zero (0.01 to 0.09; Table 4).

Estimates of the genetic correlations of milk production traits (MY, FY and PY) with urination traits were between -0.20 and 0.43; the phenotypic correlations were low and near zero for these same traits (Table 4).

Table 1 Number (N) of records, N of cows, unadjusted phenotypic mean and standard deviation (SD) of urination and milk traits in grazing dairy cows.

Trait	N records	N cows	Mean	SD
UN (g/d)	1,075	339	203.8	78.9
UNconc (g/L)	1,075	339	6.07	1.71
Uvol (L/d)	1,121	343	34.4	11.9
Unum (count)	1,121	343	12.3	4.2
VolEvent (L/event)	1,121	343	2.87	0.70
MY (L/d)	1,090	343	18.8	4.9
FY (kg/d)	1,089	343	0.91	0.22
PY (kg/d)	1,079	340	0.70	0.17
MUNY (g/d)	1,079	340	2.09	0.74
MUN (mg/dL)	1,079	340	11.18	3.12

daily urinary nitrogen (UN), mean UN concentration (UNconc), urination volume (Uvol), urination number (Unum), mean volume per urination event (VolEvent), milk yield (MY), milk fat yield (FY), milk protein yield (PY), milk urea nitrogen yield (MUNY) and milk urea nitrogen concentration (MUN).

Discussion

Measuring urination phenotypes of lactating cattle grazing outdoors is logistically challenging and expensive and, therefore, restricted for research purposes. Few studies have summarised whole-day urination traits in grazing dairy cattle (Shepherd et al. 2017; Bryant et al. 2018; Mangwe et al. 2019), and none have quantified genetic variation. Our estimates of heritability for the urination traits were moderate and greater than the heritability of urinary creatinine (0.05) or phosphorus concentrations (0.12) in a Danish study (Løvendahl & Sehested 2016). That same study reported repeatabilities for urinary creatinine (0.38) and phosphorus concentrations (0.21), comparable to the

Table 2 Posterior means with 95% credibility intervals (95% CI; presented in brackets) of the heritability, repeatability and the difference between cows that had High or Low MUN breeding values for urination and milk traits in grazing dairy cows.

Trait	Heritability	95% CI	Repeatability	95% CI	High – Low MUNBV	95% CI
UN (g/d)	0.17	(0.07, 0.26)	0.31	(0.24, 0.38)	6.6	(-9.3, 22.4)
UNconc (g/L)	0.35	(0.18, 0.49)	0.58	(0.53, 0.64)	0.24	(-0.17, 0.66)
Uvol (L/d)	0.32	(0.14, 0.45)	0.50	(0.43, 0.55)	0.4	(-2.7, 3.5)
Unum (count)	0.28	(0.11, 0.44)	0.50	(0.44, 0.56)	0.2	(-0.9, 1.3)
VolEvent (L/event)	0.18	(0.08, 0.32)	0.56	(0.49, 0.61)	0.03	(-0.18, 0.23)
MY (L/d)	0.51	(0.31, 0.67)	0.75	(0.71, 0.78)	-0.6	(-1.5, 0.3)
FY (kg/d)	0.27	(0.10, 0.49)	0.56	(0.49, 0.62)	0.00	(-0.04, 0.04)
PY (kg/d)	0.44	(0.21, 0.59)	0.66	(0.61, 0.71)	-0.01	(-0.04, 0.02)
MUNY (g/d)	0.17	(0.06, 0.35)	0.72	(0.67, 0.76)	-0.04	(-0.19, 0.12)
MUN (mg/dL)	0.29	(0.11, 0.52)	0.73	(0.68, 0.77)	0.17	(-0.49, 0.81)

daily urinary nitrogen (UN), mean UN concentration (UNconc), urination volume (Uvol), urination number (Unum), mean volume per urination event (VolEvent), milk yield (MY), milk fat yield (FY), milk protein yield (PY), milk urea nitrogen yield (MUNY) and milk urea nitrogen concentration (MUN).

Table 3 Posterior means of the phenotypic (above the diagonal) and genetic (below the diagonal) correlations among urination traits in grazing dairy cows with 95% credibility intervals (presented in brackets).

Trait	UN	UNconc	Uvol	Unum	VolEvent
UN	-	0.28	0.66	0.55	0.14
	-	(0.21, 0.35)	(0.61, 0.70)	(0.49, 0.60)	(0.07, 0.22)
UNconc	0.32	-	-0.45	-0.24	-0.29
	(-0.07, 0.62)	-	(-0.51, -0.39)	(-0.32, -0.17)	(-0.36, -0.21)
Uvol	0.43	-0.67	-	0.73	0.33
	(-0.19, 0.75)	(-0.81, -0.41)	-	(0.69, 0.77)	(0.26, 0.40)
Unum	0.67	-0.29	0.79	-	-0.33
	(0.44, 0.84)	(-0.55, 0.22)	(0.64, 0.89)	-	(-0.40, -0.26)
VolEvent	-0.29	-0.51	0.40	-0.21	-
	(-0.61, 0.24)	(-0.74, -0.19)	(0.03, 0.68)	(-0.63, 0.12)	-

daily urinary nitrogen (UN), mean UN concentration (UNconc), urination volume (Uvol), urination number (Unum), mean volume per urination event (VolEvent)

Table 4 Posterior means of the genetic and phenotypic correlations between milk traits and urination traits in grazing dairy cows with 95% credibility intervals (presented in brackets).

	UN	UNconc	Uvol	Unum	VolEvent
Genetic correlation					
MY	0.36 (0.01, 0.68)	-0.06 (-0.34, 0.25)	0.34 (0.02, 0.68)	0.32 (-0.03, 0.57)	-0.08 (-0.50, 0.33)
FY	0.43 (-0.29, 0.85)	0.12 (-0.37, 0.46)	0.24 (-0.31, 0.68)	0.32 (-0.04, 0.63)	-0.10(-0.69, 0.57)
PY	0.25 (-0.20, 0.63)	-0.12 (-0.39, 0.2)	0.40 (0.04, 0.69)	0.41 (0.05, 0.76)	-0.20 (-0.68, 0.29)
MUNY	0.24 (-0.29, 0.72)	0.35 (-0.01, 0.67)	-0.06 (-0.49, 0.42)	0.21 (-0.42, 0.69)	-0.23 (-0.64, 0.23)
MUN	0.09 (-0.64, 0.75)	0.41 (-0.02, 0.72)	-0.39 (-0.80, 0.22)	-0.31 (-0.68, 0.12)	-0.27 (-0.74, 0.36)
Phenotypic correlation					
MY	0.14 (0.06, 0.21)	-0.04 (-0.13, 0.04)	0.16 (0.08, 0.24)	0.10 (0.02, 0.18)	0.08 (0.00, 0.17)
FY	0.14 (0.07, 0.22)	-0.06 (-0.14, 0.02)	0.17 (0.10, 0.25)	0.08 (0.00, 0.16)	0.14 (0.06, 0.22)
PY	0.15 (0.08, 0.23)	-0.08 (-0.16, 0)	0.21 (0.13, 0.28)	0.14 (0.06, 0.22)	0.09 (0.01, 0.17)
MUNY	0.09 (0.01, 0.17)	0.04 (-0.05, 0.12)	0.04 (-0.04, 0.12)	0.06 (-0.02, 0.14)	0.01 (-0.08, 0.10)
MUN	0.01 (-0.07, 0.09)	0.11 (0.02, 0.20)	-0.09 (-0.17, -0.01)	-0.02 (-0.10, 0.06)	-0.08 (-0.16, 0.01)

daily urinary nitrogen (UN), mean UN concentration (UNconc), urination volume (Uvol), urination number (Unum), mean volume per urination event (VolEvent), milk yield (MY), milk fat yield (FY), milk protein yield (PY), milk urea nitrogen yield (MUNY) and milk urea nitrogen concentration (MUN).

repeatability estimate for UN in the current study, but less than for the other urination traits. Repeatability estimates for the urination traits measured across four consecutive days ranged from 0.31 to 0.58, as these estimates are not near unity, they are indicative that reliable prediction of urination traits would benefit from measurements being repeated over several days, potentially spread over lactations and stages of lactation.

The heritability estimate for MUN of 0.29 in the current study is in the range reported in New Zealand cattle of 0.22 to 0.49 (Lopez-Villalobos et al. 2018; Ariyarathne et al. 2019; Beatson et al. 2019) and in overseas Holsteins from 0.14 to 0.59 (Wood et al. 2003; Mitchell et al. 2005; Hossein-Zadeh & Ardalan 2011). The heritability of MUNY was comparable to the value of 0.19 reported by Lopez-Villalobos et al. (2018). The estimates of repeatability for MUN and MUNY were greater than the range of 0.35 to 0.55 reported in New Zealand studies (Lopez-Villalobos et al. 2018; Ariyarathne et al. 2019), likely due to the data in the present study being collected on any one cow over just four consecutive days rather than on four days spread over early, mid and late lactation.

It has been suggested that MUN measured using MIR from individual herd tests could be used as an estimator of per cow UN excretion (Beatson et al. 2019). The genetic correlations of daily MUN with UN was 0.09 and of MUNY with UN was 0.24, i.e., low and near zero. The 95% credibility intervals around the genetic correlations were wide, thus, these preliminary estimates based on 343 cows should be viewed as indicative, and collection of further data from a larger number of cows might be warranted.

The phenotypic correlations between MUN and UN was 0.01 and for MUNY and UN was 0.09. These near-zero phenotypic correlations indicate that cows that have lower MUN or MUNY are not likely to have lower UN excretion compared with cows that have higher MUN or MUNY. It has been reported internationally that MUN is related linearly to UN excretion at the phenotypic level (Kauffman & St-Pierre 2001; Spek et al. 2013). However, this relationship appears to be due to diet differences, i.e., if a cow is fed a higher-protein diet, their MUN and UN increases linearly (Kauffman & St-Pierre 2001; Spek et al. 2013). Based on the results from the current study it appears that the phenotypic relationship between MUN and UN from controlled feeding studies does not hold in a within-herd comparison of grazing cows with higher MUN compared to those with lower MUN when they were offered the same diet. Although grazing cattle may be offered the same quantity of pasture, their voluntary feed intake can differ considerably. Therefore, it is difficult to identify which cows are consuming more feed and, hence, more N than others and what impact this has on their subsequent MUN or UN excretion.

A 0.67 g/L decrease in urinary urea N concentration per unit decrease in MUNBV was reported in a study of 48 lactating Holstein-Friesian cattle divergently selected for MUNBV (Marshall et al. 2020). Urinary urea N was not measured in the current study, but it is a component of UN. There were small positive genetic and phenotypic correlations between MUN and UNconc, but the 95% credibility intervals for the genetic correlation were wide (-0.02, 0.72), leading to uncertainty whether or not selection for decreased MUN would result in cows that had markedly decreased UNconc. Furthermore, UNconc is only one factor influencing N leaching, with VolEvent and Unum being other important animal factors to consider (Kennett et al. 2020). The genetic and phenotypic correlations between UNconc and VolEvent were negative in the current study indicating that selection for lower UNconc would likely be associated with an increase in volume per urination event.

Marshall et al. (2020) reported an R^2 of 0.67 for the regression of MUN phenotype on MUNBV. The R^2 for the regression of phenotype on true breeding value is the definition of heritability (Bourdon 2000; Mrode 2005). The value of 0.67 reported by Marshall et al. (2020) is outside the 0.14 - 0.59 range of estimates for the heritability of MUN (Wood et al. 2003; Mitchell et al. 2005; Hossein-Zadeh & Ardalan 2011; Lopez-Villalobos et al. 2018; Ariyarathne et al. 2019; Beatson et al. 2019). The higher than expected R^2 indicates that the MUNBV was not a parent average estimate but was likely calculated by including individual MUN phenotypes of the cows studied by Marshall et al. (2020). Although using the cow's own MUN phenotype will increase the accuracy of the estimated MUNBV, it introduces a non-genetic covariance between MUNBV and any subsequent MUN phenotype measured on the individual, hence, it was not simply a genetic relationship being studied by Marshall et al. (2020) but a repeatability relationship. Therefore, their conclusion that selecting dairy cows with low MUNBV can reduce urinary urea N deposition onto pasture may not hold in practice and is not supported by the results from the current study on a larger dataset.

A recent study modelled the likely change in UN excretion when including MUN in the New Zealand national selection index (Breeding Worth; BW) and concluded that there was unlikely to be much environmental benefit from including selection for lower MUN (Ariyarathne et al. 2021). The authors attributed this to whole-herd N leaching being more a function of stocking rate rather than individual cow UN excretion. They predicted that after 10 years of selection for lower MUN with 20% emphasis in BW, UN excretion per hectare would be 154.5 kg, only 3.8 kg less than if MUN was not included in BW at all.

Overall, the results from the current study indicate that urination traits may be changed through direct selection. However, the low and near-zero estimates of genetic correlations between MUN and urination traits suggest that MUN may not be a useful estimator of individual UN excretion in pasture-fed dairy cattle that vary in voluntary feed intake from cow to cow.

Acknowledgements

Low N Livestock is a DairyNZ-led collaborative research programme delivering science for better farming and environmental outcomes. The main funders are DairyNZ and the Ministry of Business, Innovation and Employment, with co-funding from CRV-Ambreed and Fonterra. The field work was conducted with the assistance of the farm, technical and academic staff at Ashley Dene, Lincoln University, Massey University No.4 Dairy, AgResearch Tokanui Dairy Research farm, and Scott Farm, DairyNZ. Olivia Jordan, Emma Masterton, Addam Knight, Charissa Thomas, Fiona Sharland and Hamish Doohan had leading roles.

References

Ariyarathne HBPC, Correa-Luna M, Blair H, Garrick D, Lopez-Villalobos N 2021. Can nitrogen excretion of dairy cows be reduced by genetic selection for low milk urea nitrogen concentration? Animals 11: 1-18.

Ariyarathne HBPC, Correa-Luna M, Blair HT, Garrick DJ, Lopez-Villalobos N 2019. Estimation of heritabilities for milk urea and efficiency of crude protein utilization by day of lactation in a New Zealand dairy cow herd. New Zealand Journal of Animal Science and Production 79: 183-187.

Beatson PR, Meier S, Cullen NG, Eding H 2019. Genetic variation in milk urea nitrogen concentration of dairy cattle and its implications for reducing urinary nitrogen excretion. Animal 13: 2164-2171.

Betteridge K, Costall DA, Li FY, Luo D, Ganesh S 2013. Why we need to know what and where cows are urinating – a urine sensor to improve nitrogen models. Proceedings of the New Zealand Grassland Association 75: 33-38.

Bourdon RM 2000. Heritability and Repeatability. 2nd ed. New Jersey, United States: Pearson New International.

Bryant RH, Welten BG, Costall D, Shorten PR, Edwards GR 2018. Milk yield and urinary-nitrogen excretion of dairy cows grazing forb pasture mixtures designed to reduce nitrogen leaching. Livestock Science 209: 46-53.

Cheng H, Fernando RL, Garrick DJ 2018. JWAS: Julia implementation of whole-genome analysis software. Proceedings of the World Congress on Genetics Applied to Livestock Production 11: 859.

Dickerson GE 1973. Inbreeding and heterosis in animals. Journal of Animal Science (Suppl): 54-77.

Hossein-Zadeh NG, Ardalan M 2011. Estimation of genetic parameters for milk urea nitrogen and its relationship with milk constituents in Iranian Holsteins. Livestock Science 135: 274-281.

Kauffman AJ, St-Pierre NR 2001. The Relationship of Milk Urea Nitrogen to Urine Nitrogen Excretion in Holstein and Jersey Cows. Journal of Dairy Science 84: 2284-2294.

Kennett J, Romera A, Beukes P, Amer P 2020. A model to assess the impact of potential genetic selection of dairy cows for volume per urination and total urine volume per day on nitrogen leaching. New Zealand Journal of Animal Science and Production 80: 60-64.

Kolver ES, Muller LD 1998. Performance and Nutrient Intake of High Producing Holstein Cows Consuming Pasture or a Total Mixed Ration. Journal of Dairy Science 81: 1403-1411.

Lopez-Villalobos N, Correa Luna M, Burke J, Sneddon N, Schultz M, Donaghy D, Kemp P 2018. Genetic parameters for milk urea concentration and milk traits in New Zealand grazing dairy cattle. New Zealand Journal of Animal Science and Production 78: 56-61.

Løvendahl P, Sehested J 2016. Short communication: Individual cow variation in urinary excretion of phosphorus. Journal of Dairy Science 99: 4580-4585.

Mangwe MC, Bryant RH, Beck MR, Beale N, Bunt C, Gregorini P 2019. Forage herbs as an alternative to ryegrass-white clover to alter urination patterns in grazing dairy systems. Animal Feed Science and Technology 252: 11-22.

Marshall CJ, Beck MR, Garrett K, Barrell GK, Al-Marashdeh O, Gregorini P 2020. Grazing dairy cows with low milk urea nitrogen breeding values excrete less urinary urea nitrogen. Science of the Total Environment 739: 139994.

Mitchell RG, Rogers GW, Dechow CD, Vallimont JE, Cooper JB, Sander-Nielsen U, Clay JS 2005. Milk urea nitrogen concentration: Heritability and genetic correlations with reproductive performance and disease. Journal of Dairy Science 88: 4434-4440.

Mrode RA 2005. Genetic Evaluation with Different Sources of Records. 2nd ed. Oxfordshire, United Kingdom: CABI Publishing.

Sargolzaei M, Chesnais JP, Schenkel FS 2014. A new approach for efficient genotype imputation using information from relatives. BMC Genomics 15: 478.

Selbie DR, Buckthought LE, Shepherd MA 2015. The Challenge of the Urine Patch for Managing Nitrogen in Grazed Pasture Systems. Advances in Agronomy 129: 229-292.

Shepherd MA, Shorten P, Costall D, Macdonald KA 2017a. Evaluation of urine excretion from dairy cows under two farm systems using urine sensors. Agriculture, Ecosystems and Environment 236: 285-294.

Shepherd MA, Welten BG, Costall D, Cosgrove GP, Pirie M, Betteridge K 2017b. Evaluation of refractive index for measuring urinary nitrogen concentration in a sensor worn by grazing female cattle. New Zealand Journal of Agricultural Research 60: 23-31.

Spek JW, Dijkstra J, Van Duinkerken G, Bannink A 2013. A review of factors influencing milk urea concentration and its relationship with urinary urea excretion in lactating dairy cattle. Journal of Agricultural Science 151: 407-423.

Wood GM, Boettcher PJ, Jamrozik J, Jansen GB, Kelton DF 2003. Estimation of genetic parameters for concentrations of milk urea nitrogen. Journal of Dairy Science 86: 2462-2469.

Evaluation of existing equations to rank methane yields of sheep fed a range of fresh forages

A Jonker[a*], M Eugène[bc], C de Klein[d], D Pacheco[a] and R Vibart[a]

[a]*Grasslands Research Centre, AgResearch Ltd., Private Bag 11008, Palmerston North, New Zealand;* [b]*INRAE, UMR Herbivores, F-63122, Saint-Genès-Champanelle, France;* [c]*Clermont Université, VetAgro Sup, UMR1213 Herbivores, BP 10448, F-63000, Clermont-Ferrand, France;* [d]*Invermay Agricultural Research Centre, AgResearch Ltd., Private Bag 50034, Mosgiel, New Zealand*

[*]*Corresponding author.* Email: arjan.jonker@agresearch.co.nz

Abstract

The objective of the current study was to evaluate the performance of published equations to predict CH_4 yield (g CH_4/kg dry matter intake; DMI) in NZ sheep fed a range of fresh forages. A database with 182 NZ sheep CH_4-yield treatment means was used to validate 17 published CH_4-yield prediction equations. The CH_4-yield predictions were evaluated using the square root of mean-square prediction error (RMSPE) and Lin's concordance correlation coefficient (CCC). The three prediction equations with the smallest error (<22% of RMSPE) and the only ones with moderate CCC with observed CH_4 yield were all based on data from animals fed forage only. These three models predicted CH_4 yield based on DMI plus dietary neutral detergent fibre, and animal body weight, feed crude protein, or feed organic matter digestibility; and they performed better than the DMI only equations that are commonly used in national greenhouse gas inventory models and on-farm accounting models.

Keywords: Meta-analysis; enteric methane; model validation; fodder crops

Introduction

Ruminant enteric methane (CH_4) emissions in national greenhouse gas (GHG) inventories and on-farm accounting models are, in most cases, predicted as a function of dry-matter intake (DMI) only, without considering the effect of diet composition and animal characteristics on CH_4 output (Vibart et al. 2021). For example, enteric methane output of cattle in the NZ GHG inventory and the farm model Overseer is calculated as DMI (kg/d) × 21.6 g CH_4/kg DMI (i.e., CH_4 yield). These equations do not capture the effects of diets that reduce emissions per unit of DMI. For example, sheep fed forage rape emitted consistently less CH_4 per unit of DMI than did sheep fed pasture in multiple trials (Sun et al. 2016) and sheep fed a high-sugar diploid perennial ryegrass cultivar had a lower CH_4 yield than did sheep fed a conventional diploid ryegrass cultivar (Jonker et al. 2018), but these effects are not captured in the national GHG inventory or Overseer (Vibart et al. 2021). Outside NZ, more-comprehensive CH_4-prediction equations that include dietary chemical composition [e.g., starch, neutral detergent fibre (NDF), crude protein (CP)] have been developed and validated with animal CH_4-production measurement data (Benaouda et al. 2019). However, to our knowledge, no validation of CH_4-prediction equations has been performed with CH_4 production measurement data from animals fed largely fresh herbage, either from pastures or forage crops. Furthermore, most validations have been performed with CH_4 expressed as g/d, not as g/kg DMI. Methane production measurements data of NZ sheep published in literature were collected as part of an international project, Capturing the Effects of Diet on Emissions from Ruminant Systems (CEDERS; https://www.eragas.eu/en/eragas/Researchprojects/CEDERS-1.htm). The objective of the current study was to evaluate the performance of published equations in predicting CH_4 yield in NZ sheep fed a range of fresh forages.

Materials and methods

A database with 182 treatment means was generated from 27 publications reporting CH_4 production measurements conducted in NZ with sheep. Information extracted included animal age and body weight, DMI, CH_4 production, CH_4-measurement technique used [i.e., respiration chambers (63% of data) and SF_6 tracer technique (37% of data)], diet fed (i.e., ryegrass/white clover-based pasture (66% of data) and other forages (34% of data), including forage brassicas, legumes, herbs, fodder beet and maize silage) and feed composition (ash, CP, NDF, lipids) and DM and OM digestibility (DMD, OMD; *in vivo*, *in vitro* and NIRS digestibility) (Table 1). Missing data on chemical composition was estimated based on relationships with data in the database or based on published equations (INRA 2018) and non-fibre carbohydrates (NFC) were calculated as 1000 − (ash + lipids + CP + NDF). Missing data on diet DMD was estimated using a faecal DM output equation based on NZ data of sheep fed a range of fresh forages (van Lingen et al. 2021). Conversions between DMD, OMD, OMD on DM basis (DOMD), and digestible energy (DE) were performed according to equations of CSIRO (1990), and OMD at maintenance feeding level (OMDm) was calculated as described by Ramin & Huhtanen (2013). Dietary metabolisable energy (ME; MJ/kg DM) was calculated from diet DOMD (CSIRO 1990).

Prediction equation evaluation

A literature search was performed using 'Google Scholar' to find CH_4 production (g/d) prediction equations generated from data of sheep and of multiple species, including sheep. Seventeen CH_4 production prediction equations were extracted from 11 publications. Of these, 12 were generated using sheep data only and five from multiple species (including sheep); nine of the equations were generated using data of animals fed 100% forage and eight equations from animals fed a wide range of feeds with variable forage-to-concentrate ratios. All 17 equations contained intake (DMI or feeding level) as explanatory variable. Only results of the five best-performing equations (see criteria below) are presented in Table 2, plus two additional equations of Zhao et al. (2016) to aid the discussion. Three equations had only DMI as a predictor variable in the model: 21.6 × DMI (IPCC 2006) and MS_2 and YZ_1 in Table 2. The other equations evaluated (explanatory variables in equations between brackets), but not presented, include eq. 14.1 (OMD, feeding level and concentrate proportion in the diet) of Sauvant et al. (2018); eq. 1f (DMI, N) and eq. 11 (DMI, N, NDF) of Zhao et al. (2016); eq. 17 (feeding level, OMD, lipids, NDF, NFC) of Ramin & Huhtanen (2013); eq. 10 (NDF intake, lipid intake, NFC intake, CP intake), eq. 11 (DMI, OMD), eq. 12 (DMI, feeding level, NDF), and eq. 13 (DMI, feeding level, OMD) of Patra et al. (2016); and eq. 4 (OMD, lipid, feeding level) of Bell et al. (2017).

Methane production (g/d), observed and predicted, were first converted to CH_4 yield [CH_4 (g/d) / DMI (kg/d)] before evaluation of the equations using the database described above. This conversion allows evaluation of the CH_4 predictions independent of differences in DMI, the main driver of CH_4 production (Swainson et al. 2018). Because DMI is the same for observed and predicted values, any deviation in CH_4 yield is due to errors in CH_4 prediction, and DMI and CH_4 measurement errors in the sheep trials.

Errors in the CH_4-yield predictions were evaluated as detailed by Benaouda et al. (2019) determining the mean-square prediction error [MSPE; i.e., sum of (observed – predicted values)^2] and partitioning MSPE into mean bias, slope bias, and random error (Bibby & Toutenburg 1977). The square root of MSPE (RMSPE) is given in the same unit as observed values. Smaller RMSPE, mean bias and slope bias values, and greater random error as percentage of MSPE indicate better accuracy of the CH_4-yield prediction equation. Lin's concordance correlation coefficient (CCC), which is the product of Pearson correlation (r) and a bias correction factor (C_b), between observed and predicted values were determined using GenStat (19th edition; VSN International, Hemel Hempstead, UK). In addition, Spearman correlation (r_s) between observed and predicted values was determined using GenStat, which enables the evaluation of precision of ranking of treatments independent of the accuracy of values predicted. Greater CCC, r, C_b, and r_s values (range from 0.0 to 1.0) indicate better precision of the CH_4-yield prediction equation.

Results

The mean observed CH_4 yield in sheep fed a range of forages was 21.6 (±4.8) g/kg DMI (Table 1), predicted mean CH_4 yield ranged from 18.0 to 27.2 g/kg DMI across the 17 equations (data not shown), and 20.2 to 23.0 g/kg DMI for the seven prediction equations presented in Table 2. The RMSPE, expressed as a percentage of mean CH_4 yield observed, ranged from 21.2 to 39.2% (4.4 to 8.1 g/kg DMI) and random error ranged from 100 to 34.9 % of MSPE across the 17 prediction equations. The CCC and r were moderate (>0.4) only for three prediction equations and r_s was moderate (>0.4) only for five equations (Table 2), while all other equations had weak correlations (<0.28) with observed CH_4 yields. The three best-performing equations across the whole sheep CH_4 yield database were also the best performing equations for predicting CH_4 yield of sheep in respiration chamber trials only, sheep in SF_6 trials only, sheep fed ryegrass based-pasture only, and of sheep fed legumes and forage crops only (data not shown).

Discussion

The three prediction equations with the smallest error (RMSPE) and the only ones with moderate correlations (>0.4; CCC, r, r_s) with observed CH_4 yield were the two equations of van Lingen et al. (2021) (VL_1 and VL_2) and the equation of Sauvant et al. (2018) for animals fed forage

Table 1 Mean, standard deviation (Stdev), minimum (min), and maximum (max) of animal characteristics, intake, forage composition, and methane emissions of the database with published treatment means from sheep fed a range of fresh forages (n = 182).

Parameter	Mean	Stdev	min	max
Age (months)	14.3	9.1	3.3	48.0
Body weight (BW; kg)	45.8	8.3	32.5	77.0
Dry matter intake (DMI; kg/d)	0.98	0.32	0.35	2.30
DMI/BW (g/kg)	21.7	6.5	8.4	48.8
Crude protein (g/kg DM)	177	56	58	328
Ether extract (g/kg DM)	31	10	11	57
Ash (g/kg DM)	97	28	11	196
Neutral detergent fibre (g/kg DM)	425	120	127	662
Non-fibre carbohydrate (g/kg DM)	271	114	74	586
OM digestibility (fraction 0-1)[1]	0.76	0.07	0.59	0.96
Metabolisable energy (MJ/kg DM)	11.3	1.0	8.7	13.8
CH_4 production (g/d)	20.6	6.1	8.1	39.5
CH_4 yield (g/kg DMI)	21.6	4.8	8.3	43.9

[1]OM digestibility, organic matter digestibility at actual intake level, includes measured values and estimated values based on DMD calculated according to van Lingen et al. (2021) and converting DMD to OMD according to CSIRO (1990).

Table 2 Evaluation of the performance of equations[1] to predict CH_4 yield [g/kg dry matter intake (DMI)] of sheep fed a range of fresh forages

	Equation[1]						
Parameter[2]	VL_1	VL_2	MS_2	NS_1	YZ_1	YZ_4	YZ_5
Mean predicted CH_4 yield	20.4	21.0	23.0	22.6	20.2	19.2	22.3
Standard deviation	3.02	3.25	2.74	1.47	1.16	3.13	2.07
RMSPE (g/d)	4.36	4.38	4.41	4.56	4.74	5.71	5.21
RMSPE (% of mean CH_4 yield)	21.2	21.3	21.4	22.2	23.0	27.8	25.3
Decomposition of error							
ECT (mean bias; % of RMSPE)	7.7	1.9	11.3	4.7	8.6	17.3	2.0
ER (slope bias; % of RMSPE)	1.8	4.8	0.4	0.8	1.4	14.5	12.9
ED (random; % of RMSPE)	90.6	93.3	88.3	94.5	90.1	68.2	85.1
Correlations							
CCC	0.44	0.44	0.41	0.21	0.15	0.16	0.03
Pearson *r*	0.51	0.48	0.51	0.39	0.36	0.20	0.04
C_b (slope bias)	0.86	0.92	0.80	0.54	0.42	0.78	0.71
Spearman *r*	0.61	0.57	0.50	0.43	0.43	0.27	0.03

[1]**VL_1**: CH_4 (g/kg DMI) = (-4.00 + 12.9 × DMI + 0.121 × (NDF/10) + 0.132 × BW)/DMI, **VL_2**: CH_4 (g/kg DMI) = (1.98 + 15.1 × DMI + 0.117 × (NDF/10) – 0.138 × (CP/10))/DMI (van Lingen et al. 2021); **MS_2**: CH_4 (g/kg DMI) = (DMI × 0.001 × (OMD × 1000) × (34.95 – 4.05 × (DMI/BW/10) + 0.27 × (NDF/10) – 0.01 × (OMD × 1000)))/DMI (Sauvant et al. 2018): **NS_1**, CH_4 (g/kg DMI) = (exp(0.763 × ln_DMI + 3.039))/DMI (Swainson et al. 2018), **YZ_1**: CH_4 (g/kg DMI) = (16.7 × DMI + 3.1)/DMI, **YZ_4**: CH_4 (g/kg DMI) = (17.9 × DMI -84 × (OM/1000) + 18.7 × (NDF/1000) + 69.0)/DMI, **YZ_5**: CH_4 (g/kg DMI) = (18.8 × DMI + 5.0 × DE – 4.9 × ME – 9.9)/DMI (Zhao et al. 2016); where, DMI, dry matter intake (kg/d); NDF, neutral detergent fibre (g/kg DM); BW, body weight (kg); CP, crude protein (g/kg DM); OMD, organic matter digestibility at feeding level (fraction 0-1); DMI/BW (g/kg); OM, organic matter (g/kg DM); DE, digestible energy (MJ/kg DM); ME, metabolizable energy (MJ/kg DM).

[2]RMSPE, square root of the mean square prediction error; ECT, error due to central tendency; ER error due to deviation from the regression slope; ED, error due to disturbance; CCC, concordance correlation coefficient; *r*, correlation; C_b, a bias correction factor

only (MS_2). These three models predicted CH_4 production based on DMI plus dietary NDF and BW, CP, or OMD (see footnote Table 2). This suggests that equations that include dietary variables would be better at capturing the CH_4-mitigating effects of forages and forage crops with variable composition fed to sheep than equations that only use DMI to predict CH_4 production (NS_1 and YZ_1) (Zhao et al. 2016; Swainson et al. 2018). Both DMI-only based prediction equations were developed from respiration-chamber trials with sheep fed fresh ryegrass-based pasture only and both performed especially poorly with data of sheep fed legumes and forage crops (data not shown).

The equations of van Lingen et al. (2021) were generated from individual sheep CH_4 respiration-chamber data on ryegrass, clover, and fresh forage crops, and the Sauvant et al. (2018) equation for ruminants fed forage only was generated from 592 dietary-treatment means, including data from cattle, sheep, and goats fed mainly grass (C3 and C4) or legume based forages (fresh and conserved). The range in forage NDF concentration reported by van Lingen et al. (2021) was much greater with 166 to 660 g/kg DM and those reported by Sauvant et al. (2018) with 127 to 845 g/kg DM compared to the NDF concentration across ryegrass batches of 427 to 594 g/kg DM in Zhao et al. (2016). This likely explains why NDF concentration is an important predictor in the equations of van Lingen et al. (2021) and Sauvant et al. (2018), but not in those of Zhao et al. (2016)

In contrast to the above, across the equations of Zhao et al. (2016) evaluated in the current analysis (YZ_1, YZ_4, and YZ5 plus two others not shown), the equation that predicted CH_4 production from DMI alone had a lower prediction error and stronger correlations than did equations that included DMI plus additional nutritional parameters like NDF, OM, DE and ME. However, after a closer inspection of the CCC components, it is apparent that the two DMI-only equations have a much lower C_b (i.e., greater slope bias) than any of the equations that include nutritional components. When plotting observed *vs* predicted data (for selected equations), it becomes apparent that the CH_4 yield predicted by the DMI-only equation of Swainson et al. (2018) (and of Zhao et al. 2016, not shown) was on average close to the observed CH_4 yield, and with similar precision for the relationship (i.e., *r*; Fig. 1; Table 2), however, there was a large slope bias visible, which was identified using C_b, but not when using ER% of MSPE. Similar observations were made by Ellis et al. (2010) when evaluating equations to predict CH_4 production of dairy cows.

The prediction equations that were generated from databases that included CH_4-yield data of ruminants fed forage plus different levels of concentrates, starch, or oil supplementation (Ramin & Huhtanen 2013; Bell et al. 2016; Patra et al. 2016; Sauvant et al. 2018) performed poorly at predicting CH_4 yield for the current database of sheep fed fresh forages in NZ (data not shown). These equations

Figure 1 Predicted *vs.* observed methane (CH_4) yield of sheep fed a range of fresh forages, for three prediction equations evaluated (Sauvant et al. 2018; Swainson et al. 2018; van Lingen et al. 2021).

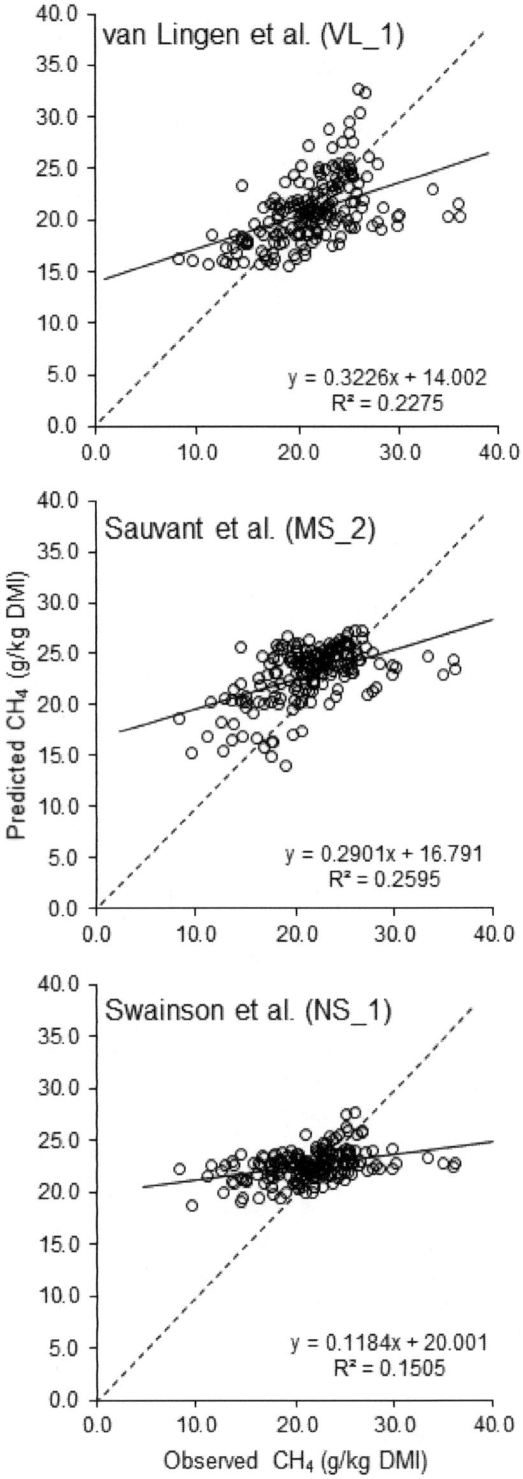

were tested because forage crops fed to sheep in NZ have a nutrient composition more similar to concentrates (e.g., high non-fibre carbohydrates) than to grass-based herbage. But apparently, the physical form of concentrates (dry and small particles) results in a different CH_4-yield response when fed to animals, than does feeding forage crops.

Collectively, these findings indicate that equations based on data from animals fed forage only, and that include dietary variables, would be better at capturing the CH_4-mitigating effects of forages and forage crops with variable composition fed to sheep in NZ than equations that only use DMI to predict CH_4 yield. Therefore, replacing the DMI only based CH_4 yield prediction equations currently used in on-farm accounting models with the more complex equations, that include animal and feed characteristics, would improve CH_4 yield predictions for sheep fed a range for fresh forages. However, the evaluation also revealed that there remains scope to develop improved CH_4-yield prediction equations for sheep fed a range of fresh forages.

Acknowledgments

The collation of the database used in the current analysis was funded by the New Zealand Government in support of the objectives of the Livestock Research Group of the Global Research Alliance on Agricultural Greenhouse Gasses and to contribute to the ERA-GAS project Capturing the Effects of Diet on Emissions from Ruminant Systems (CEDERS; https://www.eragas.eu/en/eragas/Researchprojects/CEDERS-1.htm).

References

Bell M, Eckard R, Moate P, Yan T 2016. Modelling the effect of diet composition on enteric methane emissions across sheep, beef cattle and dairy cows. Animals 6: 54.

Benaouda M, Martin C, Li X, Kebreab E, Hristov AN, Yu Z, Yáñez-Ruiz DR, Reynolds CK, Crompton LA, Dijkstra J, Bannink A, Schwarm A, Kreuzer M, McGee M, Lund P, Hellwing ALF, Weisbjerg MR, Moate PJ, Bayat AR, Shingfield KJ, Peiren N, Eugène M 2019. Evaluation of the performance of existing mathematical models predicting enteric methane emissions from ruminants: Animal categories and dietary mitigation strategies. Animal Feed Science and Technology 255: 114207.

Bibby JM, Toutenburg H 1977. Prediction and improved estimation in linear models. Chichester, NY, USA, Wiley.

CSIRO 1990. Feeding standards for Australian livestock. Ruminants. Melbourne, Victoria, Australia, CSIRO Publications.

Ellis JL, Bannink A, France J, Kebreab E, Dijkstra J 2010. Evaluation of enteric methane prediction equations for dairy cows used in whole farm models. Global Change Biology 16: 3246-3256.

INRA 2018. INRA Feeding System for Ruminants. Wageningen, the Netherlands, National Institute for Agricultural Research. Wageningen Academic Publishers.

IPCC 2006. Intergovernmental Panel on Climate Change. Chapter 10 Emissions from livestock and manure management. Available at http://www.ipcc-nggip.iges.or.jp/public/2006gl/pdf/4_Volume4/V4_10_Ch10_Livestock.pdf [Accessed 28/09/2021].

Jonker A, Molano G, Sandoval E, Taylor PS, Antwi C, Olinga S, Cosgrove GP 2018. Methane emissions differ between sheep offered a conventional diploid, a high-sugar diploid or a tetraploid perennial ryegrass cultivar at two allowances at three times of the year. Animal Production Science 58: 1043-1048.

Patra AK, Lalhriatpuii M, Debnath BC 2016. Predicting enteric methane emission in sheep using linear and non-linear statistical models from dietary variables. Animal Production Science 56: 574-584.

Ramin M, Huhtanen P 2013. Development of equations for predicting methane emissions from ruminants. Journal of Dairy Science 96: 2476-2493.

Sauvant D, Giger-Reverdin S, Eugène M 2018. Chapter 14. Enteric methane emissions. In INRA Feeding System for Ruminants. Eds P Nozière, D Sauvant, L Delaby. Wageningen Academic Publisher, Wageningen, the Netherlands, pp. 209-212.

Sun X, Pacheco D, Luo D 2016. Forage brassica: a feed to mitigate enteric methane emissions? Animal Production Science 56: 451-456.

Swainson N, Muetzel S, Clark H 2018. Updated predictions of enteric methane emissions from sheep suitable for use in the New Zealand national greenhouse gas inventory. Animal Production Science 58: 973-979.

van Lingen HJ, Jonker A, Kebreab E, Pacheco D 2021. Quantitative joint evaluation of sheep enteric methane emissions and faecal dry matter and nitrogen excretion. Agriculture, Ecosystems & Environment 305: 107116.

Vibart R, de Klein C, Jonker A, van der Weerden T, Bannink A, Bayat AR, Crompton L, Durand A, Eugène M, Klumpp K, Kuhla B, Lanigan G, Lund P, Ramin M, Salazar F 2021. Challenges and opportunities to capture dietary effects in on-farm greenhouse gas emissions models of ruminant systems. Science of The Total Environment 769: 144989.

Zhao YG, O'Connell NE, Yan T 2016. Prediction of enteric methane emissions from sheep offered fresh perennial ryegrass (*Lolium perenne*) using data measured in indirect open-circuit respiration chambers. Journal of Animal Science 94: 2425-2435.

TECHNICAL NOTE: Emissions improbable: how inappropriate use of the Global Livestock Environmental Assessment Model (GLEAM) can overestimate the emissions associated with dairy-production chains

AJ Fletcher[a,b,c]*, NW Sneddon[c] and JP Hill[a,b,c]

[a]*Riddet Institute, Massey University, Palmerston North, New Zealand;* [b]*Sustainable Nutrition Initiative, Riddet Institute, Massey University, Palmerston North, New Zealand;* [c]*Fonterra Research and Development Centre, Palmerston North, New Zealand*

**Corresponding author:* Email Andrew.Fletcher@fonterra.com

Abstract

In this paper we calculate the key differences between emissions estimates made using the Food and Agricultural Organisation of the United Nations (FAO) Global Livestock Environmental Assessment Model (GLEAM) and the New Zealand-specific analysis performed for the New Zealand Greenhouse Gas Inventory and the Fonterra Farm Gate Milk life-cycle analysis (LCA) that provides a proxy for all New Zealand milk production. We discovered that a wide range of factors can be responsible for more than a doubling of emissions calculated using GLEAM compared with emissions calculated using New Zealand-specific data. Such inappropriate use of GLEAM and the resulting conclusions have been used to undermine the efficiency of New Zealand production systems in publications such as Emissions Impossible.

Keywords: Life cycle analysis; modelling; dairy processing

Background

The Global Livestock Environmental Assessment Model (GLEAM) produced and maintained by the Food and Agricultural Organisation of the United Nations (FAO) provides a fantastic resource in a world of increasing scrutiny on environmental impacts of farming. This model, as its name indicates, is a global assessment model and that is where its strengths ultimately lie: at a global level it is predicted to have an error of ±18%. This is an acceptable level of accuracy at a global level, given that it has to balance the highly industrialised farming of Northern America and Western Europe with the extensive/subsistence farming practices of Africa, Asia and South America. The accuracy of the model changes by production and animal species, with dairy cattle emissions estimated to be 30% greater in GLEAM than comparable Life Cycle Analysis (LCA) studies (MacLeod et al. 2018). The data used in the model is extensive and well-curated, however, the data held in the model is only current to 2010. Any model can be used in an incorrect manner if defaults are not properly understood, leading to greater variances in some regions or nations.

The use of greenhouse-gas modelling is becoming a battlefield for the promotion or deriding of industry impacts. Recent examples of this have been in the publications by GRAIN and the Institute for Agriculture and Trade Policy (IATP). The publications titled *"Emissions impossible: How big meat and dairy are heating up the planet"* in 2018 (GRAIN & IATP 2018) and *"Milking the Planet: How big dairy is heating up the planet and hollowing rural communities"* in 2020 (IATP 2020) are two examples of how to poorly apply an emissions intensity factor. These two studies claim to have used GLEAM to estimate the emissions associated with various animal-sourced nutrition companies, including Fonterra and Dairy Farmers of America.

External estimation of Fonterra's emissions footprint

There are two publicly available sources of data from GLEAM, both of which are based on 2010 reference data from the FAO.

These are: **GLEAM_Data_public_release.xls** – an Excel file containing a set of outputs from GLEAM at a regional level (e.g., Oceania) grouped by major commodity class (e.g., milk) and **GLEAM i** – a cut down version of the GLEAM model that enables users to explore the impact of changing farming system parameters on production and emissions at a country and sector level.

It appears that both the GRAIN and IATP studies have used regional data (for Oceania) from the public data release in generating their estimates (Table 1) by multiplying an emissions factor of 1.88 by an assumed milk production intake for Fonterra (between 22 and 24 billion litres). This produces estimates of over 40 million tonnes of CO_2 equivalents, which are significantly greater than the total of all New Zealand agricultural emissions in 2010 (36 million tonnes CO_2eq) or 2018 (38 million tonnes CO_2eq). By contrast a bottom-up estimate from full LCA studies gives a figure of 19 million tonnes CO_2eq.

The top-down approach is a grossly inaccurate method to determine emissions, as it does not include country-specific emissions, factors resulting from animal diet, production standards, or other environmental factors. Even in a bottom up LCA, which starts with a farm level and works up to industry as opposed to industry level down, there can be bias introduced using assumed or default values. An example of this is the study recently published by the World Resources Institute (Wirsenius et al. 2020), where two assumptions on replacement rate and milk production increase emissions intensity by 55%.

Table 1 Total emissions and emission intensity estimations from different sources.

Measure	Grain & IATP 2018	IATP 2020	Wirenius 2020	Ledgard 2020	MFE 2020
Total (millions of tonnes of CO_2)	41.5	44.0	34.2*	19.0	18.85‡
Intensity (kg CO_2 per kg FPCM¥)	1.88	1.88	1.40	0.78	NA

*Calculated from Fonterra 2020 data ((1.40/0.91)*22.2 billion litres FPCM; ¥FPCM (Fat and Protein Corrected Milk)

‡ Data extrapolated from Dairy being 50% of New Zealand agricultural emissions.

Differences in emission estimates

The following are some of the key differences between the emissions estimates from GLEAM and the Fonterra LCA. There is not a single key point on which the models differ, due in part to the use of generated data vs current data, rather it is the accumulation of a number of differences – each of these are of the order of 20% in the respective areas of the model – that contributes to an eventual figure from GLEAM that is more than double the estimates from the Fonterra LCA or the MFE (2020) greenhouse gas inventory.

System Boundary

The GLEAM model includes a provision for emissions – including transport, processing and packaging, that occur after the farm gate. These CO_2 emissions, which comprise 15.1% of the total emissions in the GLEAM result are not included in the Fonterra LCA and account for 0.28 kg CO_2eq/kg FPCM.

Dairy herd models

GLEAM has a well-constructed herd model that considers the various sub-populations required as part of a dairy herd, including milking cows, growing replacement females, bulls, replacement males, and surplus male and female animals. There is a built-in assumption that all surplus animals are raised for beef, and the model allocates feed to these animals. Although this does not appear to directly influence the GHG calculations for the milk-producing component of the herd, an indirect result is that GLEAM assumes that all cattle in New Zealand come from the dairy herd. This effectively folds the beef industry into the dairy industry masking any beef-industry-specific emissions.

Milk / meat allocation model

The FAO has a heavy focus on protein as a nutrient, and emissions intensity results from GLEAM are presented as kg CO_2eq/kg protein. The emissions factor of 1.88 used by GRAIN/IATP is obtained by multiplying the value given in the GLEAM public data set by a 3.3% protein concentration.

This emphasis on protein also connects through into the method used to allocate emissions between the milk and meat outputs of the core dairy herd (milking cows, bulls, and replacements). The result of this is that approximately 96% of the emissions are allocated to milk production. The Fonterra LCA uses the IDF/ISO (International Dairy Federation/International Organization for Standardization) allocation method, which for 2018 resulted in allocation of 85% of the emissions to milk, and the remainder to meat. This difference accounts for a change of 0.19 kg CO_2eq/kg FPCM.

Global warming potential factors for methane

GLEAM uses a value of 34 kg CO_2eq/kg CH_4 for the 100-year warming potential of methane (FAO 2018). The Fonterra LCA uses a biogenic methane factor of 27.75 kg CO_2eq/kg CH_4 (Ledgard et al. 2012). This accounts for 0.146 kg CO_2eq/kg FPCM between the models. The New Zealand GHG Inventory currently uses a methane factor of 25 kg CO_2eq/kg CH_4 (Foster et al. 2007) and the latest IPCC AR-6 estimate is 27.2 kg CO_2eq/kg CH_4.

Global warming potential factors for nitrous oxide

GLEAM uses a value of 298 kg CO_2eq/kg N_2O for the 100-year warming potential of nitrous oxide (FAO 2018). The Fonterra LCA uses 265 kg CO_2eq/kg N_2O accounting for 0.048 kg CO_2eq/kg FPCM.

Feed quality and enteric emissions

The quality of feed has a major impact on the enteric emissions of ruminants. Methane production is broadly proportional to the dry-matter intake, regardless of how much energy the animal can extract from this feed. GLEAM uses global values for the energy and nitrogen content, and digestibility, of grass – the primary source of nutrition for New Zealand dairy cattle – of 18 MJ/kg dry matter (DM), 22 g N/kg DM and 66% digestibility. The New Zealand inventory is based on specific values that vary by both region and throughout the year, with average values of 11.2 MJ/kg DM, 35 g N/ kg DM, and 76.6% digestibility.

After adjustment to use the IDF allocation method the GLEAM results come down to 0.587 kg CO_2eq/kg FPCM. This is 0.056 kg CO_2eq/kg FPCM higher than the Fonterra LCA.

Country-specific emissions factors for nitrous oxide emissions

New Zealand has developed country-specific factors for nitrous oxide emissions from agriculture that have been internationally peer reviewed and accepted by the Interim Climate Change Committee for use within the national greenhouse-gas inventory. The use of these factors results in N_2O emissions being 59% lower than those calculated with the global defaults given the same amount of nitrogen in faeces and urine. This change in emissions factor causes a difference of 0.253 kg CO_2eq/kg FPCM.

Emissions associated with electricity use

New Zealand has a greater proportion of renewable electricity generation compared with most countries which manifests in a very low contribution from electricity to the Fonterra LCA (0.010 kg CO_2eq/kg FPCM). However, the GLEAM model assumes an emissions factor of ~0.067 kg CO_2eq/kg FPCM in New Zealand and ~0.098 kg CO_2eq/kg FPCM in Australia. The greater-than-average protein content of New Zealand milk leads to the final GLEAM figure being 0.054 kg CO_2eq/kg FPCM greater than the IDF/ISO allocation method, a difference of 0.044 kg CO_2eq/kg FPCM.

Different assumptions on effluent treatment

Both GLEAM and the New Zealand GHG Inventory assume that approximately 5% of the manure from the dairy herd is captured and stored in anaerobic lagoons where it decays to release methane at a much greater rate than the manure deposited on paddocks (Marrow & Gibbs 2021). The Fonterra LCA uses additional information on the range of effluent-treatment systems (weeping wall, lagoon, covered lagoon, etc.,) that results in a much lower figure (note - The data set on which these assumptions is based is now ageing and needs to be updated). This is a difference of 0.055 kg CO_2eq/kg FPCM.

Summary of differences

Table 2 contains a summary of the differences identified above, and the extent to which they contribute to the overall difference between the results. The first four factors comprising approximately 65% of the accounted differences relate to generic parameters of the model that impact on the analysis for all countries – although there is a small effect of herd characteristics on the milk/meat allocation. The remaining 35% of the difference reflects country specific factors particular to New Zealand dairy farming.

Table 2 Summary of factor differences between GLEAM model defaults and New Zealand Specific emission factors.

Factor	Impact (kg CO_2 per kg FPCM*)	% of difference
GRAIN/IATP use of GLEAM	1.880	
System boundary	0.280	27.3%
Milk/meat allocation model	0.190	18.5%
Global warming potential factor for methane	0.146	14.2%
Global warming potential factor for nitrous oxide	0.048	4.7%
Feed quality and enteric emissions	0.056	5.4%
Country specific emissions factors for nitrous oxide	0.210	20.4%
Emissions associated with electricity use	0.044	4.2%
Effluent treatment	0.055	5.3%
Value after accounting for known causes	0.854	
Fonterra life-cycle analysis	0.776	
Unexplained difference	0.078	

*FPCM – Fat and Protein Corrected Milk

Conclusion

From the comparisons undertaken in this review the accuracy of a global model can be concluded to be just that, accurate at a global level to the extent that the available or used data allows. When comparisons are to be conducted at a regional, country or industry level, the broad accuracy of a global model introduces broad inaccuracies causing significant differences between analyses. This reinforces the requirement to use a specific tool or methodology for specific analyses. Inappropriate use of models may distort views of the role of various animal-based supply chains in the provision of sustainable nutrition.

References

FAO. 2018. Global Livestock Environmental Assessment Model Version 2.0. http://www.fao.org/fileadmin/user_upload/gleam/docs/GLEAM_2.0_Model_description.pdf [accessed 6/7/2020]

Fonterra 2019. Fonterra Sustainability Report: year ending 31 July 2019. https://view.publitas.com/fonterra/sustainability-report-2019/page/1 [accessed 6/7/2020].

Forster P, Ramaswamy V, Artaxo P, Berntsen T, Betts R, Fahey DW, Haywood J, Lean J, Lowe DC, Myhre G, Nganga J, Prinn R, Raga G, Schulz M and Van Dorland R, 2007: Changes in Atmospheric Constituents and in Radiative Forcing. In: Climate Change 2007: The Physical Science Basis. Contribution of Working Group I to the Fourth Assessment Report of the Intergovernmental Panel on Climate Change [Solomon S, Qin D, Manning M, Chen Z, Marquis M, Averyt KB, Tignor M, Miller HL (eds.)]. Cambridge University Press, Cambridge, United Kingdom and New York, NY, USA.

GRAIN and IATP. 2018. Emissions Impossible: How big meat and dairy are heating up the planet. https://www.grain.org/article/entries/5976-emissions-impossible-how-big-meat-and-dairy-are-heating-up-the-planet [accessed 6/7/2020]

IATP 2020. Milking the planet: How big dairy is heating up the world and hollowing rural communities. https://www.iatp.org/milking-planet [accessed 6/7/2020]

Ledgard SF, Falconer SJ, Abercrombie R, Philip G, Hill JP 2020. Temporal, spatial and management variability in the carbon footprint of New Zealand milk. Journal of Dairy Science 103: 1031-1046.

MacLeod MJ, Vellinga T, Opio C, Falcucci A, Tempio G, Henderson B, Makkar H, Mottet A, Robinson T, Steinfeld H, Gerber PJ 2018. Invited review: A position on the Global Livestock Environmental Assessment Model (GLEAM). Animal 12: 383-397.

Marrow B, Gibbs J 2021. Methodology for calculation of New Zealands agricultural greenhouse gas emissions – Version 7. Ministry for Primary Industries. https://www.mpi.govt.nz/dmsdocument/13906-Detailed-methodologies-for-agricultural-greenhouse-gas-emission-calculation- [accessed 6/8/2020]

Ministry for the Environment (MFE) 2020. New Zealands Greenhouse Gas Inventory. https://www.mfe.govt.nz/sites/default/files/media/Climate%20Change/new-zealands-greenhouse-gas-inventory-1990-2018-vol-1.pdf [accessed 6/7/2020]

Wirsenius S, Searchinger T, Zionts J, Peng L, Beringer T, Dumas P 2020. Comparing the life cycle greenhouse gas emissions of dairy and pork systems across countries using land-use carbon opportunity costs. World Resources Institute. https://www.wri.org/publication/comparing-life-cycle-greenhouse-gas-emissions-dairy-pork-systems [accessed 6/7/2020]

Volatile fatty acids (VFA) in rumen samples collected pre-feeding had a stronger relationship with methane emissions in lambs fed graded levels of forage rape than VFAs in post-feeding samples

MM Della Rosa*, PH Janssen, P Reid, E Sandoval, D Luo, D Pacheco and A Jonker

Grasslands Research Centre, AgResearch Ltd, Tennent Drive, Private Bag 11008, Palmerston North 4442, New Zealand

**Corresponding author:* Email: milagros.dellarosa@agresearch.co.nz

Abstract

To evaluate the effect of sampling time on volatile fatty acid (VFA) concentrations and their ability to predict methane yield (g/kg DMI), wethers (n=14/treatment) were fed a ryegrass-based pasture substituted with 0, 25, 50, 75 and 100% forage rape (FR) on a dry-matter basis. Rumen samples were collected before morning feeding (pre-feeding) and 2 h after feeding (post-feeding) and VFA concentrations were measured. Methane emissions and dry-matter intake (DMI) were measured in respiration chambers. Two VFA ratios were calculated: **R1**= [acetate + butyrate] / [propionate + valerate], **R2** = [acetate + butyrate + caproate] / [propionate + valerate + (3 × propanol)].

The relationship of VFA-ratios and methane yield was determined using linear and quadratic regression. VFAs were statistically analysed including dietary treatment as fixed effect and sampling timing as repeated measurement. Pre-feeding R1 included in a quadratic regression was a better and simpler methane-yield predictor (RMSE 1.83) than was R2. In pre-feeding samples, acetate proportion was greater than that in post-feeding samples in all diets except in FR25% ($P<0.01$). Butyrate proportion before feeding was lower than the proportion after feeding in diets with 50% or more FR while propionate was lower in FR25% and higher in FR100% than in post-feeding ($P<0.01$). Considering sampling time can improve the predictive power of a methane proxy.

Keywords: brassica; methane yield; rumen fermentation

Introduction

Volatile fatty acid (VFA) concentrations in rumen can be used as a proxy to infer changes in rumen fermentation related to methane (CH_4) emissions (Williams et al. 2019) and rank individual sheep by their emissions (Jonker et al. 2021). VFAs from rumen samples collected pre- or post-feeding have been used indistinctly in several studies to make inferences about CH_4 emission as well as individual VFAs or ratios of different VFAs (Jonker et al. 2016; Williams et al. 2019). However, because VFA proportions change during the day (van Lingen et al. 2017), their value as a CH_4 proxy may depend on sampling timing relative to feeding.

Post-feeding rumen conditions are more dynamic over time, but production of more-reduced VFAs increases rapidly due to changes in rumen-liquid dissolved-hydrogen (H_2) pressure (van Lingen et al. 2016; van Lingen et al. 2017). Some of these electron sinks, such as caproate and alcohols, are not commonly included in stochiometric "H_2" balances to predict CH_4 emissions from rumen samples. Changes in rumen fermentation occur after feeding regardless of diet type because of an increase in fermentable substrates, but the extent of those changes is diet related. Some changes in VFA concentrations are related to microbial responses to excess H_2 (Janssen 2010).

Sheep fed forage rape were consistently found to produce less CH_4 compared to those fed ryegrass pasture (Sun et al. 2012; Sun et al. 2015a; Sun et al. 2015b), and in an earlier experiment, forage rape decreased methane emission in a linear fashion (Sun et al. 2015b). It is expected that feeding graded levels of forage rape will result in changes in VFA proportions and consequently CH_4 emissions, which would enable exploring the effect of time of feeding on VFA as proxies. The first aim of this study was to compare VFA proportions, including minor end-products, in rumen samples collected pre-feeding and post-feeding, and determine their relationship with CH_4 yield. The second aim was to evaluate the variation of individual end-products across diets and sampling times.

Materials and methods

Animals and diets

Seventy 10-month old Romney wether lambs (38.8±1.8 kg), and spare animals, were randomly assigned to one of five dietary treatments: (on dry-matter (DM) basis) 100% ryegrass-based pasture (FR0%), 75% ryegrass + 25% forage rape (FR25%), 50% ryegrass + 50% forage rape (FR50%), 25% ryegrass + 75% forage rape (FR75%) and 100% forage rape (FR100%). The animals were transitioned onto and adapted to grazing forage rape (*Brassica napus cv. Titan*) at AgResearch's Aorangi farm over a three-week period. The lambs were then transported to AgResearch Grasslands Research Centre (Palmerston North), for adaptation to indoor group housing and feeding of cut forages for at least three weeks prior to the measurement phase. The experiment was reviewed and approved by the Grasslands Animal Ethics Committee (Palmerston North, NZ; approval number 14990), and animals were cared for according to the AgResearch Code of Ethical Conduct.

Dry-matter intake and CH_4 emission measurements

The measurement phase consisted of two days of housing in individual crates for acclimatisation and two days in individual respiration chambers for CH_4 emission measurements. Dry-matter intake (DMI) was determined during the two days of measurements in the respiration chambers by calculating the difference in the dry matter of feed offered and refused.

Twenty-four respiration chambers, connected to three SERVOMEX 4900 (Servomex Group Ltd., East Sussex, UK) analysers, i.e., eight chambers per cluster, were available for CH_4 measurements. The 70 trial animals were allocated to three batches of 24, 24 and 22 animals. The three batches went into chambers in three consecutive weeks. Full details of chambers and procedures were described by Jonker *et al.* (2016). Methane emission per day was divided by unit of DMI to calculate CH_4 yield.

Rumen sampling

Two rumen samples were collected by oesophageal stomach tubing, the first one approximately 2 h after morning feeding on the day before lambs entered the respiration chambers, and the second sample was collected two days later before morning feeding when the lambs left the respiration chambers. A subsample of 1.8 mL was placed in ice, taken back to the laboratory within ~1 hour, centrifuged at 21,000 × *g* for 10 min at 4°C, before 900 mL of the supernatant was transferred to a new tube containing 100 μL of internal standard (19.87 mM 2-ethylbutyric acid in phosphoric acid 20%, vol/vol) and stored at −20°C until VFA and alcohol analysis. After at least 24 h storage, samples containing the internal standard were thawed and centrifuged at 21,000 × *g* for 10 min at 4°C. Molar concentrations of VFAs (acetate, propionate, butyrate, valerate and caproate) and alcohols (ethanol and propanol) were determined by gas chromatography (Tavendale et al. 2005).

Data calculations

The proportion of each individual VFA and alcohol were expressed as mole percentage of total mol fermented hexoses (FH) equivalent. The mole hexose equivalents for each product are acetate, 0.5; propionate, 0.5; butyrate, 1.0; valerate, 1.0; caproate, 1.5; ethanol, 0.5; propanol, 0.5. Additionally, two different VFA ratios were calculated from the molar concentrations of the products:

R1= [acetate + butyrate] / [propionate + valerate],

R2 = [acetate + butyrate + caproate] / [propionate + valerate + (3 × propanol)].

The ratios represent electron ("H_2") production/incorporation for each fermentation product, according to principles of rumen stoichiometry (Wolin 1960). The coefficients for acetate, butyrate and caproate are all 1.0, in accordance with long-established practice where every four electrons (two × "H_2") is given a value of 1.0 in the numerator, based on the product formation from hexoses. The coefficients for end products that result in net electron incorporation 1.0 for propionate and valerate, and 3.0 for propanol, again in accordance with established practice where every two electrons (one × "H_2") is given a value of 1.0 in the denominator. Ethanol formation does not result in net electron production/release, so was not included in the VFA ratio.

Statistical analyses

Data analyses were performed using the 'predictmeans', 'lme4' packages in statistical software R 4.0.3 (R Core Team, 2020). Methane yield (g/kg DMI) was analysed with dietary treatment as fixed effect, batch and chamber nested within cluster as random effects. Fermentation end-product data were analysed using a repeated-measurements model that included dietary treatment, time, and their interaction as fixed effects, and lamb ID nested within batch as a random effect. A multiple comparison was performed on the modelling results with P values adjusted by the "BH" method (Benjamini & Hochberg 1995). Significant mean differences were declared at $P<0.05$.

To evaluate relationship between R1or R2 calculated for pre-and post-feeding samples and CH_4 yield separately, a polynomial (up to quadratic level) regression with batch and chamber nested within cluster as random effects was used in the analysis. In this case, the CH_4 yield was averaged over two days of measurements.

Results

Relationship between end-product hydrogen balances and CH_4 yield

R1 ratios were slightly greater than those calculated with R2, regardless of sampling time. Pre-feeding R1 and R2 had stronger relationship with CH_4 yield than post-feeding R1 and R2 ratios (Table 1). Adding R1 or R2 ratios as quadratic terms in the model to predict CH_4 yield further reduced the RMSE, especially for pre-feeding samples (Table 1). CH_4 yield appeared to be overestimated when the measured CH_4 yield was less than 17 g/kg DMI and underestimated when measured CH_4 yield was greater than 17 g/kg DMI regardless of the VFA ratio or sampling time chosen.

Rumen sampling timing and diet interactions

There was an interaction between diet and sampling time ($P\leq0.01$) for all fermentation end-products, except valerate ($P=0.64$) (Table 2). Total calculated fermented hexoses were greater post-feeding than pre-feeding in diets which included 50% or less forage rape ($P<0.05$), while there was no difference between pre-feeding and post-feeding total fermented hexoses in lambs fed 75 and 100% forage rape ($P>0.05$).

The post-feeding acetate proportion was less than that pre-feeding for all dietary treatments ($P<0.05$). Post- and pre-feeding butyrate proportions were similar in diets containing 0 or 25% forage rape but were greater post-feeding in diets containing 50% or more forage rape (p$P<0.05$). The post-feeding caproate proportion was greater than pre-feeding only for 100% forage rape ($P<0.05$). The post-feeding propionate proportion was greater than that obtained pre-feeding when the diet contained 25% forage rape but was less if the diet contained 100% forage rape ($P<0.05$) without differences between sampling time for the other dietary treatments. The post-feeding valerate proportion was greater than that obtained pre-feeding

Table 1 Intercept (βo), linear ($\beta 1$) and quadratic terms ($\beta 2$) (coefficients ± s.e.) in regression equations to predict CH_4 yield using volatile fatty acid ratios as independent variables.

Sample	Ratio	Term	β_0	β_1	β_2	RMSE	AIC	BIC
Pre-feeding	R1	Linear	7.55 ± 1.21	2.73 ± 0.33		2.22	327.55	341.04
Pre-feeding	R1	Linear + Quadratic	-5.82 ± 3.73	11.23 ± 2.33	-1.26 ± 0.35	1.83	317.73	333.48
Pre-feeding	R2	Linear	7.41 ± 1.21	2.81 ± 0.34		2.18	326.79	340.29
Pre-feeding	R2	Linear + Quadratic	-5.86 ± 3.73	11.36 ± 2.35	-1.29 ± 0.35	1.83	317.01	332.75
Post-feeding	R1	Linear	11.39 ± 1.75	1.78 ± 0.53		2.96	362.56	376.06
Post-feeding	R1	Linear + Quadratic	-6.76 ± 6.62	12.85 ± 3.94	-1.61 ± 0.57	2.80	356.18	371.92
Post-feeding	R2	Linear	10.99 ± 1.77	1.93 ± 0.55		2.93	361.37	374.86
Post-feeding	R2	Linear + Quadratic	-8.23 ± 6.90	13.91 ± 4.21	-1.79 ± 0.62	2.77	354.60	370.34

RMSE: root mean square error, AIC: Akaike information criterion, BIC: Bayesian information criterion
R1: [acetate + butyrate]/ [propionate + valerate]
R2: [acetate + butyrate + caproate] / [propionate + valerate + (3 × propanol)]

Table 2 Mean methane and hydrogen yield (g/kg dry matter intake), and volatile fatty acids, ethanol and propanol as a proportion of total fermented hexoses (FH) in rumen samples collected pre-and post- feeding in sheep fed ryegrass-based pasture substituted with 0, 25, 50, 75 and 100% of forage rape (FR0%, FR25%, FR50%, FR75%, FR100%)

Measure[1]	FR0%	FR25%	FR50%	FR75%	FR100%	SED	diet	time	diet × time	*r* pre vs. post	*r* CH_4 yield vs end products
Dry matter intake [kg/d]	0.98^{c}	1.01^{bc}	1.08^{ab}	1.10^{a}	1.06^{abc}	0.03					
CH_4 yield [g/kg dry matter intake]	19.18^{a}	18.65^{a}	18.09^{ab}	17.02^{b}	12.67^{c}	0.66	<0.01				
Total hexoses fermented (FH) [mM][2]											
pre-feeding	26.00^{e}	29.08^{e}	34.18^{de}	40.97^{abc}	51.52^{a}	2.73	<0.01	<0.01	0.01	0.37	-0.41
post-feeding	38.73^{cd}	43.71^{abc}	47.68^{ab}	45.46^{abc}	51.44^{a}						-0.34
Acetate [% of FH]											
pre-feeding	62.89^{a}	59.07^{b}	56.07^{b}	53.27^{d}	44.69^{f}	1.10	<0.01	<0.01	<0.01	0.80	0.61
post-feeding	59.71^{b}	54.23^{cd}	48.54^{e}	45.76^{f}	41.97^{g}						0.66
Propionate [% of FH]											
pre-feeding	15.61^{e}	17.16^{de}	19.41^{bcd}	18.80^{bcd}	22.28^{a}	0.81	<0.01	0.89	<0.01	0.23	-0.71
post-feeding	17.08^{cde}	19.42^{bc}	20.28^{ab}	20.09^{ab}	16.08^{e}						-0.03
Butyrate [% of FH]											
pre-feeding	18.67^{f}	20.21^{ef}	21.37^{def}	24.08^{cd}	26.30^{bc}	0.99	<0.01	<0.01	0.01	0.61	-0.14
post-feeding	19.78^{ef}	22.64^{de}	27.18^{b}	28.89^{b}	32.22^{a}						-0.44
Valerate [% of FH]											
pre-feeding	1.77^{d}	2.09^{d}	1.94^{d}	2.14^{d}	3.97^{b}	0.33	<0.01	<0.01	0.64	0.54	-0.63
post-feeding	2.49^{cd}	2.41^{cd}	2.68^{cd}	3.25^{bc}	4.94^{a}						-0.68
Caproate [% of FH]											
pre-feeding	0.69^{cd}	0.92^{bcd}	0.87^{bcd}	1.37^{bc}	1.54^{b}	0.23	<0.01	0.48	<0.01	0.56	-0.15
post-feeding	0.59^{d}	0.81^{bcd}	0.77^{cd}	1.08^{bcd}	2.51^{a}						-0.42
Ethanol [% of FH]											
pre-feeding	0.31^{d}	0.40^{cd}	0.27^{d}	0.29^{d}	1.06^{b}	0.12	<0.01	<0.01	<0.01	0.72	-0.44
post-feeding	0.26^{d}	0.35^{d}	0.47^{cd}	0.73^{bc}	2.10^{a}						-0.53
Propanol [% of FH]											
pre-feeding	0.09^{e}	0.10^{de}	0.09^{e}	0.08^{e}	0.19^{b}	0.01	<0.01	<0.01	<0.01	0.36	-0.48
post-feeding	0.07^{e}	0.09^{e}	0.14^{cd}	0.16^{bc}	0.22^{a}						-0.56
R1											
pre-feeding	4.40^{a}	3.85^{b}	3.32^{df}	3.36^{cde}	2.59^{h}	0.18	<0.01	<0.01	<0.01	0.50	0.70
post-feeding	3.81^{bc}	3.21^{defg}	2.93^{egh}	2.86^{fgh}	3.33^{defg}						0.38
R2											
pre-feeding	4.33^{a}	3.81^{b}	3.30^{d}	3.34^{cde}	2.55^{g}	0.17	<0.01	<0.01	<0.01	0.51	0.73
post-feeding	3.78^{bc}	3.19^{def}	2.89^{efg}	2.81^{fg}	3.25^{def}						0.67

SED: average standard error of the differences
[1] **R1**: [acetate + butyrate]/ [propionate + valerate], **R2**: [acetate + butyrate + caproate] / [propionate + valerate + (3 × propanol)],
[2]FH = 0.5 × acetate + 0.5 × propionate + butyrate + valerate + 1.5 × caproate + 0.5 × ethanol + 0.5 × propanol
r pre vs. post: correlation coefficient between sampling time within each rumen fermentation end-product.
r CH_4 y vs end products: correlation coefficient between CH_4 yield and each fermentation end-product measured at each sampling time
[abcdefgh] letter-based representation of pairwise comparisons at significant level '0.05'

($P<0.05$) only when diet contained 100% forage rape. Post-feeding proportions of ethanol were greater than that obtained pre-feeding in diets containing >75% forage rape and for propanol when >50% forage rape in the diet. The pre-feeding VFA ratio calculated either with the R1 or R2 equation was less than the post-feeding VFA ratio in diets containing up to 75% of forage rape ($P<0.05$). However, the post-feeding VFA ratio was greater than the pre-feeding value ($P<0.05$) when the diet contained 100% forage rape.

Acetate, butyrate and ethanol proportions had medium-high ($r \sim 0.70$) correlations between pre- and post-feeding samples, while propionate and propanol had weak ($r < 0.40$) correlations (Table 2).

Discussion

The main finding of the current study was that VFA ratios (R1 and R2) in rumen liquid collected pre-feeding were a stronger predictor of CH_4 yield in lambs fed graded levels of forage rape than were VFA ratios (R1 and R2) from rumen samples collected post-feeding. Propionate proportion behaved unpredictably in post-feeding rumen samples, while the other VFAs changed in the direction that was reported in previous studies in sheep fed forage rape (Sun et al. 2012; Sun et al. 2015a). The finding that the pre-feeding VFA ratio was a stronger predictor of CH_4 yield than in post-feeding samples is consistent with findings of a recent meta-analysis of data from sheep fed with a range of brassica crops in New Zealand (He et al. 2020). This meta-analysis showed that pre-feeding VFA ratio explained 42% of variability in CH_4 yield, while post-feeding VFA ratio was not selected in the model to explain CH_4 yield.

The low predictive value of post-feeding VFA ratios may be a consequence of a greater variability in VFA concentrations after feeding due to rapid changes in fermentation rate and consequently rumen environment after feeding (van Lingen et al. 2017). The main VFA produced in all diets was acetate, as its production is thermodynamically more efficient than production of other fermentation end-products (van Lingen et al. 2016). However, H_2 concentrations in the rumen increase after feeding due to rapid feed fermentation. This results in an increased concentration of soluble hydrogen that forces alternative electron-accepting pathways to maintain re-oxidation of reduced electron carriers (van Lingen et al. 2016). Propionate is generally the main electron sink, but also other electro-accepting pathways increase when CH_4 yield decreases, resulting in formation of products such as butyrate, valerate, caproate, ethanol and propanol (van Lingen et al. 2017; Ungerfeld, 2020). Propionate was the most variable VFA across the two sampling times and diets which could indicate that the threshold for switching to this less-energy-yielding pathway is relatively higher and a more-permanent switch than the condensation of existing VFAs to produce butyrate, caproate and valerate to remove some of the hydrogen created. Feeding FR100% resulted in the lowest CH_4 yield across diets and in the greatest propionate proportion in pre-feeding samples. However, lambs fed FR100% had the lowest propionate proportion post-feeding, which greatly affected the electron balances (as R1 and R2 VFA ratios) calculated post-feeding. These results were different from previous findings, where propionate proportion increased post-feeding in sheep fed forage rape (Sun et al. 2012; Sun et al. 2015a). The other electron sinks that appear in the R1 and R2 VFA-ratios calculations, valerate and propanol, represent together roughly between 2 and 6% of fermented hexoses equivalent across the diets, with the greatest level for FR100%. Including propanol in the R2 electron balance (VFA ratio) calculation generated numerically smaller values, but this change did not result in a better prediction of CH_4 yield.

Inclusion of VFA ratio as quadratic term in the model did improve the CH_4 yield prediction accuracy in pre- and post-feeding samples, but pre-feeding samples were still more accurate than post-feeding samples. In a previous study, the addition of VFA ratio as linear and quadratic terms to explain CH_4 yield in sheep fed lucerne silage substituted with graded level of maize grain and silage (Jonker et al. 2016) was also stronger than only the linear term, indicating the factors other than simple diet composition are playing a major role in the hydrogen balance of the rumen. In both our study and the study of Jonker et al. (2016), CH_4 yield changed non-linearly with increasing substitution of one feed with another feed, which might explain the improvement in CH_4 yield prediction when the VFA ratio is added as a quadratic term.

Conclusions

Pre-feeding VFA ratios had a stronger relationship with CH_4 yield in lambs fed graded levels of forage rape than did VFA ratios in post-feeding rumen samples. Propionate behaved unpredictably in post-feeding rumen samples with the smallest proportion in lambs fed 100%FR, while being greatest pre-feeding. Including caproate, ethanol and propanol in R2 did not improve its relationship with CH_4 yield compared to R1. Time of sampling relative to feeding and type of diet evaluated is a key factor to consider for developing CH_4 proxies from rumen metabolite concentrations.

Acknowledgements

This research was funded by the New Zealand Agricultural Greenhouse Gas Research Centre (NZAGRC) and Maria Della Rosa was financially supported by the New Zealand Government through the Global Research Alliance Livestock Emissions Network (LEARN) awards programme.

References

Benjamini Y, Hochberg Y 1995. Controlling the false discovery rate: a practical and powerful approach to multiple testing. Journal of the Royal statistical society: series B (Methodological) 57: 289-300.

He Y, Sun X, You P 2020. Animal, feed and rumen fermentation attributes associated with methane emissions from sheep fed brassica crops. Journal of Animal Physiology and Animal Nutrition 105: 210-218.

Janssen PH 2010. Influence of hydrogen on rumen methane formation and fermentation balances through microbial growth kinetics and fermentation thermodynamics. Animal Feed Science and Technology 160: 1-22.

Jonker A, Hickey S, Boma P, Woyimo Woju C, Sandoval E, MacLean S, García Rendón Calzada M, Yu W, Lewis S, Janssen PH, McEwan JC, Rowe S 2021. Individual-level correlations of rumen volatile fatty acids with enteric methane emissions for ranking methane yield in sheep fed fresh pasture. Animal Production Science 61: 300-305.

Jonker A, Lowe K, Kittelmann S, Janssen PH, Ledgard S, Pacheco D 2016. Methane emissions changed nonlinearly with graded substitution of alfalfa silage with corn silage and corn grain in the diet of sheep and relation with rumen fermentation characteristics in vivo and in vitro. Journal of Animal Science 94: 3464-3475.

R Core Team (2020). R: A language and environment for statistical computing. R Foundation for Statistical Computing, Vienna, Austria. URL https://www.R-project.org/.

Sun X, Henderson G, Cox F, Molano G, Harrison SJ, Luo D, Janssen PH, Pacheco D 2015a. Lambs fed fresh winter forage rape (Brassica napus L.) emit less methane than those fed perennial ryegrass (Lolium perenne L.), and possible mechanisms behind the difference. PloS one 10: e0119697.

Sun, X, Sandoval E, Pacheco D 2015b. BRIEF COMMUNICATION: Substitution of perennial ryegrass with forage rape reduces methane emissions from sheep. Proceedings of the New Zealand Society of Animal Production 75: 64-66.

Sun, X, Waghorn G, Hoskin S, Harrison SJ, Muetzel S, Pacheco D 2012. Methane emissions from sheep fed fresh brassicas (Brassica spp.) compared to perennial ryegrass (Lolium perenne). Animal Feed Science and Technology 176: 107-116.

Tavendale MH, Meagher LP, Pacheco D, Walker N, Attwood GT, Sivakumaran S 2005. Methane production from in vitro rumen incubations with Lotus pedunculatus and Medicago sativa, and effects of extractable condensed tannin fractions on methanogenesis. Animal Feed Science and Technology 123: 403-419.

Ungerfeld EM 2020. Metabolic hydrogen flows in rumen fermentation: principles and possibilities of interventions. Frontiers in Microbiology 11: 589.

van Lingen HJ, Edwards JE, Vaidya JD, van Gastelen S, Saccenti E, van den Bogert B, Bannink A, Smidt H, Plugge CM, Dijkstra J 2017. Diurnal Dynamics of Gaseous and Dissolved Metabolites and microbiota composition in the bovine rumen. Frontiers in Microbiology 8: 425.

van Lingen HJ, Plugge CM, Fadel JG, Kebreab E, Bannink A, Dijkstra J 2016. Thermodynamic driving force of hydrogen on rumen microbial metabolism: A theoretical investigation. PLoS One 11: e0161362.

Williams SRO, Hannah M, Jacobs J L, Wales WJ, Moate PJ 2019. Volatile fatty acids in ruminal fluid can be used to predict methane yield of dairy cows. Animals 9: 1006.

Wolin MJ 1960. A theoretical rumen fermentation balance. Journal of Dairy Science 43: 1452-1459.

Sheep divergently selected for methane yield showed differences in meat fatty acid composition

PL Johnson[1*], M Agnew[2], A Jonker[2], SM Hickey[3], JC McEwan[1], B Bryson[4], KG Dodds[1] and SJ Rowe[1]

[1]*Invermay Agricultural Centre, AgResearch Ltd, Private Bag 50034, Mosgiel, New Zealand;* [2]*AgResearch Grasslands Research Centre, Private Bag 11008, Palmerston North, New Zealand;* [3]*Ruakura Research Centre, AgResearch Ltd, Private Bag 3115, Hamilton, New Zealand;* [4]*Woodlands Research Station, AgResearch Ltd, Woodlands RD1, New Zealand*

**Corresponding author:* Email: tricia.johnson@agresearch.co.nz

Abstract

The purpose of this work was to investigate whether 18-month-old rams from the methane selection lines differed in their levels of soft tissue depth over the 12th rib (GR, as a proxy of carcass fatness) intramuscular fat (IMF), and the fatty acid composition of IMF. The rumen is the site of microbial biohydrogenation of consumed plant fatty acids and synthesis of unique microbial body fatty acids. It is known that there are differences in the microbiome and fermentation characteristics of sheep selected for and against methane production (methane selection lines). When comparing the high and low lines several differences were found. Rams from low-methane-yield-selection lines tended to have lower GR ($P<0.06$) than those from high line rams, but there was no evidence for a difference in levels of IMF between the lines ($P=0.77$). There were, however, many significant differences in the fatty acid composition (as a % of total fatty acids) of the IMF between the lines. The IMF from the low-methane-yield line had a significantly ($P<0.05$) higher proportion of branched-chain fatty acids, and fatty acids that are associated with the early stages of the biohydrogenation pathways in the rumen, specifically conjugated linoleic acid (CLA) and C18:1 t11 but with less C18:0 stearic acid (the end product of biohydrogenation) and a significantly higher proportion of C16:1 t9 which is produced through β-oxidation of C18:1 t11. These results provide insight into how differences in sheep from the methane-yield selection line may influence the fatty acid composition of their fat.

Keywords: Methane; sheep; intramuscular fat; fatty acid composition

Introduction

The outputs of New Zealand sheep-production systems are predominately meat and wool, although there are now several dairy-sheep operations. There is increasing consumer expectations of high eating and nutritional quality in the products that they consume, and that the environmental impact of the production system is sustainable. With meat quality in mind, there is increased emphasis on ensuring that the eating quality of lamb is consistently high, with recent breeding emphasis starting to place pressure on reducing levels of carcass subcutaneous fat, whilst maintaining or increasing levels of intramuscular fat (Johnson et al. 2018). Traditionally, there has not been an emphasis on intramuscular fat in New Zealand lamb but overseas research is increasingly demonstrating that minimum levels are required to ensure consumer acceptance of the product (Hopkins et al. 2006; Lambe et al. 2017; Realini et al. 2021).

With the increasing emphasis on environmental impacts, there is an initiative in New Zealand to breed sheep that produce less methane per unit of dry-matter intake (i.e., methane yield) and, therefore, reduce the contribution of sheep production to global warming (Pinares-Patiño et al. 2013). In the first instance, selection lines for high and low-methane-yield have been established (Pinares-Patiño et al. 2013). Published data from the selection lines has demonstrated that they tend to possess different rumen microbiomes (Kittleman et al. 2014). Within low-methane-yield sheep, there is evidence of two predominant ruminotypes, with one of the ruminotypes harbouring higher proportions of the bacteria *Quinella ovalis,* which is involved in propionate production, and correspondingly, these sheep have lower rumen acetate-to-propionate ratios (Kittleman et al. 2014). Other attributes of the low-methane-yield line are that the sheep appear to have lower subcutaneous fat levels compared with the high-methane-yield line (Elmes et al. 2014).

To date, there has been no investigation as to whether the sheep methane-selection lines differ in intramuscular fat. Since the 1980s it has been hypothesised that intramuscular fat preferentially utilises glucose (with propionate a pre-cursor) for its carbon source, whilst subcutaneous fat preferentially utilises acetate (Smith & Crouse 1984). This hypothesis is supported by the selection lines described above where at least one low-methane-yield ruminotype produces more propionate and is associated with less subcutaneous fat (Elmes et al. 2014; Jonker et al. 2020). It is, therefore, hypothesised that the low-methane sheep with their higher propionate will have higher levels of intramuscular fat. An additional consideration is whether the fatty acid profile of the intramuscular fat differs between the selection lines. The possibility of these differences is supported by results from a trial undertaken in dairy cattle (Rico et al. 2016) where variation in the fatty acid composition of the milk was shown to be significantly correlated to differences in methane production, with a model weighted for values of *cis* and *trans* isomers of C14:1, C17:1, C18:1 and C18:2 the most predictive.

The objectives of the current study were to investigate whether there are any differences in the total amount and composition of intramuscular fat or other carcass traits between 18-month-old rams sourced from the New Zealand methane-yield selection lines.

Materials and methods

The experiment adhered to the guidelines of the 1999 New Zealand Animal Welfare Act and AgResearch Code of Ethical Conduct. The current study was approved by the AgResearch Grasslands (Palmerston North, NZ) Animal Ethics committee.

Animals

The rams used in this study were from the methane-selection lines as described by Pinares-Patiño et al. (2013), with further descriptions of the specific sheep (Exp. 1) used reported by Jonker et al. (2021). The sheep were 2016-born rams that had been used in a series of trials to investigate the repeatability of methane measurements over time and on different forages, but always had undergone the same feeding and treatment conditions (Jonker et al. 2021). The series of trials concluded in December 2017. From that point up to slaughter in early March 2018, the sheep were grazed together with *ad libitum* access to a leafy perennial-ryegrass white-clover sward. A total of 65 rams were available made up of 33 from the high-methane-yield line (progeny from four sires) and 32 from the low-methane-yield line (progeny from five sires). These numbers are lower than those reported by Jonker et al. (2021) due to some sheep being removed for animal-health reasons. Those remaining were assessed to be healthy. Genotyping of the rams included assessment of a known mutation with the growth differentiation factor 8 gene (GFD8) which segregates within the population and is associated with carcass traits (Johnson et al. 2003).

Measurements

Two days before slaughter, the sheep were weighed immediately off pasture to obtain a full body weight (FBW). Approximately 24 hours before slaughter the rams were removed from feed and weighed 12 hours later, immediately prior to trucking to provide an empty body weight (EBW). The sheep were processed following standard protocols at Alliance Ltd Levin. In the chiller, immediately post-slaughter, carcass weight, depth of soft tissue at the GR site over the 12th rib at 110 mm from mid-line (CGRM, mm) and carcass measurement of buttock circumference (CBUTT, cm) were recorded. The day following slaughter, the right-side *M. longissimus lumborum* was collected and chilled to -1°C. In the research laboratory, a 2-cm slice from the posterior end of the *M. longissimus lumborum* was collected before freezing the sample for subsequent fatty acid analysis as described below.

M. longissimus lumborum fatty acid analysis

Before fatty acid composition analysis, all *M. longissimus lumborum* samples were freeze-dried (Cuddon Freeze Drier, Blenheim, New Zealand) and ground into a fine powder. Fatty acid concentrations were measured using trans-methylation of the fatty acids and quantification by gas chromatography (GC), as described by Agnew et al. (2019). The peak areas obtained from the GC were integrated using Shimadzu Lab-solution software (version 4.20, Shimadzu, Kyoto, Japan) in a post-run analysis. Peaks were identified by comparison of their retention times with those of commercial standard mixtures (FAME mix 37 components from Supelco Inc, Bellefont, PA, USA) and quantified by using the internal standard (C11:0) and theoretical FID response factors (Agnew et al. 2019). All known peaks were identified, and other peaks were each given a unique identifier. The quantity of individual fatty acids within the wet muscle was calculated using the freeze-dried factor as described by Agnew et al. (2019).

Analysis

Univariate linear mixed models (LMMs) were fitted for each trait using ASREML v4.1 (Gilmour et al. 2015). All models included fixed effects of GDF8 mutation status (double copy, single copy, no copy of the mutation) and methane yield line (high versus low). Carcass weight was fitted as a covariate for all traits (except for the analysis of carcass weight itself), and total intramuscular fat was fitted as a covariate for fatty acid composition traits (a model not fitting intramuscular fat as a covariate was investigated but the results did not change and are not presented). Wald test statistics were used to compare the phenotypes of the high-methane-yield rams with those of the low-methane-yield rams.

Results and discussion

The trait averages are provided in Table 1. Of note, the sheep used in this study are atypical of New Zealand sheepmeat destined for export. The sheep were chosen for this study on the basis that if differences in intramuscular fat were to occur, they would be more-evident in older sheep as marbling (visible intramuscular fat) is a late-maturing fat depot (Pethick et al. 2006). The percentage of intramuscular fat reported for young lambs slaughtered in New Zealand (Craigie et al. 2017) are approximately half those observed in this study.

Carcass traits

The results show that at the same carcass weight there were no significant differences in gross carcass traits between methane selection lines, but there was suggestive evidence ($P<0.10$) that the carcasses from the high-methane-yield sheep had higher GR values (soft tissue depth as an estimate of carcass fatness) and a higher DO% when using empty liveweight as the liveweight-ratio trait. The results for GR are consistent with those of Elmes et al. (2014) which showed low-methane-line ram lambs to have lower carcass fatness, but the difference for DO% is inconsistent with the literature where it is suggested that the low-methane-line animals have higher DO% due to having smaller rumen volumes (Bain et al. 2014).

Intramuscular fat levels and composition

The results show that at the same carcass weight there were no significant differences in intramuscular fat % between the lines.

The analysis of the *M. longissimus* data expressed each fatty acid as a percentage of total fat and fitted total fat as a covariate to account for differences in relationships

Table 1 Differences in carcass composition traits and fatty acid composition of intramuscular fat between lines of rams divergently selected for high- (n=33) or low- (n=32) methane-yield. The models for the fatty acid composition traits included total intramuscular fat (%) as a covariate. The local of the unknown fatty acids occur in sequence as to where in the chromatograph they were located.

Variable	Alternate Name	Mean	(Low – High) *100/ Low	SED	P value
Carcass weight (kg)		35.8	-5.0	0.95	0.06
GR (mm)		17.7	-11.2	1.03	0.06
Buttock circumference (cm)		76.4	-0.3	0.36	0.54
Full live weight (kg)		86.4	-3.1	1.86	0.16
Empty live weight (kg)		77.9	-2.4	1.65	0.27
Full dressing out (%)		41.4	-0.7	0.37	0.44
Empty dressing out (%)		45.9	-1.5	0.37	0.07
Intramuscular Fat (%)		4.6	-2.4		0.77
Fatty acid Composition (%)					
C10:0		0.12	-7.1	0.002	0.07
artifact 1		0.12	1.0	0.011	0.92
C12:0		0.07	2.7	0.004	0.63
iso C14		0.03	10.7	0.002	0.23
C14:0	Myristic acid	1.67	-0.1	0.053	0.97
iso C15		0.14	6.1	0.006	0.14
anteiso C15		0.16	12.8	0.008	0.02
C14:1		0.03	10.0	0.002	0.18
C15:0		0.33	7.2	0.013	0.07
artifact 2		0.81	5.7	0.044	0.29
iso C16		0.13	8.6	0.005	0.02
unknown 3		0.05	19.0	0.005	0.06
unknown 4		0.07	12.4	0.004	0.06
C16:0	Palmitic acid	22.91	-0.7	0.281	0.56
unknown 5		0.09	10.4	0.004	0.04
unknown 6		0.03	6.0	0.004	0.63
unknown 7		0.03	19.8	0.004	0.10
C16:1t		0.11	32.6	0.011	<0.01
iso C17		0.43	4.0	0.009	0.07
unknown 8		0.35	2.7	0.007	0.17
C16:1t9	Palmitoleic acid	0.91	6.5	0.023	0.01
anteiso C17		0.49	6.4	0.014	0.03
unknown 9		0.04	-8.3	0.004	0.41
C17:0	Heptadecanoic acid	0.90	2.1	0.016	0.22
artifact 3		0.88	-2.2	0.049	0.69
unknown 10		0.14	3.2	0.003	0.20
unknown 11		0.10	3.5	0.003	0.27
C17:1	Heptadecenoic acid	0.40	4.8	0.009	0.05
unknown 12		0.05	27.8	0.007	0.06
unknown 13		0.11	-0.3	0.007	0.97
C18:0	Stearic acid	18.53	-4.6	0.295	0.01
C18:1 t6-8		0.13	8.1	0.004	0.02
C18:1 t9		0.20	7.2	0.005	0.01
C18:1 t11	Vaccenic acid	3.05	21.2	0.239	0.01
C18:1c9	Oleic acid	36.50	-3.0	0.421	0.01
C18:1 c11	Trans-vaccenic acid	0.87	8.0	0.014	<0.001
C18:1 c12		0.15	14.7	0.005	<0.001
C18:1 c13		0.06	10.8	0.004	0.13
C18:1 c14/t16		0.34	9.3	0.004	0.00
unknown 14		0.14	7.3	0.010	0.02
unknown 15		0.09	1.2	0.004	0.80
C18:2tt n6		0.31	13.1	0.011	<0.001
unknown 16		0.17	12.9	0.004	<0.001
unknown 17		0.02	8.3	0.002	0.28
C18:2 t11/c15		0.28	22.5	0.001	<0.001
C18:2 n6	Linoleic acid	2.21	14.6	0.087	<0.001
C18:2 c9/c15		0.06	9.3	0.392	0.16
unknown 18		0.02	4.2	0.001	0.36
unknown 19		0.04	1.7	0.004	0.88
unknown 20		0.03	-2.3	0.002	0.76
C20:0		0.06	-17.5	0.008	0.21
C18:3 n3	Alpha linolenic acid	1.76	12.4	0.061	<0.01
C18:2c9, t11	Conjugated linoleic acid	1.08	22.6	0.075	0.00
unknown 21		0.03	19.8	0.003	0.06
unknown 22		0.03	5.6	0.002	0.53
unknown 23		0.02	2.3	0.002	0.73
unknown 24		0.20	21.0	0.009	<0.001
unknown 25		0.18	-20.2	0.008	<0.001
unknown 26		0.08	-4.9	0.004	0.16
C20:4 n6	Arachidonic acid	0.56	2.7	0.002	0.53
C22:1		0.02	-7.0	0.002	0.32
unknown 27		0.04	-23.3	0.010	0.34
unknown 28		0.03	-15.1	0.005	0.76
unknown 29		0.03	-11.6	0.003	0.66
unknown 30		0.05	-15.9	0.008	0.71
C20:5 n3	Eicosapentaenoic acid (EPA)	0.59	-5.3	0.031	0.52
C22:2		0.05	-14.5	0.007	0.05
unknown 31		0.02	6.3	0.002	0.45
C22:5 n3	Docosapentaenoic acid (DPA)	0.49	0.2	0.020	0.96
C22:6 n3	Docosahexaenoic acid (DHA)	0.19	-6.6	0.012	0.35
Saturated FA		44.6	-1.4	0.35	0.18
Monounsaturated FA		41.8	-1.0	0.45	0.12
Polyunsaturated FA		7.24	12.5	0.207	<0.001
Branched-chain FA		1.36	2.8	0.034	0.02

across the range of total fatty acids weights (long chain polyunsaturated fatty acids make up a higher proportion of total fatty acid weight at low levels of total intramuscular fat (Wood et al. 2008)). Although no differences in total intramuscular fat levels between the lines were observed, there were many significant differences between the lines in fatty acid composition expressed as percentages of the total intramuscular fat. Assignment of specific peaks within the GC profile relies on the peak being identified within the standards used. The standards commonly used by the laboratory to measure fatty acid profiles capture the major fatty acids, however several additional peaks were also generated, most likely isomers of other reported fatty acids with low levels. These were identified as unknown in this study but were retained for analysis, as Rico et al. (2016) had found many isomers that had relatively low concentrations but were significantly correlated to methane production. The analysis of the individual fatty acids resulted in many significant results. Of the known fatty acid peaks, the most significant ($P<0.01$) were C16:1t9, C18:1t9, C18:1t11, C18:1c11, C18:2tt n6, C18:2n6, C18:3n3 and CLA, all of which had higher levels in the low-methane-line sheep and C18:0 which was higher in the high-methane-line sheep. A series of the branched-chain fatty acids (anteiso C15, iso C16, anteiso C17) were significantly higher in the low-methane-line sheep, with iso C17 higher (but not significantly ($0.05<P<0.10$) in the low-methane-line sheep. There were several significant differences ($P<0.05$) between isomers of C18:1 and C18:2 that are less commonly reported in the literature including C18:1c12, C18:1c14/t16 and C18:2t11c15 and unknown compounds 30 and 31 which lie between conjugated linoleic acid (CLA) and C20:4n6. Other fatty acids significant at the $P<0.05$ level included C15:0, C18:1c9. Of the composite traits, the percentage of polyunsaturated fatty acids was significantly ($P<0.001$) higher in the low-methane-line sheep and the levels of branched-chain fatty acids were also significantly higher ($P<0.05$) in the low-methane sheep, whilst saturated fatty acid levels were significantly higher in the high-methane-line sheep.

The cluster of C16:1t9, C18:1t11 and CLA has previously been reported by Heck et al. (2012) who investigated the clustering of fatty acids in milk at the genetic and herd (environment) level. In their work, they suggested that C16:1t9 is formed during rumen conditions which also promote concentrations of the other two fatty acids in the cluster and as proposed by Luna et al. (2009) that C16:1t9 in milk could originate from β-oxidation of C18:1t11. Additionally, Kadegowda et al. (2013) reported that CLA can be endogenously synthesised from C16:1t9, with C18:1t11 as an intermediate, indicating that there is potentially a multi-directional relationship among this trio of fatty acids.

The major fatty acids in pastures are C16:1, C18:0, C18:1, C18:2 and C18:3. As reviewed by Kalac & Samkova (2010), the relative levels of the different fatty acids differ between pasture species, but that additionally, even within pasture species, relative levels are impacted by the stage of growth and level of maturity of the plant. Consistently the fatty acid with the greatest concentration is C18:3, which commonly makes up 50-70% of the total fatty acids in pasture species (Kalac & Samkova 2010). However, these fatty acids are then processed in the rumen. Both C18:2 and C18:3 are essential fatty acids in that they cannot be synthesised by sheep (human or ruminants) and can, therefore, only be sourced from the diet. Despite the high level of C18:3 in plant material, the resulting levels in sheep (muscle, fat and milk) are much lower as both C18:2 and C18:3 are further processed by the population of microbes in the rumen. The major pathways that feed-derived C18:2 and C18:3 enter are the biohydrogenation pathways, which if taken to completion, yield C18:0, however, this conversion is not always completed, and intermediaries of the process include CLAs and C18:1 isomers (Vahmani et al. 2015). The greater levels of polyunsaturated fatty acids in the low-methane-line indicate that the rumen environment in these sheep was less favourable to biohydrogenation.

The fatty acids obtained through the diet undergo a series of reactions in the rumen. The first is lipolysis which is the release of free fatty acids from the ester form in which they were consumed. The second is the process of biohydrogenation which is carried out by some, but not all, species of ruminal bacteria and involves a reduction in the number of double bonds in the carbon chain. The literature is continuing to expand as to the role that specific bacteria have in the biohydrogenation pathway. An early review by Jenkins et al. (2008) reported that of the 26-predominant species of ruminal bacteria, eleven metabolise C18:2 substantially, but through different mechanisms, with C18:1 t11 produced by 3 *Butyrivibrio* strains and *Clostridium proteoclasticum* strains, with C18:0 only produced by the later. They also summarised work that showed that *B. fibrisolvens* bacteria that produced CLA and C18:1 t11 were butyrate producers, but that the reverse relationship was not true, with all butyrate producers such as *Eubacterium* and *Clostridium* spp. producing substantial amounts of CLA and C18:1 t11. More recent studies have demonstrated that *B. hungatei* is also capable of converting C18:1 t11 to C18:0.

Conclusion

Overall, these results demonstrate that differences between the sheep methane-yield selection lines have resulted in the fat precursors outputs of the rumen processes being different. The results broadly fit with the known biohydrogenation pathway and fatty acids specifically synthesised in microbial bodies. Further research is needed to specifically link the results to the individual animal microbiome versus the summarised differences between the lines. However, if the results can be further validated it would suggest that reducing the methane yield of lambs could also improve the fatty acid profile, in terms of increased levels of polyunsaturated fatty acids and branched chain fatty acids, of the meat that they produce.

Acknowledgements

The methane yield selection line sheep used in this study were funded by the New Zealand Agricultural Greenhouse Gas Research Centre (NZAGRC) and the Pastoral Greenhouse Gas Research Consortium (PGgRc). Funding for measurements undertaken in this study was provided by AgResearch strategic funding.

References

Agnew MP, Craigie CR, Weralupitiya G, Reis MM, Johnson PL, Reis MG 2019. Comprehensive Evaluation of Parameters Affecting One-Step Method for Quantitative Analysis of Fatty Acids in Meat. Metabolites 18: 189.

Bain WE, Bezuidenhout L, Jopson NB, Pinares-Patiño CS, McEwan JC 2014. Rumen differences between sheep identified as being low or high-methane emitters. Proceedings of the Association of the Advancement of Animal Breeding and Genetics 20: 376–378.

Craigie C, Johnson P, Shorten P, Charteris A, Maclennan G, Tate M, Agnew MP, Taukiri KR, Stuart AD, Reis M 2017. Application of Hyperspectral imaging to predict the pH, intramuscular fatty acid content and composition of lamb *M. longissimus lumborum* at 24 h post mortem. Meat Science. 132: 19-28.

Elmes S, Bain W, Greer G, Hickey S, Young E, Pickering N, Rowe SJ, Knowler KJ, Pinares-Patiño CS, McEwan J 2014. BRIEF COMMUNICATION: An exploratory investigation of the effects of selection for divergence in methane emissions on rumen digesta and carcass traits in 8-month old sheep. Proceedings of the New Zealand Society of Animal Production 74: 142-144.

Gilmour AR, Gogel BJ, Cullis BR, Thompson R 2015. ASReml User Guide Release 4.1. VSN International Ltd, Hemel Hempstead, Hertfordshire, UK.

Heck JML, van Valenberg HJF, Bovenhuis H, Dijkstra J, van Hooijdonk ACM 2012. Characterization of milk fatty acids based on genetic and herd parameters. Journal of Dairy Research 79: 39-46.

Hopkins D, Hegarty R, Walker P, Pethick D 2006: Relationship between animal age, intramuscular fat, cooking loss, pH, shear force and eating quality of aged meat from sheep. Australian Journal of Experimental Agriculture 467: 879-884.

Jenkins T, Wallace R, Moate P, Mosley E 2008. BOARD-INVITED REVIEW: Recent advances in biohydrogenation of unsaturated fatty acids within the rumen microbial ecosystem. Journal of Animal Science 86: 397-412.

Johnson PL, Agnew M, Shorten PR, Reis M.M, Cullen N. Craigie CR, 2018. Preliminary genetic parameter estimates for intramuscular fat percentage in New Zealand lamb. Proceedings of the World Congress on Genetics Applied to Livestock Production. 11: 611.

Jonker A, Hickey S, Boma P, Woyimo WC, Sandoval E, MacLean S, García Rendón Calzada M, Yu W, Lewis S, Janssen PH, McEwan JC, Rowe S 2020. Individual-level correlations of rumen volatile fatty acids with enteric methane emissions for ranking methane yield in sheep fed fresh pasture. Animal Production Science 61: 300-305.

Kadegowda AK, Burns TA, Miller MC, Duckett SK. 2013. Cis-9, trans-11 conjugated linoleic acid is endogenously synthesized from palmitelaidic (C16:1 trans-9) acid in bovine adipocytes. Journal of Animal Science 91: 1614-23.

Kalac P, Samkova E 2010. The effects of feeding various forages on fatty acid composition of bovine milk fat: A review. Czech Journal of Animal Science 12: 521-537.

Kittelmann S, Pinares-Patiño C, Seedorf H, Kirk M, Ganesh S, McEwan J, Janssen P 2014. Two Different Bacterial Community Types Are Linked with the Low-Methane Emission Trait in Sheep. PLoS ONE 97: e103171.

Lambe N, McLean K, Gordon J, Evans D, Clelland N, Bunger L 2017. Prediction of intramuscular fat content using CT scanning of packaged lamb cuts and relationships with meat eating quality. Meat Science 123: 112-119.

Luna P, Rodríguez-Pino V, de la Fuente M 2009. Occurrence of C16:1 isomers in milk fats from ewes fed with different dietary lipid supplements. Food Chemistry 1172: 248-253.

Pethick, D, Harper, G, Hocquette, J, Wang, Y 2006. Marbling biology-what do we know about getting fat into muscle? Australian beef the leader; 7-8 March 2006; University of New England, Armidale, NSW.

Pinares-Patiño CS, Hickey SM, Young EA, Dodds KG, MacLean S, Molano G, Sandoval E, Kjestrup H, Harland R, Hunt C, Pickering NK, McEwan JC 2013. Heritability estimates of methane emissions from sheep. Animal 7: 316-321.

Realini CE, Pavana E, Johnson PL, Font-i-Furnols M, Jacob N, Moon C, Agnew M, Bermingham EN, Craigie CR 2021. Role of intramuscular fat level and composition on consumer linking scores of M. longissimus lumborum from pasture finished lambs. Meat Science 173: 108380.

Rico D, Chouinard P, Hassanat F, Benchaar C, Gervais R 2016. Prediction of enteric methane emissions from Holstein dairy cows fed various forage sources. Animal 10: 203-211.

Smith S, Crouse J 1984: Relative Contributions of Acetate, Lactate and Glucose to Lipogenesis in Bovine Intramuscular and Subcutaneous Adipose Tissue. The Journal of Nutrition 1144: 792-800.

Vahmani P, Mapiye C, Prieto N, Rolland D, McAllister T, Aalhus J, Dugan M 2015. The scope for manipulating the polyunsaturated fatty acid content of beef: a review. Journal of Animal Science and Biotechnology 61: 29.

Wood J, Enser M, Fisher A, Nute G, Sheard P, Richardson R, Hughes S, Whittington F 2008. Fat deposition, fatty acid composition and meat quality: A review. Meat Science 784: 343-358.

The fate of calves born on New Zealand dairy farms and dairy farmer attitudes towards producing dairy-beef calves

JP Edwards[a]*, S Cuthbert[b], JB Pinxterhuis[a] and A McDermott[c]

[a]*DairyNZ Ltd, Lincoln 7608;* [b]*Cuthbert & Associates, Rototuna, Hamilton 3210;* [c]*AgriFood Strategy Ltd, Collingwood 7073.*

**Corresponding author:* Email: Paul.Edwards@dairynz.co.nz

Abstract

Dairy calves that are not required as herd replacements provide an opportunity for beef production with a relatively low environmental footprint. A telephone survey of 262 New Zealand dairy farmers was conducted in October 2020 to gain insight into current mating and calving practices, and attitudes to dairy-beef breeding and rearing calves for beef. This number of respondents provided a margin of error of ±6%. On average, per herd, 28% of calves were kept as dairy replacements, 12% were reared on-farm for beef, 15% sold for beef rearing, 35% sold to process as 'bobby' calves, 5% euthanised and 4% born dead or died. The largest barriers to producing more beef calves were uncertainty around their value, followed by calving difficulties with beef bulls. The main barrier to rearing more calves for beef was higher earnings for milk than for beef. However, 60% of respondents agreed that the sector should aim to decrease the number of 'bobby' calves. This survey provides data for the development of cross-sector initiatives to increase dairy-beef integration and the utilisation of calves. These results provide a rich description of the current state of the sector and they suggest that at least 74% of New Zealand beef production originates in the dairy industry.

Keywords: dairy calf; dairy-beef; bobby calf; calf rearing; mating

Introduction

The New Zealand dairy sector is guided by a vision and strategy for its future, the Dairy Tomorrow Strategy (DairyNZ et al. 2017). One commitment in this strategy is a high level of animal care that provides all animals a life worth living. Consumer and public attention for the wellbeing of animals and the ethics of current animal production systems is growing, and it is expected that concerns about processing of young animals will increase (Bolton & Von Keyserlingk, 2021). Thus, the dairy sector and partners are developing pathways to reduce the number of dairy calves slaughtered at a young age, known as bobby calves.

Economic modelling of the beef value chain has indicated that increasing the number of dairy-beef-cross progeny would increase net value to the beef industry (McDermott et al. 2005). Value could potentially be shared by the various participants along the dairy-beef value chain, including dairy farmers, calf rearers, finishers and processors (McDermott et al. 2005). Modelling of environmental impact has demonstrated the potential of further dairy-beef integration to reduce greenhouse gas emissions by -22% (of the kg CO_2e per kg carcass weight) through substituting all suckler-beef production with dairy-beef production (van Selm et al. 2021).

Research also demonstrated the potential of animals of dairy origin for beef production (Muir et al. 2001; Pike et al. 2019), lending further weight to increasing dairy-beef production as a strategy to reduce the number of bobby calves. According to a 2004 survey, 13% of calf rearers and 41% of beef finishers used only dairy-beef animals in their systems; 41% of calf rearers and 82% of beef finishers indicated they wanted to move to or remain dairy-beef only (Oliver & McDermott 2005). However, it is not clear that dairy-beef numbers have increased since this time. Rearers and finishers indicated least satisfaction with their ability to source high-quality four-day old calves or weaner calves, respectively (both scoring 6.8 out of 10). Calf rearers selected calves based on breed, live weight, general health, colour, and sufficient colostrum. For beef finishers, breed was almost twice as frequently cited as purchase criterion for selecting weaner cattle than any other. However, breed by itself may not be a good indicator of the genetic potential of the animal for growth (Martín et al. 2020). Level of satisfaction of calf rearers with calves growing rapidly after weaning was 7.9 and with calf growth rates achieved 7.6 (out of 10). Beef finishers scored their satisfaction with growth rates as 7.5.

To achieve a significant change to the number of dairy-origin animals being finished for beef production, these 2004 survey results suggest more dairy-beef calves need to be available that will satisfy performance expectations. The start of this system is the dairy farmer producing calves that are desirable for beef production. Therefore, our study sought to gain insight into current mating and calving practices by New Zealand dairy farmers, and their attitudes to dairy-beef breeding and rearing calves for beef. Future surveys are planned amongst calf rearers and beef finishers to update the results of the 2004 survey, as well as research into consumer expectations and public perception. This information can be used for the development of effective cross-sector initiatives and helps to describe the current state of the sector, providing a platform for evaluating change.

Materials and methods

A telephone survey of dairy farmers was completed between 29 September and 27 October 2020. Questions related to the fate of calves born alive (reared for dairy replacement, sold for beef rearing, reared on farm for beef, euthanised on farm, or sold as a bobby calf) and mating period plans for 2020. Euthanasia refers to any calf born alive but killed on farm, irrespective of reason for euthanising. Pilot testing of the questionnaire indicated that the contents of the survey were complex and required respondents to pause to consider their answers. As a result, the survey was adapted to give participants the option to give the researchers permission to access data on semen use for artificial insemination stored by LIC (Hamilton, New Zealand), thereby reducing the number of questions asked by telephone. In total 82 respondents took this option. If permission was not given the full suite of questions was asked by telephone.

A random sample of dairy farmers from each region of New Zealand who had not been surveyed in the previous 12 months, was sourced from DairyNZ's customer relationships management database. Only one contact per farm was supplied. The number of contacts for each region was proportional to the number of farms located in that region. Telephone interviews were conducted by independent interviewers contracted by Cuthbert and Associates (Hamilton, NZ). A total of 382 calls were made. The response rate was 69%, the remainder unable to be contacted (21%), refused to participate (7%), or were ineligible (4%; e.g. had left the dairy sector). The average call duration was 19 minutes. The number of respondents provided a margin of error of ±6% on all answers.

Survey questions

The survey covered demographics of the respondents, their mating practices, the fate of calves born, and barriers to use of proven beef genetics and rearing more calves for beef. Questions were developed based on the earlier survey from Oliver and McDermott (2005) and the authors' knowledge of current practices and opinions regarding dairy-beef. The introductory section collected information about the number of farms the respondent owned, sharemilked or managed (they were asked to answer questions about the farm they were most familiar with), the size of farm, their experience in the dairy industry, their use of support land, calving pattern (spring, autumn, split) and herd breed. The second section related to mating practices, including farmer use of artificial insemination (AI) and length of mating, and choice of natural mating bulls and breeds. Potential factors that could influence farmer decisions for producing more calves from proven beef genetics or rearing more calves for beef were evaluated. For these questions, a scale of 1-7 was used, with 1 being the factor having no influence whatsoever on the decision and 7 the factor being extremely influential. In the third section, farmers were asked questions about the fate of calves born on their farm; the number of calves that were reared as dairy replacements to keep or sell, sold for beef rearing, reared for beef (to keep or sell), died on farm, euthanised on farm, and sold as bobby calves for processing were recorded. Finally, participants were asked about their attitudes (on a scale of 1-7, strongly disagree to strongly agree) to a series of statements about the direction the industry should take regarding bobby calves.

Data analysis

The results were analysed and reported for all survey answers, with sub-group or cluster analysis reported where applicable. The number of responses (n) varied slightly among questions due to omitting obvious typographical errors, or variations to the questionnaire in response to pilot feedback (e.g., adding a question). Not all percentages sum to one hundred because of rounding or the question was multi-response.

A cluster analysis was performed to examine possible groupings of farmer behaviour based on what they do with calves born on-farm. The characteristics of clusters were compared. Significance testing was conducted between total results and the results for key demographic sub-groups. Results are based on two-sided tests assuming equal variances with a confidence level $P<0.05$.

Results

Demographic information

The mean herd size was 473 cows (393 in the North Island and 622 in the South Island) with an effective milking platform size of 167 ha (144 in the North Island and 207 in the South Island). In terms of herd breed, 28% were Friesian, 34% were Friesian-cross, 16% Friesian-Jersey-cross (i.e., major breed fraction was seven to nine parts Jersey or Friesian), 6% Jersey-cross, 12% Jersey and 4% other. Respondents had an average 25 years' experience in dairying. Seventy-four percent operated one farm and 67% owned or shared a support block, mean size of 92 ha (median 67 ha). Although support-block ownership was similar between the North and South Islands (69% and 68%, respectively), support-block size was larger in the South (median 101 ha compared with 50 ha in the North Island, in line with the effective platform size). Fifty-eight percent of respondents wintered some cows off the milking platform; mean 199 cows, median 100 cows. Cows were more likely to be wintered off the milking platform in the South Island than the North Island (74% and 49%, respectively) and 45% of farmers in the South Island wintered off the whole herd, compared with 10% of farmers in the North Island. Replacement young stock were grazed at a support block for 50% of respondents, 34% contracted them out, and 12% kept them on the milking platform. The remaining respondents had other arrangements, for example a combination of the above or not rearing young stock themselves.

Mating practices

Eighty-five percent of respondents operated a spring-calving system, 3% autumn calving and 12% operated a split spring-autumn calving system. The mean mating length was seven weeks for autumn-calving herds (autumn only and split), of which AI was used for four weeks (median six weeks mating and five weeks AI). For spring-calving herds (spring only and split), the mean and median mating length was 10 weeks, with six weeks of AI. Of the milking herds that were mated in spring, 17% were mated with AI only.

The length of AI was more likely to be longer in the South Island than the North (6.8 and 5.6 weeks, respectively).

Ninety-nine percent of respondents used AI with semen from dairy breeds, 72% used semen from beef breeds (including short-gestation length Hereford), and 36% used short-gestation-length (SGL) semen (e.g., at the tail end of mating). Forty-two percent of respondents used AI mating on rising two-year-old (R2) heifers. For this group, the average milking herd size was 593 and the mean number of R2s mated by AI was 118.

For cows that were pregnant to AI, an average of 83% of calves born were from dairy breeds, 13% from beef breeds and 4% from SGL. A subset of respondents (n=144) answered a version of the survey that related to mating outcome of all cows in the herd. On average 57% of the cows were pregnant to AI dairy sires, 3% to AI SGL sires, 21% to natural-mating bulls, 8% to AI beef sires, and 11% were not pregnant.

Four percent of respondents did not use any natural-mating bulls for either their milking herd or R2s. For those that used natural-mating bulls (for cows and/or R2s), 27% used dairy sires only, 30% used beef sires only and 44% used a mixture of both. The most common breed of sires was Jersey (35%), followed by Hereford (24%), Angus (13%), Friesian (8%) Jersey × Friesian cross (6%) and 'Other beef breeds' (14%). The majority (73%) of natural-mating bulls were unrecorded (i.e., the genetic merit of the animal was unknown).

Fate of calves

On average, 28% of all calves born on-farm were reared as dairy replacements (to keep or sell), 27% were for the beef industry (reared, kept or sold), 35% were sold to process as bobby calves, 4% were born dead or died, 5% were euthanised, and 1% were categorised as 'other' (e.g., bulls kept for breeding). The median number of animals finished for beef was 36, equivalent to 8% of the milking herd size.

Fifty-nine percent of respondents finished animals for beef on their milking platform or support block. This includes 19% of the respondents who kept five animals or fewer to finish. This group could be assumed to produce meat for their own consumption, instead of producing beef for the general market.

Barriers to use of proven beef genetics and rearing more calves for beef

Factors influential in preventing the respondents producing more dairy-beef calves of proven beef genetics are presented in Fig. 1. The most influential barriers were 'the uncertain value of dairy-beef calves' and 'more calving difficulties with beef bulls compared with dairy bulls' (40% of respondents scoring these barriers a 5-7, with 7 being extremely influential). Of least importance were 'the difficulty identifying beef calves from dairy calves' (13% scoring this 5-7) and 'the cost of beef semen relative to the cost of natural mating' (15%). Other reasons mentioned that were not listed in the survey, mostly related to the limited market for calves with dairy genetics (46% of the responses to 'Other reasons').

Barriers to rearing most of the calves born on farm are listed in Fig. 2. The statement that respondents agreed

Figure 1 Responses to the question "On a scale of 1-7, where 1 means no influence whatsoever and 7 means extremely influential, how much does [statement] prevent you from producing more proven* beef calves on your farm?" n=221. Percentage of respondents that 'agreed' (scored 5-7) labelled at the end of the bar.

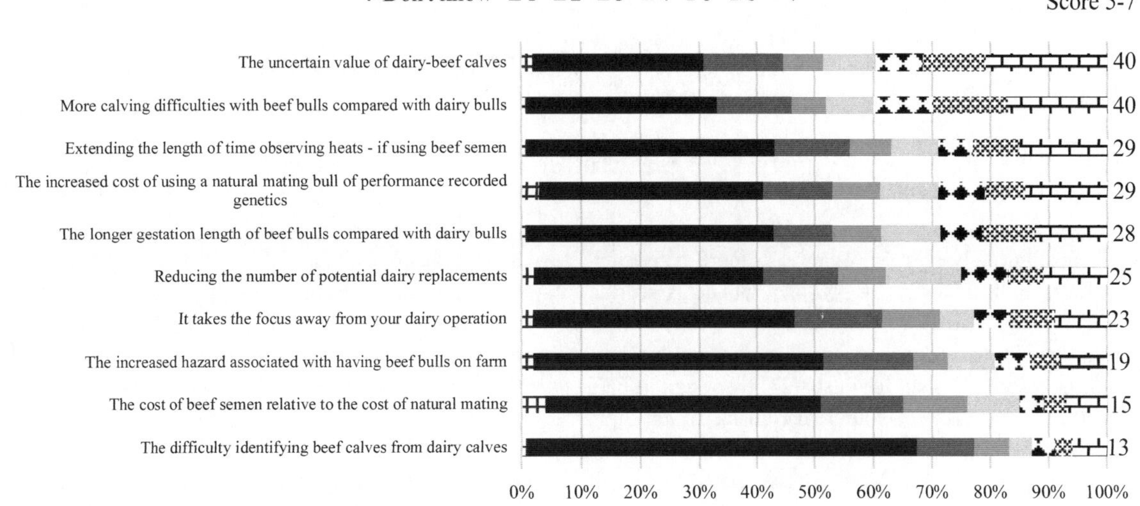

*A definition of the word proven was not given to respondents unless asked, in that case proven was considered having beef genetics that has been proven to perform well by an artificial breeding company or beef stud.

Figure 2 Responses on a scale of 1-7 (strongly disagree to strongly agree) to the question "how much do you agree that [statement] prevents you from rearing all or most of the calves born on your farm?" n=235. Percentage of respondents that 'agreed' (scored 5-7) labelled at the end of the bar.

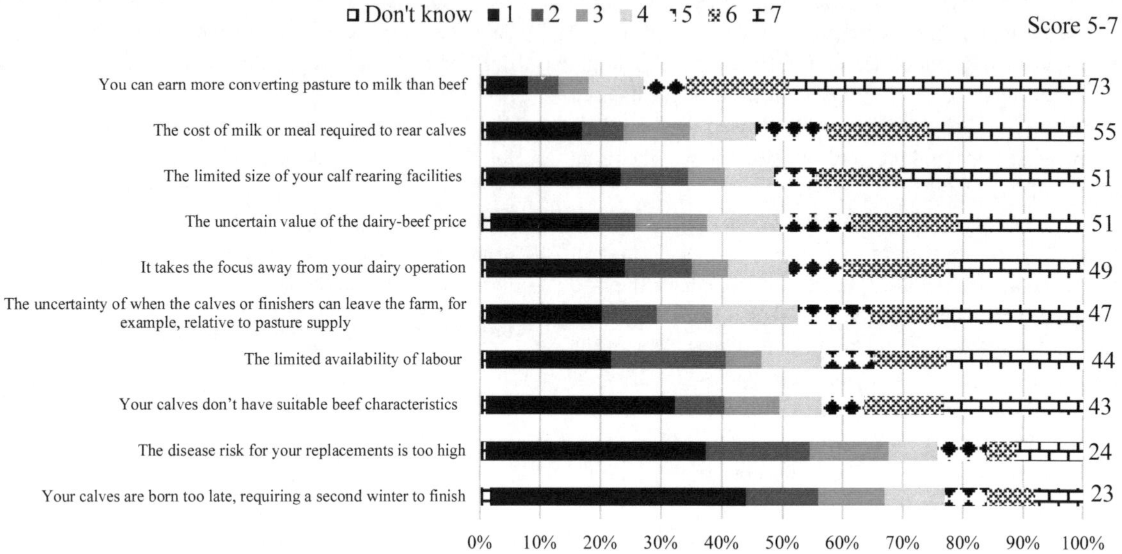

Figure 3 Responses to the question "using a 1 to 7 scale, where 1 means strongly disagree and 7 means strongly agree, how much do you disagree or agree with the following statements?" [about bobby calf practices] n=237. Percentage of respondents that 'agreed' (scored 5-7) labelled at the end of the bar.

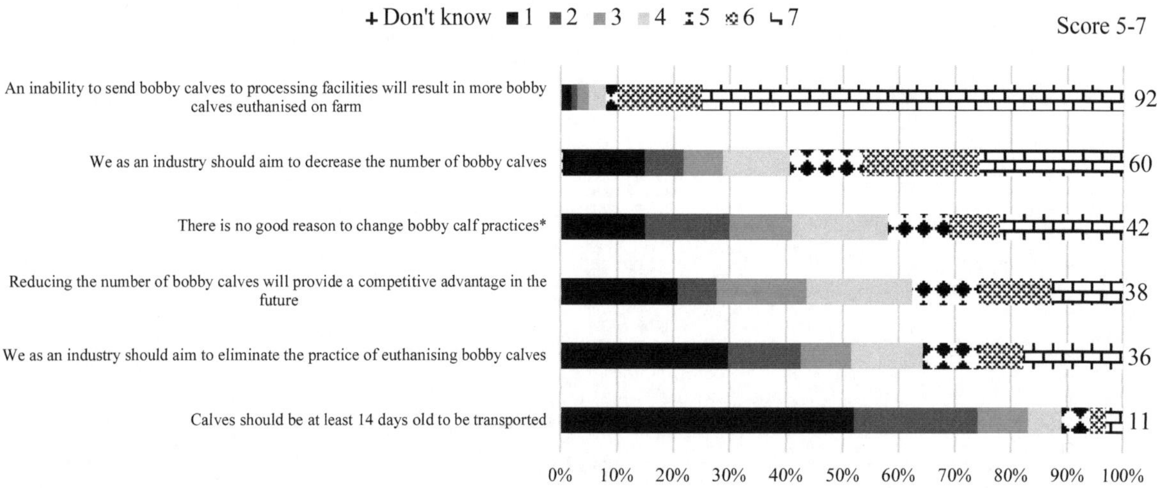

*This statement was changed part-way through data collection. Earlier version: "dairy farmers are being forced to change bobby calf practices for no good reason", n=103. The answers to this earlier version of the question are included in the figure.

with most strongly was 'you can earn more converting pasture to milk than beef' (73% of respondents scoring this 5-7). Two statements were considered considerably less important than others: 'your calves are born too late, requiring a second winter to finish' (23% scoring this 5-7) and 'the disease risk for your replacements is too high' (24%). Other reasons mentioned that were not listed in the survey were diverse, but 28% of these related to lack of market.

Nearly all respondents agreed that 'an inability to send bobby calves to processing facilities will result in more calves euthanised on farm' (92% scoring 5-7) and disagreed that 'calves should be at least 14 days old to be transported' (83% scoring 1-3; Figure 3). The majority (60%) agreed (scoring 5-7) that 'we as an industry should aim to decrease the number of bobby calves'; 28% disagreed (scoring 1-3). There was less agreement (36% scoring 5-7 and 52% scoring 1-3) that 'we as an industry should aim to eliminate the practice of selling bobby calves for processing'. Verbal comments from respondents illustrated that, while in general they agreed with the aim to decrease the number of bobby calves, they do not yet see a way to make this work at scale.

"I understand there may be a bobby calf issue but until somebody comes up with a solution (alternative) there is nothing I can do about it. I will not keep more calves on farm as they will drink the milk I want to sell."

"I struggle with the industry saying they want to eliminate or decrease bobby calves. It's nice to say, however, the everyday reality and practicalities of it makes it just not practical in everyday life."

One respondent provided a clear driver for change, echoed by others and related to the main barrier of the uncertain value of the calves: the price you can get for the animals. Note the average lamb price has not reached $200/head in the last decade (Beef+Lamb NZ 2021) and, therefore, would be an attractive proposition:

"Bobby calves are an essential part of dairying farming - but they are worth nothing, so nobody keeps them. If you could get a reasonable price (compare lambs at $200) then we would all find a way to make it work. Until the value of a young calf goes up nothing will change."

Distinct clusters regarding fate of calves

Four distinct clusters were identified based on respondents' estimates of the fate of calves born in their herds. Cluster 1 was 'euthanise on farm', Cluster 2 'sell bobby calves', Cluster 3 'sell for beef rearing' and Cluster 4 'rear for beef'. Significant differences in demographics and attitudes were apparent among the clusters (Table 1). The respondents that sold relatively more calves for beef or reared for beef had mostly Friesian or Friesian-cross herds. Those that reared for beef used less SGL semen than Clusters 1 and 2, and more AI beef semen than Cluster 2. Most barriers to rearing all or most calves born on farm were felt significantly less for Clusters 3 and 4 relative to Cluster 1 and 2. Respondents in Cluster 1 clearly considered labour to be a major barrier. The limited opportunity to earn income from beef calves was an important barrier for all groups, with average scores above 4 for the statement that they could earn more converting pasture to milk, rather than beef. The barriers of least importance were calves being born too late and disease risk for replacements.

Table 1 Mean percentage of calves by fate for four distinct clusters (n=261) and a description of the mean characteristics of each segment. Barriers were scored on a scale from 1 (no barrier at all) to 7 (severe barrier); values given are the mean of the scores.

	Cluster 1: Euthanise on farm	Cluster 2: Sell as bobby calves	Cluster 3: Sell for beef rearing	Cluster 4: Rear for beef
Number and percentage of respondents	n=12*	n=126	n=75	n=48
	5%	48%	29%	18%
Fate of calf, %#				
Kept as dairy replacement	27%	27%	30%	30%
Sold for beef rearing	3%	7%	**34%**	4%
Reared for beef	9%	6%	9%	**45%**
Euthanised on farm	**58%**	2%	2%	2%
Sold to process as bobby calf	0%	**55%**	20%	15%
Description of segment^				
Demographics#				
Years working in the dairy industry	19 b	24 b	27 a	23 ab
Herd size (number of cows)	620	531	503	438
Friesian or Friesian cross	36% b	51% b	81% a	82% a
Friesian-Jersey	27%	24%	8%	9%
Jersey or Jersey cross	36% ab	25% a	11% b	9% ab
Herd pregnant to AI by type, %#				
AI Short Gestation Length	6.6% a	5.0% a	3.0% ab	1.7% b
AI Dairy	79.6% ab	85.0% a	82.2% ab	79.0% b
AI Beef	13.8% ab	12.2% a	15.0% ab	19.4% b
Barriers to rearing all or most calves born on farm (mean on scale of 1-7)				
The cost of milk or meal	5.6 a	5.0 a	4.1 b	3.6 b
Uncertainty of the dairy beef price	5.3 a	4.7 a	4.4 a	3.2 b
Limited availability of labour	6.0 a	4.3 b	3.2 c	3.4 c
Limited size of calf rearing facilities	5.2 ab	4.8 a	3.7 bc	3.6 c
Earning more converting pasture to milk	5.6 ab	5.9 a	5.4 ab	4.7 b
Unsuitable beef characteristics of calves	4.1 ab	4.6 a	3.0 b	2.7 b
Calves are born too late	3	3.1	2.8	2.7
The disease risk for replacements is too high	3.9 a	3.1 ab	2.5 b	3.2 ab
Uncertainty of when calves or finishers can leave the farm	4.7 ab	4.5 a	4.2 ab	3.5 b
Takes the focus away from their dairy operation	5.2 a	4.5 a	4.0 a	2.9 b

*Caution is advised due to low n number

#Percentages sum within cluster

^Means across a row followed by a common letter are not significantly different by the LSD test at the 0.05 level of significance

Respondents' comments illustrate that currently there is a lack of market opportunities, especially for calves with Jersey genetics:

"*Our herd is Jersey - people don't want a Jersey beef cross.*"

"*We are a Jersey herd, so people don't want beef calves out of our Jersey calves. There is no market for them.*"

"*We're actually going back to Jerseys this year as we couldn't sell enough beef cross calves. Had about 700 and only sold about 450, had to put the rest on the bobby and didn't want to do this. Just not a ready market for them, so disappointed.*"

Other respondents experienced that crossbreeds with Jersey genetics were suitable for beef:

"*We are a Kiwi Cross herd. The beef calf, we have to grow them through ourselves as they are all sorts. A week-old calf is worth nothing however to grow them through is of significant value.*"

Another barrier mentioned was the existence of *Mycoplasma bovis* infection (recorded in verbatim comments 13 times):

"*Bovis has stuffed everything for the bobby-calf market and we can't sell anything. No one in the South Island wants a dairy cross.*"

"*Gone to Jersey due to Bovis and to be more self-contained.*"

However, *M. bovis* was also mentioned as the reason to produce more dairy-beef calves from natural mating:

"*Have Hereford bulls as can't trust Friesian bulls anymore due to Bovis. We buy bulls that come from up in the hills with no neighbours and are from a closed herd.*"

Discussion

The results of this survey provide an opportunity to improve current estimates of the extent of integration between the dairy and beef industries, and potentially account for a large number of animals that previously have had an unknown fate (van Selm et al. 2021). The present survey results relate to the 2020/21 dairy season; the latest dairy cow population data available from the New Zealand Dairy Statistics are for 2019/20, when the size of the dairy herd was 4,921,548 (LIC and DairyNZ 2020). There were 1.88 million calves processed in the year to December 2020 (StatsNZ 2020), representing produce from 38% of dairy cows. This is within the ±6% margin of error for the survey, lending validity to the results and suggesting that currently 1.33±0.30 million dairy calves are used for beef production, which is much higher than a previous estimate of 0.82 million (van Selm et al. 2021). Further, using a conservative estimate of the number of dairy calves reared for beef (1.03 million), combined with the 0.12 million dairy heifers (likely not-pregnant R2 animals) and 0.81 million cull dairy cows reported (StatsNZ 2020), these results indicate that around 74% of the 2.66 million adult cattle slaughtered in 2019/20 (StatsNZ 2020) originated in the dairy industry. This illustrates that the dairy sector already has a major role in beef production in New Zealand and that further increasing this proportion may be difficult. For example, beef farmers often mention the importance of beef cows for managing pasture quality in marginal areas such as hill country. It is unclear to us how many beef cows are currently used for this purpose and to what extent other cattle classes could fulfil this same function, such as carry-over dairy cows (non-pregnant and non-lactating) or growing beef animals.

The cluster analysis based on the fate of dairy calves indicates that the 47% of respondents classified in the 'rear for beef' or 'sell for beef' clusters are responsible for producing 73% of the dairy calves used for beef production and were only responsible for 23% of bobby calves. This indicates that to achieve further dairy-beef integration and a meaningful reduction in the number of bobby calves, efforts should focus on the 48% of respondents in the 'sell as bobby calves' cluster. However, whilst small, the 'euthanise on farm' cluster should not be overlooked, although their availability of labour barrier has to be addressed. Herds in these two clusters were less likely to be Friesian or Friesian cross. Despite evidence that animals with Jersey genetics can be reared successfully for beef (Muir et al. 2001), having Jersey or Kiwicross herds appears to still be a major barrier for rearers buying calves. These breeds are well-suited to New Zealand spring-calving and pasture-focused dairy systems (Edwards et al. 2019), aligning with this cluster's view about the ability to earn more converting pasture to milk. Therefore, it is unlikely that these farmers will make a large swing to Friesian or dual-purpose breeds to produce a different type of calf altogether. Future work, including demonstration of successful supply chains, will need to increase acceptance of Jersey genetics for beef production and facilitate mating strategies that produce more calves with beef sires.

Two factors that were key barriers to using improved genetics for beef production were certainty about the value of dairy-beef calves and concern about calving difficulties using beef bulls. These were the two most-important factors in a 2004 survey of dairy farmers (Oliver & McDermott 2005). Concern around calving difficulty may relate to the high proportion of unrecorded sires being used. Recent research highlights that appropriately selected beef sires have low rates of assisted calvings for adult dairy cows (Coleman 2020). Given that a dairy-beef calf can be processed as a bobby calf, the concern about their uncertain value implies that respondents believe that they are more costly to produce than dairy calves, or that their existing culling rate and reproductive performance dictate that their mating strategy is solely focused on generating replacement dairy animals. Potentially this has some validity, if their current system does not produce surplus replacements (e.g., 60% six-week in-calf rate × 50% female × 90% suitable for rearing = 24% available for replacements), then the most straightforward method for increasing the amount of dairy-beef would be via a mating programme with sexed semen (~$63/straw), which is considerably more expensive than conventional semen (~$22/straw) and is not sufficiently offset by the lower cost of beef semen (~$18/straw). Under this system, to maintain profit, the price paid for the resulting dairy-beef animal would need to be higher than that received for a bobby calf (Oliver & McDermott 2005). Modelling could determine what this premium would need to be, or conversely, how much cheaper sexed semen would need to be, to influence semen choice based on costs and calf prices only. Oliver and McDermott (2005) indicated a premium of $30/calf would motivate dairy farmers to use beef semen, and/or that beef semen would need to be priced appropriately to encourage use. The risk of not achieving this premium would also need to be considered. Despite this risk, since this 2004 survey, the number of dairy farmers using beef semen has approximately doubled. In a supply-chain context, the resulting calf would also need to be worth more than the current bobby-calf price to a calf rearer or beef finisher. Supply-chain contracts where calves are bred to a clear specification could be one way of achieving this, although this may also require market development. Another potential avenue could be decreasing rearer and finisher uncertainty in the growth potential of the animal. This could be achieved through the use of herd recording data, where the parentage of the animal and growth potential are available to each part of the supply chain. If these options prove to be insufficient to attract more buyers of dairy-beef calves, dairy farmers may need to sell quality dairy-beef calves at bobby-calf prices or rear more calves themselves to reduce the number of bobby calves slaughtered.

Conclusion

Generally, dairy farmers surveyed agreed that their sector should aim to reduce the number of bobby calves slaughtered, but there was little support for eliminating the practice altogether. Cross-sectoral initiatives to increase dairy-beef integration and reduce the number of bobby calves should focus on the 48% of the dairy sector represented by the cluster 'sell as bobby calves'. Potential areas for initiatives to address include the real or perceived lack of value for Jersey and Jersey-cross animals and increasing the confidence of calf rearers and beef finishers in the growth potential of dairy-beef animals. Collecting data on genetics, liveweight gain and slaughter quality would assist with these aspects. Supply chain and market initiatives would also help to increase the demand for these animals.

Acknowledgements

This project was funded by the dairy farmers of New Zealand through DairyNZ Inc, contract AC2002. We would like to thank the farmers who participated in the survey for their time and LIC for extracting herd data.

References

Beef+Lamb NZ. 2021. All grades lamb $ per head. https://beeflambnz.com/sites/default/files/data/files/All%20grades%20lamb%20%24%20per%20head.pdf [Accessed 19 Jul 2021].

Coleman L. 2020. The use of high genetic merit Angus and Hereford bulls in a New Zealand dairy herd: a thesis presented in partial fulfilment of the requirements for the degree of Doctor of Philosophy (PhD), Massey University.

DairyNZ, Federated Farmers, DCANZ, Dairy Women's Network. 2017. Dairy Tomorrow: The future of New Zealand dairying. https://www.dairytomorrow.co.nz/ [Accessed 24 Feb 2021].

Edwards JP, Spaans OK, Neal M, Macdonald KA 2019. *Short communication:* Milk fat payment affects the relative profitability of Jersey and Holstein-Friesian cows at optimal comparative stocking rate. Journal of Dairy Science 102: 9463–9467. doi 10.3168/jds.2018-16152.

LIC, DairyNZ. 2020. New Zealand dairy statistics 2019-20. https://www.dairynz.co.nz/media/5794073/nz-dairy-statistics-2019-20-dnz.pdf [Accessed 24 Feb 2021].

Martín N, Schreurs N, Morris S, López-Villalobos N, McDade J, Hickson R 2020. Sire Effects on Post-Weaning Growth of Beef-Cross-Dairy Cattle: A Case Study in New Zealand. Animals 10: doi 10.3390/ani10122313.

McDermott A, Dooley AE, Smeaton DC 2005. Identifying opportunities to add value to the New Zealand beef industry: a modelling approach. Proceedings of the New Zealand Society of Animal Production 65: 252-255.

Muir PD, Fugle CJ, Smith NB, Ormond AWA 2001. A comparison of bull beef production from Friesian type and selected Jersey type calves. Proceedings of the New Zealand Grassland Association 63: 203-207.

Oliver L, McDermott A 2005. More beef calves from the dairy industry: a survey. Proceedings of the New Zealand Grassland Association 67: 73-79.

Pike SJ, Schreurs NM, Hickson RE, Hunt JJ, Kenyon PR, Garrick DJ, Blair HT, Morris ST 2019. BRIEF COMMUNICATION: Meat quality of light-weight, yearling steers of dairy origin. New Zealand Journal of Animal Science and Production 79: 156-158.

StatsNZ. 2020. InfoShare. http://infoshare.stats.govt.nz/ [Accessed 26 Feb 2021].

van Selm B, de Boer IJM, Ledgard SF, van Middelaar CE 2021. Reducing greenhouse gas emissions of New Zealand beef through better integration of dairy and beef production. Agricultural Systems 186: 102936. doi 10.1016/j.agsy.2020.102936.

Is there sire variation in seasonal liveweight gain of rising-yearling red deer?

JF Ward[a*], NR Cox[a], BR Thompson[a], CD McMahon[b], DR Stevens[a] and GW Asher[a]

[a]*AgResearch Ltd, Invermay Agricultural Centre, Private Bag 50034, Mosgiel 9053, New Zealand;* [b]*AgResearch Ruakura Research Centre, Private Bag 3123, Hamilton 3240, New Zealand*

**Corresponding author:* Email: jamie.ward@agresearch.co.nz

Abstract

Red deer exhibit the greatest seasonal control of nutritional physiology of farmed ruminants in New Zealand, highlighted by seasonal inappetence during winter, present in all sex and age classes. Four years of data from 2048 DNA-pedigreed mixed-sex progeny of 112 sires and 1069 dams from a red deer breeding stud were interrogated to investigate the hypothesis of sire variation in growth across seasons in rising yearlings. Analysis used 17145 live weights from progeny with ≥ 4 live weights to a minimum of 300 days of age. Estimates of the weights of all animals at the same age were calculated by fitting a restricted maximum likelihood model to adjust for multiple variables and create sire best linear unbiased predictions of rising-yearling growth. Liveweight gain was examined at 11 key seasonal ages. While overall post-weaning variation in liveweight gain was low (CV = 3.6-4.7%), for seasonal periods it was higher (CV 9.2-15.5%), with the greatest variability during the winter period (CV = 15.5%). Ranking sires by the predicted proportion of gained live weight of their offspring over seasonal periods showed sire variation across different seasons. Such variation may be beneficial allowing genetic selection of animals to fit d ferent farm systems.

Keywords: red deer; *Cervus elaphus*; seasonality; growth; live weight

Introduction

Red deer sub-species (*Cervus elaphus*), which originate from temperate Northern Hemisphere latitudes and climates, are ruminant herbivores that are opportunistic grazers and browsers (Kay & Staines 1981). This dietary plasticity, along with seasonal synchronicity of growth, reproduction and activity are key characteristics of cervids (Klein 1984). Kay (1981) characterised four three-monthly periods of growth for young red deer, birth to three months of age, three to six months of age, six to nine months of age and nine to fifteen months of age, noting that appetite falls to low levels during winter (6-9 months). Fennessy et al. (1981b) demonstrated that post-weaning liveweight gain (LWG) was low from 3 to 9 months of age, followed by high LWG from 9 to 15 months of age. Noting different peak weights between progeny of two sires, Fennessy (1981a) speculated that there may be potential for genetic selection for growth rate.

Genotype differences in growth rate of rising-one-year-old (R1) eastern European red deer *vs.* elk-crossbreds during winter and of New Zealand red deer *vs.* eastern European red deer and elk-crossbreds during spring, were described by Stevens et al. (2000). During the Deer Progeny Test (Ward et al. 2014) it was observed that offspring cohorts of individual sires with very similar genetic merit for 12-month live weight followed different pathways of growth to achieve similar sire mean live weights at 11-13 months of age. These pathways were measured at quarterly live weights at approximately 3, 6, and 9 months of age as R1s. If such genetic variation existed, lines of deer could be better fitted to individual farm systems or vice-versa.

It was hypothesised that sire variation in the proportion of liveweight gain of farmed red deer, with equivalent genetic merit for yearling live weight, exists during discrete seasonal periods to twelve months of age.

Methods

Animal data were sourced from a large deer stud breeding company Deer Improvement Limited (now trading as Melior) (Gudex et al. 2013; 2018), from animals born and raised on their breeding farm in Balfour (northern Southland, New Zealand). The progeny represented in these data were born in four different years; 2008, 2009, 2011 and 2012. Date of birth was calculated for all animals using either a known conception date from artificial breeding (either artificial insemination (AI), or embryo transfer (ET)), or estimated based on real-time ultrasonographic fetal ageing (White et al., 1989) if naturally mated (NM). Their pedigrees, (i.e., sire and dam), were determined by DNA methods (Gudex et al., 2013). Live weights of the R1 deer were available from January (approximately 60 days of age) prior to weaning (approximately 100 days of age) and then approximately monthly thereafter until a maximum of 375 days of age. Animals were raised outside on pastures of predominantly perennial ryegrass (*Lolium perenne*), containing white clover (*Trifolium repens*) and herb species including plantain (*Plantago lanceolata*) and chicory (*Cichorium intybus*). During the winter (June-September inclusive) the deer were grazed on fodder crops (forage brassica (e.g., swedes (*Brassica napus napobrassica*), rape (*Brassica napus biennis*) or kale (*Brassica oleracea acephala*) varieties), or fodder beet (*Beta vulgaris mangelwurzel*), and supplemented with ensiled pasture.

Live weights were measured with an electronic scale unit using load cells with precisions of 0.1 kg (<50 kg) and 0.5 kg (>50 kg). All individuals were weighed directly off pasture or crop (i.e., non-fasted live weight). Within the live weight dataset, not all individuals were weighed at all periods, and there were differences in the weighing dates for different mobs, among months and years. Data were excluded for animals with no pedigree, <4 live weights, or no liveweight data beyond 300 days of age (most culling was after this age). The final dataset for analysis contained 17145 live weights on 2048 R1 red deer, from 112 different sires and 1069 different dams. Many sires and most dams were used in more than one year. There were 7731 and 9414 live weights for female and male R1 deer, respectively.

The R1 deer had varying numbers and dates (or age) of weighing. To get valid comparisons of animals, estimates of the weights for all animals at the same age were calculated by fitting a restricted maximum likelihood (REML) model (Genstat v13.2 (VSN International, Hemel Hempstead UK). The model included fixed effects for; year born, sex, age (linear), interactions of these, and a linear correction for birth day (day of year; this latter term effectively gives a correction for the fact that animals reach particular ages at different times of the year), random terms for; sire, dam, animal, sire by linear age (slopes), dam by age, animal by age, and spline (or shape) terms for age, age by year born (different shape for each year), sire by age, dam by age, animal by age (Table 1). A less complex model which excluded fixed effects for sex, and year born interacting with sex, was used to produce growth curves for individual animals which was fitted independently for each sex (Table 2).

The growth curves produced live weights at 11 different ages of R1 deer: being 35, 70, 100, 145, 190, 230, 270, 310, 325 and 345 days of age. These were selected as they align with important developmental and seasonal points in the first year of life of a farmed red deer, including dam peak lactational output at Day 35, weaning at Day 100, end of autumn at Day 190, end of winter at Day 270, early and late spring at Days 310 and 345, respectively. From these 11 age points, four seasonal periods of post-weaning growth to 12 months of age were defined as: autumn between 100-190 days of age, winter between 190-270 days of age, early spring between 270-310 days of age, and late spring between 310-345 days of age. Prior to weaning, a fifth period was defined as summer between 35-100 days. Growth rates were calculated over these periods. Various data from these models were further analysed for the different seasonal growth periods, including live weight, proportional change in live weight (i.e., live weight gained over the period/live weight at the start of the period), and best linear unbiased predictors (BLUP) for the sire. Animals were ranked for each seasonal period by their proportional liveweight change over each season and outlier sires and dams were identified ranked by the proportion of their progeny in the top or bottom 5th percentile for those seasons.

Results

The REML-modelled growth splines for the R1 red deer predicted the average male in this herd to attain 30.6 kg by 35 days of age, 60.6 kg by weaning at 100 days, 78.9 kg at the end of autumn, 88.9 kg at the end of winter (270 days) and 110.2 kg in late spring at 345 days (Fig. 1). The average female was predicted to attain live weights of 28.2 kg, 55.1 kg, 69.1 kg, 75.2 kg, and 92.9 kg at the same times points, which is 8.0%, 9.2%, 12.5%, 15.4% and 15.7% less than those of the males (Figure 1).

Best unbiased linear predictions of R1 offspring live weight were produced for all sires and these are summarised in Table 1. Liveweight gains for the seasonal periods were calculated from the sire BLUP data, along with proportional changes in live weight, deviations from the mean of these seasonal liveweight changes and cumulative live weight from Day 35 (Table 2). Within this herd the overall variation of the sire liveweight BLUP at weaning (i.e., Day

Figure 1 Restricted maximum likelihood modelled mean live weights of male and female rising-yearling red deer born in 2008, 2009, 2011 and 2012 on a deer stud in Southland, New Zealand.

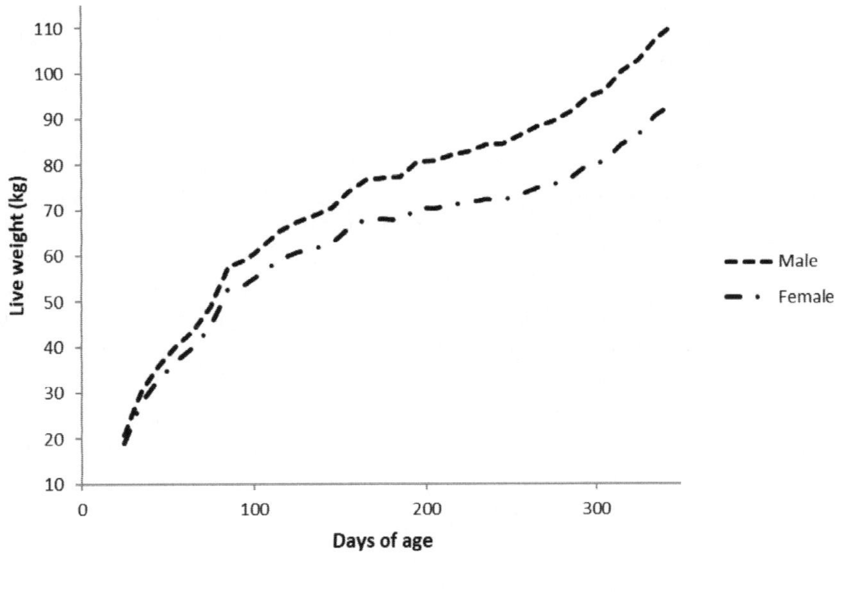

Table 1 Summary statistics, at key seasonal points from 35 to 345 days of age, of best unbiased liner predictions (BLUP) of the live weight (kg) of rising-yearling red deer sired by 112 different stags.

	Day 35	Day 100	Day 190	Day 270	Day 310	Day 345
Mean	29.4	57.5	74.0	82.0	90.2	101.6
Standard deviation	2.6	2.7	3.5	3.3	3.4	3.7
Minimum	19.4	49.7	61.6	71.3	78.9	90.1
Maximum	36.1	63.6	82.3	89.4	98.1	110.9
Coefficient of variatio	8.9%	4.7%	4.7%	4.0%	3.7%	3.6%

Table 2 Summary statistics of seasonal live weight gained (kg) and proportional change in live weight during each seasonal period (in parenthesis) from best linear unbiased predictions of the live weight of rising-yearling red deer by 112 different sires over key seasonal periods from 35 to 345 days of age (D).

	Summer D35-100	Autumn D100-190	Winter D190-270	Early spring D270-310	Late-spring D310-345
Mean	28.1 (0.97)	16.5 (0.38)	8.1 (0.11)	8.2 (0.10)	11.4 (0.13)
Standard deviation	1.8 (0.13)	1.7 (0.05)	1.3 (0.02)	1.0 (0.01)	1.0 (0.01)
Minimum	22.1 (0.73)	11.9 (0.26)	4.4 (0.05)	5.3 (0.07)	8.1 (0.09)
Maximum	31.4 (1.57)	20.1 (0.54)	12.0 (0.18)	10.5 (0.15)	13.3 (0.16)
Coefficient of variatio	6.3% (0.10)	10.2% (0.12)	15.5% (0.18)	12.8% (0.14)	9.2% (0.09)

Table 3 Best linear unbiased prediction of seasonal liveweight change in rising-yearling red deer, absolute change (kg), proportional change (in parenthesis), and seasonal haplotype (where H = high, L= low, and A = average proportional change in live weight, during summer, autumn, winter and spring respectively) for a selection of nine sires whose offspring represent population extremes for one or more season period.

Sire	Summer D35-100	Autumn D100-190	Winter D190-270	Spring D270-345	Seasonal haplotype
Bendigo	31.3 (1.18)	20.1 (0.47)	10.0 (0.13)	13.2 (0.14)	HHAL
Charles	28.1 (0.98)	18.1 (0.42)	5.2 (0.07)	14.2 (0.17)	AALA
Czar	30.1 (0.96)	19.3 (0.41)	8.0 (0.10)	12.7 (0.14)	AAAL
Ivan	27.7 (0.95)	15.4 (0.36)	8.0 (0.11)	18.8 (0.22)	AAAH
Kremlin	22.1 (0.73)	12.6 (0.30)	12.0 (0.18)	15.8 (0.19)	LLHA
Lazio	29.6 (0.95)	19.7 (0.43)	8.4 (0.10)	18.0 (0.19)	AHAA
Luciano	26.6 (0.80)	14.9 (0.32)	10.4 (0.14)	17.4 (0.19)	LLHA
Tundra	31.4 (1.05)	19.2 (0.42)	4.4 (0.05)	14.5 (0.17)	AALA
Tuscany	30.4 (0.92)	18.7 (0.38)	7.1 (0.09)	17.3 (0.18)	AALA

100) and discrete ages post-weaning was low-moderate for (CV 3.6-4.7%, Table 1), but BLUP for live weight gained over seasonal periods were much higher (CV 9.2-15.5%, Table 2) to 12 months of age.

There were sires identified whose data was at the extreme edges relative to the rest of the population for proportional liveweight gain for individual seasonal periods and in some cases for more than one seasonal period. Different representations of growth profiles for a selection of nine such sires with BLUP for proportional change in seasonal live weight over ≥1 season that were more or less than one standard deviation from the population mean, are shown in Table 3. Most of these sires were also extremes for the absolute liveweight gain in those seasonal periods. The average difference between proportional change in seasonal live weight BULP of the two most extreme selected sires, over each period, was 0.45, 0.13, 0.09 and 0.08 of liveweight gain, or 9.3 kg, 7.5 kg, 6.8 kg and 6.1 kg in absolute live weight over the summer, autumn, winter and spring periods respectively (Table 3).

These seasonal variations in growth can be represented as a haplotype to interpret the seasonal growth profile for a given individual in the first year of life as shown for the selected sires (Table 3). For the sires, the live weight BLUP at weaning and start of winter both correlated well (correlation coefficient R=0.77) with Day 345 live weight, while end-winter live weight correlated best (R=0.88) and Day 35 live weight the worst (R=0.44).

Discussion

In this analysis, the mean REML-predicted live weights produced sigmoidal curves, with greatest liveweight gain prior to weaning, lower gains in autumn, the lowest gains in winter and a sharp increase in liveweight gain in spring, as described by Kay (1981). There were BLUP estimated for individual sires that deviated markedly from these mean growth curves, with some much more linear in the post-weaning period and others a much more defined sigmoid with very low liveweight gain over winter. Key to this analysis was the proportional change in live weight over the period (i.e., liveweight gained/starting live weight), as individuals achieve very different live weights at the start of any given seasonal period. To better quantify seasonal growth potential within a herd, it is recommended to weigh R1 deer monthly to produce sufficient definition of seasonal liveweight variability, particularly across years.

While this study has identified sire variation in the seasonal growth of R1 deer, it has not considered possible mechanisms for this or factors contributing to it. These could include efficiency (e.g., digestive or metabolic (i.e., thermoregulation)) or behavioural differences (e.g., grazing behaviours or mob hierarchies). Quantification of some of these is important if genetic selection for seasonality is to be incorporated into a national breeding objective.

Identifying outlier sires within the current analysis reflects normal practice for stud breeders who are looking for freak, or "curve-bending" individuals to increase their genetic merit in particular traits or groups of traits. Within this analysis, the ability to identify sires of differing growth patterns could increase selection sophistication for growth to 12 months of age, by targeting seasonal periods. The deer production year is constrained by the seasonality of both the animal and pastures (Archer & Amer 2009). Variation of these annual cycles may offer opportunities to choose new targets or parts of the seasonal cycle to target either animals or nutrition to better suit both the animal and the farm system, better fitting the cycles of pasture production, or improving resource use efficiency.

At present there are three estimated breeding values (eBV) reported in DEERSelect for R1 live weight. Weaning weight (WWTeBV) is predicted from either a February, March or April live weight, autumn weight (AWTeBV) predicted from a May, June or July live weight, and yearling live weight (W12eBV) from an October, November or December live weight (Archer et. al. 2005). DEERSelect also has an unreported trait of spring live weight from August, September, or October measurements. If the New Zealand deer industry does want to consider selection for R1 seasonal growth, then the time period included in these traits must be more tightly defined, and a spring eBV should be reported. This would allow selection based on eBV for four key seasonal start and end points. It should be acknowledged that for breeders using selection indexes (Archer & Amer 2009), there is already a seasonal component within the indexes accounting for the seasonal variation in the cost of feeding.

Sire variation in R1 red deer growth over discrete seasonal periods offers opportunities for the New Zealand deer industry to better fit animals to farm systems, or farm systems to animals. These seasonal growth differences could be more valuable if they represent individuals that are productively more efficient (i.e., produce more kg product/$ farm system inputs). There is value in the industry further investigating the potential mechanisms of these seasonal differences (e.g., physiological and/or behavioural differences) and preparing DEERSelect to utilise or report on appropriate seasonal traits to select a better deer for future farm systems.

Acknowledgements

Deer Improvement Limited, particularly Jake Chardon and Boyd Gudex for providing the detailed Deer Improvement Balfour Research Farm data set.

References

Archer JA, Amer PR 2009. Breeding objectives for seasonal production systems: an example from New Zealand Venison systems. Proceedings of the Conference of the Association for the Advancement of Animal Breeding and Genetics 18: 207-211.

Archer JA, Ward JF, Newman SAN, Cruickshank GJ, Pearse AJ 2005. Implementing genetic evaluation in the New Zealand deer industry. Proceedings of the Conference of the Association for the Advancement of Animal Breeding and Genetics 16: 4-7.

Fennessy PF 1981a. Nutrition of red deer. Proceedings of Deer Seminar for Veterinarians, Queenstown, Deer Advisory Panel New Zealand Veterinary Association. pp. 8-16.

Fennessy PF, Moore GH, Corson ID 1981b. Energy requirements of red deer. Proceedings of the New Zealand Society of Animal Production 41: 167-173.

Gudex B, Johnson D, Benton T, Chardon J 2013. 'Deer Improvement' – Genetic selection in a recently domesticated livestock species. Proceedings of the Conference of the Association for the Advancement of Animal Breeding and Genetics 20: 336-339.

Gudex B, Chardon J, Ford D, Spelman R 2018. 'Deer Improvement' – A breeding program for farmed red deer in New Zealand. Proceedings of the World Congress on Genetics Applied to Livestock Production, Genetic Gain – Breeding Strategies 2: 66.

Kay 1981. The productive potential of domesticated red deer. Report from the Rowett Institute 37: 125-134.

Kay RNB, Staines BW 1981. The nutrition of red deer (*Cervus elaphus*). Nutritional Abstracts Review Series B 51: 601-622.

Klein 1984. In: Fennessy PF and Drew KD ed. The Biology of Deer Production. Wellington NZ, Royal Society of New Zealand, Bulletin 22. pp. 411-421.

Stevens DR Corson ID 2020. Seasonal effects of two forage-based nutrition regimens on intake and average daily gain of three genotypes of young red deer (*Cervus elaphus*). New Zealand Journal of Animal Science and Production 80: 150-155.

Ward JF, Archer JA, Asher GW, Everett-Hincks JM, Mathias-Davis HM, 2014. Design and implementation of the Deer Progeny Test (DPT). Proceedings of the New Zealand Society of Animal Production 74: 220-225.

Ward JF, McIntyre SB, Newman S-AN, 2016. DEERSelect – review of the first decade. Proceedings of the New Zealand Society of Animal Production 76: 54-58.

White IR, McKelvey WA, Busby S, Sneddon WJ, Hamilton WJ 1989. Diagnosis of pregnancy and prediction of fetal age in red deer by real-time ultrasonic scanning. Veterinary Record 124(15): 395-397.

The effects of substituting autumn nitrogen fertiliser with gibberellic acid on feed supply and milk production

ME Miller*, GR Edwards, H Hague and RH Bryant

Faculty of Agriculture and Sciences, PO Box 84, Lincoln University, Lincoln 7647, Christchurch, New Zealand

**Corresponding author.* Email: melanie.miller@lincolnuni.ac.nz

Abstract

The aim of this study was to investigate the use of gibberellic acid (GA) to improve feed supply and milk yield in late lactation. A grazing study at Lincoln used 36 lactating, mixed-age, Friesian-Jersey cross dairy cows balanced over three un-replicated treatments of perennial-ryegrass and white-clover pasture treated with nitrogen (N) fertiliser as urea (50 kg N/ha), GA as ProGibb® SG (8 g GA/ha) or untreated (Control). Treatments were applied in late March 2014 and grazed over 12 days in May. There was no effect of treatment on herbage pre-grazing pasture mass (2058±6.9 kg DM/ha, P=0.20), clover (3.6±0.86%, P=0.53), ryegrass (88.0±2.33%, P=0.24) or crude protein (CP) content (17.8±1.60%, P=0.31). All milk yield components: 14.6±0.40 L/day, 1.4±0.09 kg MS/cow/day, 5.4±0.33% fat and 4.4±0.21% protein were similar among treatment groups (P=0.71, 0.14, 0.16 and 0.06, respectively). Although faecal-N content of cows on GA pasture (3.72% N) was higher (P<0.001) than cows on N-fertilised pasture (3.45% N) there was no treatment effect (P>0.05) on urinary-N content (0.4 ± 0.02% N) or N excretion in milk (101±4.4 g N/cow/d). The lack of response to GA or N fertiliser highlights the importance of timing of growth promotants for late lactation.

Keywords: perennial ryegrass; white clover; urea; growth promotants; dairy cow; milk production

Introduction

Late-season pasture growth drives lactation length as farms consider animal condition and pasture covers leading up to winter for their drying-off decisions. New Zealand dairy pastures are typically a combination of perennial ryegrass (*Lolium perenne*) and white clover (*Trifolium repens*) that provide a primary feed source to meet seasonal feed demands. Application of N-based fertiliser is a standard practice used, throughout the season, to maximise pasture production. Over the past two decades, with increases in herd size and stocking rate, the demand for pasture has increased and, with this, the amount of N fertiliser used within New Zealand has more than doubled to a rate of over 400,000 tonnes N/year (New Zealand Fertiliser Association) of which 63% is used in the dairy industry. High rates of fertiliser N and urinary-N are important sources for N loss (Ledgard et al. 1997) and the impact of this on the environment, particularly nitrate leaching, on water quality has been of growing concern (Ledgard et al. 1999, WHO 2011). Urinary-N deposited in autumn presents a high risk of N leaching due to low plant N uptake and high drainage during winter (Selbie et al. 2014). While mitigation strategies continue to be investigated (de Klein et al. 2009; Whitehead et al. 2015; Bryant et al. 2019) a key driver to reduce N surplus is to reduce N inputs. To regulate N fertiliser use on farms, the government has imposed N-caps which will have considerable impacts on the feed supply of N-reliant farms. An alternative option for managing feed supply in the absence of N fertiliser is the use of GA. GA, which is a naturally occurring growth hormone, stimulates cell elongation, breaks dormancy and, when applied exogenously, can stimulate out-of-season dry-matter (DM) production (Matthews et al. 2009). Pasture response to low rates of exogenous GA (8-10 g GA/ha) include increased DM production in late winter, early spring (Bryant et al. 2016; Zaman et al. 2014) and autumn (van Rossum et al. 2013; Zaman et al. 2014) and reduction in herbage protein percent (Bryant et al. 2016). While these responses, which have been observed in small plot studies, offer an opportunity for farmers to reduce N intake and maintain milk yield, there has been no paddock-scale studies to test these hypotheses. This study aimed to determine the impact of using GA in late lactation on pasture production, milk yield and N-use efficiency for milk production.

Materials and method

Site and design

A grazing trial was carried out on the Lincoln University Research Dairy Farm (43°38'S, 172°28'E; 17 m above sea level) between March and May 2014 with approval from the Lincoln University Animal Ethics Committee (#567). The soil type is a Templeton fine sandy loam. The pasture is a tetraploid perennial ryegrass (cv. Bealy, AR37) and white clover (cv. Kopu II) pasture, sown in 2011. The area had been irrigated with effluent during non-grazing periods. The six-hectare experimental area consisted of 4 × 1.5 ha paddocks. The experiment was a completely randomised design consisting of three treatments: no fertiliser or GA (Control), N fertiliser (urea, Ravensdown) and GA (ProGibb® SG, Nufarm 2016).

Treatment application and pasture preparation

Four weeks prior to the experiment (27 March 2014), each paddock was grazed by dairy cows over several days, followed by mowing to a uniform compressed height of approximately 3.5 cm. Within three days of mowing, N fertiliser was applied as granules at a rate of 50 kg N/ha as urea (46% N) while the GA was applied as a foliage application at a rate of 8 g GA/ha as ProGibb® SG liquid fertiliser (40% GA). The GA was diluted with 200 L/ha water along with a surfactant, (Contact™ 0.25 ml/L) and sprayed across the treatment area. The area was left to regrow for 32 days until it reached a minimum pre-grazing compressed height of 9±0.5cm, or 20 clicks of the rising plate meter (RPM, Jenquip F150 Electronic Pasture Meter) which is equivalent to 3300 kg DM/ha based on the standard equation (kg DM/ha = 140 x RPM clicks + 500).

Thirty-six mixed-age, Friesian x Jersey cows were allocated to one of three treatment groups based on covariate milk solids (MS) production (1.4±0.01 kg MS/cow/day), milk protein (0.6±0.01 kg protein/cow/day), live weight (499±1.7 kg) and days in milk (209±0.5 days) which was measured a week prior to commencing the study. Groups were randomly assigned to one of the three pasture treatments and adapted to experimental conditions over six days from 1 May 2014. Temporary electric fencing was used to achieve a target allocation at 35 kg DM/cow/day above ground level based on pasture mass which was estimated using the RPM and winter equation (kg DM/ha = 140 x compressed height + 500). Cows had *ad lib* access to clean water in portable troughs and were milked twice a day at 0600h and 1500h and received a new allocation of pasture after each morning milking.

Herbage measurements

Throughout the study, pasture mass was measured pre- and post-grazing, by harvesting six quadrats per treatment every other day (n=66 quadrats/treatment). All plant material within 0.2 m^2 quadrats was harvested to soil level, washed and the dry weight recorded after force air oven drying to a constant weight at 60°C. Immediately prior to harvesting each quadrat the compressed height of pasture in the quadrat area was measured with the RPM to confirm the pasture mass calibration and determine apparent DM intake (kgDM/cow/day) using the disappearance method.

Calibration equations derived from regression analysis were:

Control kgDM/ha = 92*RPM (r^2=0.91, n=66),
GA kgDM/ha = 83*RPM (r^2=0.93, n=66)
N fertiliser kgDM/ha = 91*RPM (r^2=0.93, n=66).

Herbage samples for nutritive analysis were collected daily by cutting pasture to the target residual grazing height of 3.5 compressed cm at 10 random locations per treatment. The sample was mixed thoroughly and a sub-sample of approximately 20 g fresh weight (FW) was sorted into ryegrass vegetative, ryegrass reproductive, white clover, weeds and dead material. These were then oven dried at 60°C for 48 hours and dry weights recorded. These data were used for botanical composition analysis.

Species (%) = g DM of species/g DM total in the sample

Another 50g FW sub-sample was immediately frozen at 20°C and later freeze dried and ground through a 1-mm sieve (ZM200 rotor mill, Retsch Inc. Newton Pennsylvania, USA). The ground samples were then analysed using near-infrared spectroscopy (NIRSystems 500, Foss, Maryland, USA) to obtain CP, acid detergent fibre (ADF), neutral detergent fibre (NDF), water-soluble carbohydrate (WSC) and digestibility of the organic matter in the DM (DOMD). DOMD was used to calculate metabolisable energy (ME) content of the samples (Geenty et al. 1992).

Animal measurements

Milk yields for each cow were downloaded daily (Delaval Alpro Herd Management system, Hamilton, New Zealand). Milk subsamples were collected at morning and afternoon milking on days 7, 10 and 12 and used to analyse milk fat and protein content (Foss Fourier Transform infrared analyser) and milk-urea N (MUN). Milk-urea N was determined on skimmed milk after cooling and centrifuging to remove fat and analysed by an automated Modular P analyser (Roche/Hitachi). Faecal and urine samples were collected on days 7 and 10 after morning and afternoon milking. Faecal samples were collected by manually stimulating defecation. Urine samples were collected mid-stream after stimulation by rubbing the vulva by hand and acidified immediately with five drops of sulphuric acid to prevent volatilization. Faecal and urine samples were stored frozen (-20°C). Faecal samples were thawed at 4°C for 24 hours. A thoroughly mixed subsample of about 20 g FW was weighed and then oven dried at 100°C for 48 hours, then re-weighed to determine DM%. Another subsample of about 30 g FW was freeze dried for five days at 0.5 mbar, ground through a 1-mm grinder and analysed for N% using an Elemental Analyser (vario MAX CN, Analysensysteme GmbH. Hanau, Germany). Urine samples were thawed at 4°C and analysed for urea, ammonia, N% (Elemental Analyser) and creatinine (kinetic colorimetric assay, Roche Creatinine Jaffe kit).

Statistical analysis

Calibration coefficients for the RPM were derived by linear regression of compressed height and herbage mass. Herbage parameters were analysed using the general-linear-model procedure of Genstat (19th Edition, VSN International Ltd), using sampling date as replicate (since each sampling date was a different allocation) and fertiliser treatment as fixed terms in the model. Milk yield and milk composition parameters from the three herd test days (7, 10 and 12) were analysed for variance by repeated-measures analysis in Genstat using baseline values as covariates. Again, fertiliser treatment was used as the fixed term in the model and animal was the random term. An effect was considered statistically significant if the P value was less than 0.05.

Results

Climate

The cumulative rainfall during the regrowth period, to the grazing period was 142 mm. During the grazing period the cumulative rainfall was 26 mm (Table 1). During the regrowth period the average daily air temperature was

Table 1 Mean (± standard deviation, SD) weekly rainfall (mm), air temperature, and soil temperature to 10 cm below ground level (°C) from the start of the regrowth period (March 27 2014) to the last day of the trial (May 12 2014).

	Temperature				
Week commencing	Soil	SD	Air	SD	Rainfall
27/03/2014	13.2	1.72	14.6	1.55	0
3/04/2014	14.0	1.11	14.0	1.58	49.0
10/04/2014	12.7	1.49	12.3	2.02	15.4
17/04/2014	12.1	1.25	13.2	2.57	40.0
24/04/2014	9.8	1.51	10.5	2.85	56.8
1/05/2014	9.8	1.77	11.8	3.50	0
8/05/2014	8.4	0.92	9.3	0.89	6.0

Table 2 Pre-grazing pasture mass (kg DM/ha), botanical composition (% of biomass) and nutritive value (% of biomass), of untreated pasture (Control) or pasture previously treated with urea (N fertiliser) or gibberellic acid (GA).

Variate	Control	N fertiliser	GA	SEM	P val
Pasture mass	2086	2121	1968	61.6	0.20
Ryegrass	87.8	90.4	85.8	1.88	0.24
Clover	3.31	2.94	4.65	1.11	0.53
Dead	3.47	3.31	4.64	0.91	0.54
Weed	5.44	3.33	4.95	1.50	0.59
Sheath:leaf	20.8	20.3	19.5	1.92	0.89
DM	12.7	11.1	10.4	1.03	0.32
ADF	22.4	22.6	23.0	0.25	0.28
NDF	40.3	39.8	40.7	0.94	0.79
WSC	16.5	15.2	15.6	0.52	0.22
CP	18.8	20.3	19.2	0.67	0.26
WSC:CP	0.88	0.77	0.82	0.04	0.23
DOMD	80.0	79.9	78.5	0.77	0.32
ME (MJ/kg DM)	12.2	12.1	11.9	0.11	0.32

SEM = standard error of the mean, ADF = acid detergent fibre, WSC = water-soluble carbohydrate, NDF = neutral detergent fibre, CP = crude protein, ME = metabolisable energy, DM = dry matter, DOMD = digestible organic matter in the dry matter.

13.2°C with a max. of 17.0°C and a min. of 8.1°C. During the grazing period, the average daily air temperature was 10.5°C with a max. of 17.7°C and a min. of 7.5°C. During the regrowth period the average daily soil temperature to 10 cm at 9am was 12.6°C with a max. of 14.9°C and min. of 8.1°C. During the grazing period, the average daily soil temperature to 10 cm was 9.5°C with a max. of 11.9 and min. of 7.6°C.

Pastures

Due to the low calibration coefficients for the RPM, both the average pre- and post-grazing pasture mass were low, though there was no difference in herbage mass among treatments (Table 2). Similarly, pasture treatment had no effect on botanical composition whereby clover and dead material combined represented less than 10% of the biomass (Table 2). The lack of effect of treatments on herbage mass and botanical composition resulted in no difference in the nutritive value of the pasture. Average DM was low (11.4±1.14%), which influenced the pre-grazing pasture mass. Crude protein content was normal and ranged between 16 and 20% of the DM. Digestibility of organic matter in the DM was 80% resulting in ME of 12.1±0.09 MJME/kg DM.

Intake and milk production

At similar herbage allocation, GA or N fertiliser application in autumn had no effect on DM intake or milk production (Table 3). The daily pasture allocation, above ground level, was 20±1.7 kg DM/cow/day with 72% utilisation, resulting in a mean apparent intake of 14±1.1 kg DM/cow/day. Although there were no differences in milk yield (14.6±0.40 L/day, $P=0.70$) or MS yield (1.4±0.09 kg MS/cow/d, $P=0.13$) among treatments, there was a tendency ($P=0.06$) for greater milk protein content from cows in the control group from those in the GA or N fertiliser treatments. There was no effect of treatment on live weight or BCS.

Nitrogen intake and excretion in faeces, urine and milk

Nitrogen intake (453 g N/cow/day, $P=0.75$) and faecal DM content (9.1 ± 0.53%DM, $P=0.16$) were similar among treatment groups (Table 4). Cows grazing pasture treated with GA had higher ($P<0.001$) faecal-N content (3.72% N) than did the control cows (3.38% N). Urinary-N content (0.4±0.02% N, $P=0.33$) was similar among treatment groups. Treatments groups had similar ammonia, creatinine, and urea concentrations in urine. Milk-urea N (88.9±1.04 mmol/L, $P=0.97$) and milk-N excretion (101±5.7 g N/

Table 3 Allocation, dry-matter (DM) intake and milk production of cows grazing untreated pasture (Control), or pasture previously treated with urea (N fertiliser) or gibberellic acid (GA).

Variate	Control	N fertiliser	GA	SEM	P val
Allocation above ground (kg DM/cow/d)	21.9	19.9	18.6	1.85	0.45
Compressed post-graze height (cm)	3.6	3.4	3.7	0.18	0.64
Post graze pasture mass (kg DM/ha)	655	624	611	31.4	0.60
Intake (kg DM/cow/d)	15.0	12.9	13.3	1.04	0.34
ME intake	176	150	154	12.2	0.30
Milk					
Yield (kg)	15.0	14.2	14.6	0.68	0.71
Fat (%)	5.66	5.49	5.00	0.24	0.16
Fat (kg)	0.84	0.77	0.73	0.04	0.17
Protein (%)	4.62	4.39	4.21	0.12	0.06
Protein (kg)	0.69	0.62	0.62	0.03	0.12
Milk solids (kg)	1.53	1.38	1.35	0.07	0.14

SEM = standard error of the mean

Table 4 Nitrogen (N) intake, faecal dry-matter (DM) and N (%), urine ammonia, creatinine, urea and N concentration, milk-urea N (MUN), total N excretion and N use efficiency for milk (NUE) of cows grazing untreated pasture (Control) or pasture previously treated with urea (N fertiliser) or gibberellic acid (GA).

Variate	Control	N fertiliser	GA	SEM	P val
Estimated N intake (g N/c/d)	470	429	460	39.6	0.75
Faeces					
DM %	9.17	9.56	8.51	0.387	0.16
N %	3.38	3.45	3.72	0.066	<0.001
Urine					
Ammonia (mmol/l)	0.94	0.83	0.78	0.077	0.31
Creatinine (mmol/l)	1.39	1.34	1.42	0.096	0.81
Urea (mmol/l)	100.7	91.6	103.2	5.37	0.28
N concentration (g/dL)	4.04	3.73	4.10	0.24	0.33
Milk					
MUN (mmol/L)	88.1	90.1	88.6	4.89	0.95
N excretion (g N/c/d)	108	97.0	96.9	4.40	0.12
NUE for milk (%)	23.0	21.5	24.8	2.00	0.51

SEM = standard error of the mean

cow/d, P=0.12) were similar among treatment groups as was N-use efficiency for milk (23.6 ± 0.01 %, P=0.51).

Discussion

Pasture response to N or GA

There was no response in DM production to the application of N fertiliser or GA in late March through May. The lack of pasture response to either GA or N was an unexpected outcome as climate conditions were suitable for growth. Although the pre-graze herbage mass in May was low, at 2000 kg DM/ha, accumulation from a starting cover of around 690 kg DM/ha after mowing in March indicates a reasonable growth rate of 41 kg DM/ha/d. There was little dead material at the base of the sward and the absence of clover stolon or large tiller populations contribute to the low pasture mass. While this explains the low pasture mass it doesn't explain the lack of treatment effect. Factors that could have contributed to lack of response to N fertiliser and GA include: declining temperatures and day length, high soil N availability from effluent and, with respect to GA, the long regrowth interval. Although the mode of action of N fertiliser and GA differ when applied to pasture, climatic conditions will influence ryegrass growth irrespective of treatment. From application to the start of the grazing study 32 days later, soil temperatures declined from 12.6 to 9.5°C, although still above the threshold for ryegrass growth and day length from 12 to 10 hours. While it is possible to get a response to N fertiliser in autumn there is often a lower response efficiency in autumn than in spring (Cameron et al. 2005; Sun et al. 2008). Of the factors contributing to low autumn response to N, in this study it is likely that N wasn't limiting, the pasture herbage N content was >3% of the DM and the lack of clover in this instance is more likely to reflect increased competition from ryegrass under high soil mineral N from historic effluent application.

Because pasture growth rates decline naturally as daylength and temperatures decline, the response to application of GA in autumn was expected to be less than in spring, based on previous studies (van Rossum et al. 2013; Zaman et al. 2014). Nevertheless, a response was expected, though it is likely that the duration between applying GA and harvesting pasture may be an important determinant in the size of the response relative to the control. Both the observations of Ball et al. (2012) who showed a shift in partitioning from above to below ground and the yield reductions in subsequent regrowth after application reviewed by Matthew et al. (2009) imply that applying exogenous GA has a carbon cost which balances out over time. This study was conducted five to six weeks after the application of GA and, therefore, the initial response to GA may have disappeared over the extended regrowth period.

Milk production and N-use efficiency

There were limited effects of the treatments on milk yield, milk composition or N-use efficiency for milk. Although allocation was lower than target, due to the difference in plate meter calibration for pasture mass (manufacturer versus experimental) used during and after the trial, there was no difference in allocation (20 kg DM/cow/day) or apparent intake (14 kg DM/cow/day) among treatments. With similarities in botanical composition and nutritive value, the diets consumed by each treatment group were very similar. It was expected that diet changes may have affected herbage protein and N-use efficiency for milk, including changes in milk urea concentration as an indication of protein nutrition. The elevated spot faecal N% of cows in the GA group is surprising given the lack of statistical difference in N intake and milk N. From a numerical N-balance perspective, the GA cows had the most surplus N and, perhaps this is why faecal N is higher, although we question the reliability of spot samples in the absence of consistence in other variables such as milk urea, N intake and milk N.

To date, there are no published studies on the effect of GA on animal production. In a previous study (Allen 2010) we treated perennial-ryegrass and white-clover pastures with GA (8g GA/ha) or N fertiliser (50kg N/ha) in early March and, after 29 days regrowth, compared milk yield of cows grazing treated pastures. In that study, there was no effect of GA or N on pasture response or milk yield. However, applying GA in that study reduced herbage N content, increased clover content, and reduced apparent N intake, which has implications on lower urinary N excretion. A key finding of that study and the current study is that replacing N fertiliser with GA in autumn is likely to lead to similar pasture production and maintain milk yield. If soil N is not limiting pasture production, then there is little justification for late season use of N fertiliser. Provided the timing of grazing following GA application is early enough to capture the benefits then there may be value in GA use in the autumn to replace N fertiliser.

Conclusion

The timing of fertiliser application and grazing following application are important factors to consider in farm-management practices. When pasture growth rates are low due to rainfall, temperature, or daylength, it is likely that there will be a limited response in DM production to the application of GA in mid-to-late autumn. Minimal changes to DM production and botanical composition resulted in no difference in nutritive value and, in this study animal performance was not affected by the application of GA in late autumn.

References

Allen ME, 2010. The effects of gibberellic acid and time of grazing on nitrogen partitioning in dairy cows grazing perennial ryegrass pastures. B Ag Sci Hons dissertation, Lincoln University.

Ball CC, Parsons AJ, Rasmussen S, Shaw C, Rowarth JS 2012. Seasonal differences in the capacity of perennial ryegrass to respond to gibberellin explained. Proceedings of the New Zealand Grassland Association 74: 183-188.

Bryant RH, Snow VO, Shorten PR, Welten BG 2019. Can alternative forages substantially reduce N leaching? Findings from a review and associated modelling. New Zealand Journal of Agricultural Research. 63(1): 3-28.

Bryant RH, Edwards GR, Robinson B 2016. Comparing response of ryegrass-white clover pasture to gibberellic acid and nitrogen fertiliser applied in late winter and spring. New Zealand Journal of Agricultural Research 59: 18-31. doi 10.1080/00288233.2015.1119164.

Cameron KC, Di HJ, Moir J, Christie R, Pellow R 2005. Using nitrogen: what is best practice? South Island Dairy Event Proceedings. pp 1-17.

de Klein CAM, Monaghan RM, Ledgard SF, Shepherd M 2010. A systems perspective on the effectiveness of measures to mitigate the environmental impacts of nitrogen losses from pastoral dairy farming. In: Edwards, GR, Bryant RH eds. Meeting the challenges for pasture-based dairying. Proceedings of the 4th Australasian Dairy Science Symposium. Lincoln University, Christchurch, New Zealand. pp 14-28.

Geenty KG, Rattray PV 1992. The energy requirements of grazing sheep and cattle. In AM Nicol. ed. Livestock feeding on pasture. New Zealand Society of Animal Production, Hamilton, New Zealand. Occasional Publication No.10. pp 39-55.

Ledgard SF, Penno JW, Sprosen MS 1997. Nitrogen balances and losses on intensive dairy farms. Proceedings of the New Zealand Grasslands Association 59: 49-53.

Ledgard, SF, Penno JW, Sprosen MS 1999.Nitrogen inputs and losses from clover/grass pastures grazed by dairy cows, as affected by nitrogen fertilizer application. Journal of Agricultural Science 132: 215–225. doi 10.1017/S002185969800625X.

Matthew C, Hofmann WA, Osborne MA 2009. Pasture response to gibberellins: a review and recommendations. New Zealand Journal of Agricultural Research 52: 213-225. doi 10.1080/00288230909510506.

New Zealand Fertiliser Association. Fertiliser use in New Zealand. https://www.fertiliser.org.nz/site/about/fertiliser_use_in_nz.aspx [accessed 27 April 2021].

Nufarm Ltd 2016. Boost pasture production with ProGibb®-SG. https://nufarm.com/nz/product/progibb-sg/ [accessed 29 April 2021].

Selbie DR, Cameron KC, Di HJ, Moir JL, Lanigan GJ, Richards KG 2014. The fate of urine nitrogen: A grassland lysimeter study in Ireland. In: Currie LD, Christensen CL eds. http://flrc.massey.ac.nz/publications.html. Occasional Report No. 27. Fertiliser and Lime Centre, Massey University, Palmerston North, New Zealand. pp 7.

Sun T 2008. Gibberellin metabolism, perception and signalling pathways in Arabidopsis. The Arabidopsis Book No.6. pp 1-28.

van Rossum, MH 2013. Dry matter production, nutritive value and botanical composition of a perennial ryegrass white clover pasture applied with GA and nitrogen in successive periods in autumn and spring. Lincoln University, Masters Dissertation.

van Rossum MH, Bryant RH, Edwards GR 2013. Response of simple grass-white clover and multi-species pastures to gibberellic acid or nitrogen fertiliser in autumn. Proceedings of the New Zealand Grassland Association 75: 145-150 doi 10.33584/jnzg.2013.75.2935.

Whitehead D, Edwards GR 2015. Assessment of the application of gibberellins to increase productivity and reduce nitrous oxide emissions in grazed grasslands. Agriculture, Ecosystems and Environment 207: 40-50. doi. 10.1016/j.agee.2015.03.019.

World Health Organization 2011. Nitrate and nitrate in drinking-water, Back-ground document for development of WHO Guidelines for drinking water quality. Geneva: WHO Press. https://apps.who.int/iris/handle/10665/75380 [accessed 27 April 2021].

Zaman M, Ghani A, Kurepin LV, Pharis RP, Khan S, Smith TJ 2014. Improving ryegrass-clover pasture dry matter yield and urea efficiency with gibberellic acid. Journal of the Science of Food and Agriculture 94: 2521–2528 doi 10.1002/jsfa.6589.

Effect of farm system and milk urea phenotype on milk yield and milk composition of dairy cows in Canterbury

RH Bryant[a*], P O'Conner[a], H Hague[b], Y Robic[c], J Curtis[a], RC Handcock[d] and O Al-Marashdeh[a].

[a]*Faculty of Agriculture and Life Sciences, Lincoln University, PO Box 85084, Lincoln University, Lincoln 7647, Christchurch, New Zealand;* [b]*Rothamsted Research, West Common, Harpenden, Hertfordshire AL5 2JQ, United Kingdom;* [c] *Inra - UMR Pegase, 16, le clos, 35590 Saint-Gilles, France;* [d]*A.L. Rae Centre for Genetics and Breeding, Massey University, Hamilton 3214, New Zealand*

**Corresponding author*: Email: Racheal.Bryant@lincoln.ac.nz

Abstract

To investigate the effect of farm system, and cow selection for milk urea nitrogen (MUN), on milk yield and milk composition, a farmlet study was carried out between October 2018 and May 2019 in Lincoln, Canterbury. A farm system with a low stocking rate and low N fertiliser input (LSR, 2.9 cows/ha) sown with a conventional ryegrass clover and plantain diverse pastures was compared with a farm system with a moderate stocking rate and moderate N fertiliser (MSR, 3.9 cows/ha) using conventional ryegrass and white clover pastures and supplementing 3 kg DM/cow/d as crushed barley grain. Each farmlet had total herd size of 40 mixed-age HF x J spring-calving dairy cows which included six cows selected solely for a high MUN or a low MUN. There was no effect of farm system on milk fat, protein or lactose content but MUN was lower in LSR compared with MSR. Milk production was also lower for LSR (466 *vs* 429±12.4 kg MS/cow/ha, P<0.05), owing to poorer quality diet in mid lactation. Cows selected for low MUN tended to produce less milk compared with high MUN cows (4478 *vs* 3987±174 kg/cow, P<0.10) though this was partially offset by increased protein content in milk of low MUN cows. Farm system and animal selection for MUN have a greater impact on milk yield than on milk composition.

Keywords: *Plantago lanceolata*; somatic cell count; multi-species pastures;

Introduction

To meet requirements to reduce nitrogen (N) losses from intensive dairy farming systems, farmers will be expected to adopt a range of mitigation strategies. Because risk of N leaching is driven by both farm N surplus and the distribution or deposition of that surplus through urination events, many of the mitigation options proposed relate to changes in feeding and fertiliser management. Use of nutritional approaches to reduce N intake and/or reduce soil N load from urine are strategies suggested to reduce N loss at the farm and animal level. For example, feeding high-moisture feeds with lower rumen N degradability such as plantain is expected to reduce urinary N load and nitrate leaching (Bryant et al. 2020). Similarly, feeding low N supplements such as grains can also reduce N intake and improve N use efficiency in the animal.

To further improve gains made in reducing farm N losses, attention has been turned towards animal selection. Animal selection for production and production efficiency traits has been extensively explored, and limited information is available on animal selection for lower environmental footprint traits. Recently MUN has been investigated as a pathway for selecting cows with a lower N loss (Beatson et al. 2019). It has long been recognised that MUN is positively related to urinary N excretion (Spek et al. 2013). While MUN itself is a moderately heritable trait (Beatson et al. 2019), scientists have cautioned that selection for low MUN may impact other production characteristics, including milk yield (Correa-Luna et al. 2019), dry matter intake (Ariyarathne et al. 2021), and may not lead to lower UN excretion at the animal level and/or the farm level (Ariyarathne et al. 2021; Handcock et al. 2021 this Journal).

Changes to nutritional management and animal genetics will likely have implications on the composition of milk and total production which may influence product quality and profit. Consequently the purpose of this research was to compare two farm systems which combine mitigation technologies, including stocking rate, plantain and cows divergent in MUN, to determine the impact on milk yield and milk composition at the herd level.

Materials and methods

The grazing study was conducted between October 2018 and May 2019 at the Lincoln University Research Dairy Farm in Canterbury, New Zealand (43°38′S, 172°28′E). The results presented in this research are part of a longitudinal study comparing farm systems which represent different pathways to achieve environmental, social and economic sustainability. Two farming systems are compared, one which adopts a 15% increase in the average regional stocking rate (3.4 cows/ha) and the other with a 15% decrease in stocking rate. To reduce environmental impact, both farms have reduced N fertiliser inputs with the moderately stocked farm (MSR) buying in up to 0.5 t DM/cow/y as crushed grain as a low-N feed alongside perennial ryegrass and white clover (RGWC) pasture. The low stocking rate farm (LSR) has incorporated plantain diverse pastures on 45% of the farm area and aimed to winter cows on farm to reduce costs and maintain profit. Both systems were managed under irrigation. Details of the farms are presented in Table 1.

Fifty-six crossbred Holstein Friesian x Jersey dairy cows (eight primiparous and 28 multiparous) were blocked according to breeding worth, live weight and age prior to

Table 1 Farmlet details for a moderately stocked farm system supplemented with grain (MSR) and low stocked farm including plantain diverse pastures (LSR)

	MSR	LSR
Stocking rate	3.9 cows/ha	2.9 cows/ha
Farm area	10.35 ha	14.03 ha
N fertiliser use	190 kg N/ha/y to all pastures	150 to ryegrass/ clover and 50 kg N/ha/y to diverse pastures
Supplement use during lactation	0.44 t DM/cow as grain and 0.35 t DM/cow as silage	0.09 t DM/cow as silage
Pasture types	2.8 ha diploid perennial ryegrass and white clover 7.55 ha tetraploid perennial ryegrass and white clover	6 ha diploid perennial ryegrass and white clover 1.5 ha tetraploid perennial ryegrass and white clover 6.45 ha Diverse containing plantain, Italian ryegrass, red clover and white clover

Table 2 Herd details (mean±standard deviation) for a moderately stocked farm system supplemented with grain (MSR) and low stocked farm including plantain diverse pastures (LSR) farmlet with cows of high and low milk urea nitrogen (MUN) breeding values (BV).

	Farmlet		MUN group	
	MSR	LSR	High	Low
No. cows	39	39	12	12
Age (years)	4.41 (±1.8)	4.32 (±1.8)	5.2 (±0.93)	5.1 (±1.02)
BW value	98.3 (±29.0)	102 (±26.8)	72.8 (±29.3)	100 (±20.9)
PW Value	131.2 (±89.7)	126 (±84.8)	71 (±89.2)	114 (±99.2)
MUN BV	-	-	1.81 (±0.3)	-0.63 (±0.3)
Live weight (kg)	475 (±50.8)	473 (±54.6)	461 (±64.1)	476 (±63.0)

Where BW is breeding worth and PW is production worth

calving in August 2018, and randomly assigned to one of the two farm systems (Table 2). To evaluate the interaction between animal selection for MUN and farm system, an additional 12 high and 12 low MUN cows were divided into two equal groups and assigned to either the LSR or MSR farm system. Divergence in MUN was determined based on MUN breeding value (mg/dL) estimated from pedigree and the animal's own MUN records from herd-testing, as described by Beatson et al. (2019). The parent average MUN BV would have been a better indicator for the implications of selection for low MUN in the two farm systems, however, these data were not available.

Management

Over the calving period (August and September) cows were allocated to their farmlets once a minimum of 16 cows from each herd had calved. The majority of cows had calved by the beginning of October and measurements commenced once cows had reached peak milk yield at nine weeks from calving. For both farms grazing management decisions were similar which aimed to utilise pasture grown. Farm walks were carried out weekly to estimate pasture mass of each paddock using a rising plate meter. A seven day grazing plan was developed after each farm walk to maintain consistent allocation for each farmlet. Target pre grazing mass was between 2800-3000 kg DM/ha with pastures grazed to a compressed residual of approximately 4 cm. When surplus or deficits occurred, supplement was made (baleage) or fed out as required. Daily stocking density was similar for both farmlets as all paddocks were approximately 0.5 ha. Cows in the MSR system were offered 2.5 to 3.0 kg DM/day as crushed cereal grain using the in-shed feeding system (crude protein ranging between 15 and 18% and metabolisable energy (ME) content of 13-14 MJ/kg DM). Bloat prevention was by bloat oil in troughs at the first sign of bloat, which occurred in the LSR system in April 2019.

Measurements

Cows were milked twice daily from 0600 and 1430 hours and milk volume was measured at each milking through the Delpro system consisting of in-line flow meters. Milk samples from each cow were collected approximately fortnightly at consecutive afternoon and morning milkings resulting in two samplings per month except at the start and end of the study in October and May, when only one milk sampling was carried out. Twenty millilitres of milk were sampled into bronopol-containing tubes and sent to MilkTest NZ within 24 hours for analysis of fat, protein and lactose (%) concentration along with SCC and milk urea (mg/dL) using a mid-infrared (MIR) CombiFoss machine (Foss Electric, Denmark). Cow live weight was recorded twice daily after milking using walk-over weigh scales. Somatic cell count was transformed to somatic cell score (SCS, $\log_2$(SCC/1000).

Samples of herbage were collected fortnightly from paddocks which cows were grazing during milk sampling. Pluck samples were collected from random locations within each paddock and soon after were sorted into botanical components and dried at 60°C in an air force oven. The oven-dried sample was ground to pass through a 1 mm sieve and analysed for nutrient content by NIRS.

Statistical analysis

All analysis was completed using Genstat (16th edition). Summary statistics (means and standard deviations) for cow and pasture variables were collated through time for each cow or each paddock within each farmlet. To compare the effect of MUN group in the different farming systems on milk yield and milk composition, the milk variables were analysed by repeated-measures analysis of variance using farm system (LSR, MSR) and MUN group and their interaction as the fixed terms and cow replicate as the random term for each repeated herd test date (n=13 test dates). The data included only the 24 cows selected for MUN (n=12 max per MUN group). Total accumulated milk yield and milk solid yield was analysed for variance using farm system and MUN group and their interaction as the fixed terms and cow as the random term.

Results

In both MSR and LSR, the proportion of legume in both farm systems were similar, in spite of differences in N fertiliser rate (185 and 130 kg N/ha respectively). In the diverse pastures (45% of the farm) plantain content ranged from less than 10% in November 2018 to a peak of 45% in April 2019 and a seasonal average of 20% of the offered biomass in diverse pastures or 17% for the whole farm. This resulted in nearly 20% lower grass content in LSR compared with MSR (Table 3).

Table 3 Mean (and standard deviation, SD) days in milk, milk yield and milk composition, pasture cover, botanical content and pasture nutritive value for a moderately stocked farm system supplemented with grain (MSR) and a low stocked farm including plantain diverse pastures (LSR)

	MSR		LSR		
	Mean	SD	Mean	SD	% dif
Days in milk (d)	257	21.4	254	26.5	1.2%
Milk yield (L/cow)	4347	800.4	4158	876.3	4.3%
Milksolid yield (kg/cow)	466	83.9	429	70.2	7.9%
Fat (%)[1]	5.51	0.69	5.41	0.70	1.8%
Protein (%)[1]	4.16	0.38	4.16	0.40	0.1%
Fat+protein (%)[1]	9.67	1.04	9.57	1.05	1.0%
Lactose (%)[1]	4.97	0.11	4.99	0.11	-0.4%
Somatic cell score	2.05	0.44	1.98	0.51	3.3%
Urea N (mg/dL)[1]	12.7	1.6	10.4	1.8	18.7%
Fat yield (kg/d)	230	43.1	213	36.0	7.4%
Protein yield (kg/d)	236	44.2	216	37.5	8.5%
APC (kg DM/ha)	2333	126	2464	200	-5.6%
Grass (%)	79.9	0.13	64.8	0.24	18.9%
Legume (%)	8.4	7.2	10.3	10.6	-22.6%
Plantain (%)	2.7	6.4	16.7	20.9	-518.5%
Dry matter (%)	18	3.4	18	5.0	0.0%
Organic matter (%)	90.7	0.9	90.7	1.1	-0.1%
Crude protein (%)	21.9	3.25	20.4	4.17	6.8%
Neutral detergent fibre (%)	43.8	4.01	42.6	5.28	2.7%
DOMD (%)	73.5	3.37	72.9	4.11	0.7%
Metabolisable energy (MJ/kg/DM)	11.8	0.54	11.7	0.66	0.8%

Values are based on 40 cows per farmlet. Where DOMD is the in vitro digestibility of the organic matter in the dry matter and M/D is megajoules of metabolisable energy per kilogram of dry matter; APC is average pasture cover.

[1] Means are average of herd test day between October 2020 and May 2021 and are not adjusted for test day yield.

Across the season the average pasture mass was around 130 kg DM/ha greater in the LSR compared with the MSR system (Table 3). However loss of pasture control in the LSR system over mid-to-late spring (average pasture mass 2862 kg DM/ha vs 2507 kg DM/ha for LSR and MSR respectively) had a negative impact on pasture quality as frequent rain events and baleage contractor availability prevented timely harvesting of paddocks for control of surplus pasture. Carry-over effects of these impacts influenced not just feed quality with lower crude protein and M/D, but milk yield for cows in LSR was lower during the first half of the season compared to MSR.

Although there was only a 4% difference in total milk yield, the cumulative effect of volume and a 1% higher milk solid concentration resulted in 8% more milk solids in the MSR system (Table 3). Cows in both farm systems had similar fat, protein, lactose concentration and SCS. Mean MUN concentration ranged between 9 and 18 mg/dL from cows in MSR and between 5 and 18 mg/dL for cows in LSR, the latter of which was on average 17% lower than that of the MSR system.

Analysis of the subset of cows selected for divergent MUN, showed that cows selected for high MUN tended towards an 11% greater milk yield ($P<0.10$, Table 4), but lower milk protein % ($P = 0.02$) and a trend towards lower total solids % ($P=0.06$) compared with low MUN cows. Consequently, no significant difference in milk solid yield was detected between cows with high and low MUN (Table 4). Expression of MUN phenotype was evident, irrespective of farm system,

Table 4 Effect of selection for high or low milk urea nitrogen (MUN) on milk yield and milk composition

Farmlet	MSR		LSR			P value		
MUN Group	High	Low	High	Low	SEM	MUN	Farmlet	M × F
Days in milk	270	268	273	256	5.8	0.24	0.60	0.37
Milk yield (litres/cow)	4684	4066	4272	3908	174	0.06	0.26	0.61
Fat (%)	5.36	5.77	5.32	5.77	0.192	0.13	0.94	0.94
Protein (%)	4.12	4.44	4.00	4.46	0.109	0.02	0.75	0.68
Fat + protein (%)	9.5	10.2	9.3	10.2	0.288	0.06	0.86	0.83
Lactose (%)	4.94	4.95	4.99	4.98	0.030	0.97	0.33	0.88
Somatic cell score	1.88	2.17	2.10	1.86	0.137	0.91	0.84	0.18
Urea N (mg/dL)	14.2	11.4	12.5	10.3	0.38	<0.001	0.02	0.59
Fat yield (kg/cow)	237	225	217	195	8.0	0.15	0.04	0.65
Protein yield (kg/cow)	253	236	218	207	8.3	0.25	0.01	0.80
Milk solids yield (kg/cow)	489	461	435	402	15.1	0.17	0.02	0.92

Where SEM is the standard error of the mean for the main effect of MUN group with high MUN cows having nearly 20% higher MUN concentrations. No interactions between farm system type and MUN group was detected.

Discussion

In the current study, adopting a farm system with a moderately high stocking rate and supplementing with concentrate, resulted in greater milk and milk solids yield per cow than did a farm with low stocking rate incorporating plantain-diverse pastures. It is unlikely that the combination of species in the plantain-diverse pastures contributed to the lower observed milk yield as recent meta-analysis (McCarthy et al. 2019) and farm systems study (Al-Marashdeh et al. 2021) have shown that diverse pastures with plantain are able to maintain or improve milk production. The difference in milk production is probably the result of pasture management which led to a loss of quality in the LSR system in late spring, and carry-over effects of low pasture utilisation in early summer. Previous research has demonstrated the impact of low pasture utilisation on reduced milk yield for both ryegrass clover (Hoogendorn et al. 1982) and diverse (Cun et al 2018) pastures, highlighting the importance of green leaf allocation to enable high digestible nutrient intake. In earlier farm system research on the same property, lowering the stocking rate resulted in a positive effect on milk yield compared with a high stocking rate, grain-supplemented group (Chapman et al. 2021). While the study of Chapman et al. (2021) also faced challenges with pasture utilisation, the higher stocking rate of 3.5 cows/ha (compared with 2.9 cows in the present study) and increased use of mowing and topping enabled vegetative high quality pastures to be maintained.

With the exception of MUN, which was lower in LSR, there was no consistent effect of farm system on milk composition. Lack of milk composition response is in agreement with results from short-term grazing studies comparing plantain-diverse with conventional grass-clover pastures which have shown little effect on milk protein content, occasionally lower milk fat content, and regular reductions in MUN (Dodd et al. 2019; Bryant et al. 2018; Totty et al. 2013; Nkomboni et al. 2021). Nutritional manipulation to alter milk composition has been studied extensively and it is generally accepted that altering either milk protein or lactose content is difficult (Emery 1978; Sutton 1989). However, fat content along with MUN are more responsive to dietary manipulations (Sutton 1989).

From a nutritional standpoint MUN generally reflects the crude protein content of the diet, as it is positively related with N intake and blood urea (Spek et al. 2013). Although herbage CP was greater in MSR pastures, when the lower-protein grain is accounted for, it is unlikely that there were large differences in dietary crude protein content between the two systems. Rather the lower MUN concentration of individual cows in the LSR system may arise from other factors related to: timing of grazing and supplementation, greater water intake on diverse pastures, site of protein degradation, energy synchrony in the rumen - all of which are likely to play a role in the circulating concentrations of blood urea (Spek et al. 2013).

In spite of the range of dietary conditions which influence MUN, it is a relatively heritable trait (Beatson et al. 2019). Using animals selected for low MUN in the current study resulted in cows which were 7% less productive compared with cows selected for high MUN (Table 2, Figure 2). Beatson et al. (2019) reported positive genetic and phenotypic correlations between MUN and milk yield and negative genetic and phenotypic correlations between MUN and milk protein percentage, in that cows selected either at the phenotypic level or genetic level for lower MUN would be associated with lower milk yield but higher milk protein percentages. In the current study, an 11% reduction in milk yield for low-MUN cows was partially offset by greater protein composition giving a 7% reduction in milk solid yield. Marshall et al. (2020) also reported that cows selected for lower MUN had greater milk protein percentages. Similar to those in the current study, the cows used by Marshall et al. (2020) were selected based on their MUN breeding value that was calculated using ancestry, in addition to the cow's own MUN phenotype records. Both the current study and that of Marshall et al. (2020) were not designed to test the impact of genetic selection for lower MUN cattle due to incorporating the cow's own MUN phenotype in the estimation of MUN breeding values. Hojman et al. (2004), using a large database of cows in TMR systems, showed similar relationships with yield and protein as the present study, but observed that differences between high and low MUN were more pronounced at the start of lactation as differences in milk composition diminished over time. Arguably, had we used a greater number of cows, and/or full-season lactation data, milk yield differences may have been more pronounced. In the Hojman study, those authors also found a strong negative relationship between MUN and SCC which they were unable to explain and which our current research was unable to support. However, a farm system study, even at these limited animal numbers, are useful as pilot studies to indicate whether there might be obvious management interactions with genetic selection. The mode of action of herb-containing pastures on MUN versus selection for MUN did not appear to off-set each other as the low MUN phenotype was significant under both management regimes.

The purpose for selecting cows with lower MUN was based on the assumption that correlation with blood urea would lead to lower urinary urea excretion. However, recent research would suggest that due to the high inter-cow variability in MUN and urinary N excretion, selection for MUN is unlikely to reduce N loss either at the cow level (Handcock et al. 2021) or whole farm level (Ariyarathne et al. 2021). Furthermore, the results from this study highlight the importance of consideration of other important traits such as milk production which may be affected by selection on MUN alone.

Acknowledgements

This research is supported by the Low N Livestock project which is a DairyNZ-led collaborative research programme funded by DairyNZ and the Ministry of Business, Innovation and Employment, with co-funding from CRV-Ambreed and Fonterra. The farm systems study reported here is part of a wider Lincoln University initiative (Living Laboratory: Dairy Futures) which is grateful for financial support from Fonterra and seed supplied by Agricom. Finally, many thanks go to all the farm staff at LURDF and to Jasmine Tanner for statistical support.

References

Al-Marashdeh O, Cameron K, Hodge S, Gregorini P, Edwards G 2021. Integrating plantain (*Plantago lanceolata* L.) and Italian ryegrass (*Lolium multiflorum* lam.) into New Zealand grazing dairy system: the effect on farm productivity, profitability, and nitrogen losses. Animals DOI: 10.3390/ani11020376

Ariyarathne HBPC, Correa-Luna M, Blair H, Garrick D, Lopez-Villalobos N 2021. Can nitrogen excretion of dairy cows be reduced by genetic selection for low milk urea nitrogen concentration? Animals DOI: 10.3390/ani11030737

Beatson PR, Meier S, Cullen NG, Eding H, 2019. Genetic variation in milk urea nitrogen concentration of dairy cattle and its implications for reducing urinary nitrogen excretion. Animal 13:2164-2171.

Bryant RH, Snow VO, Shorten PR, Welten BG 2020. Can alternative forages substantially reduce N leaching? Findings from a review and associated modelling. New Zealand Journal of Agricultural Research 63:3-28.

Bryant RH, Welten BG, Costall D, Shorten PR, Edwards GR, 2018. Milk yield and urinary-nitrogen excretion of dairy cows grazing forb pasture mixtures designed to reduce nitrogen leaching. Livestock Science 209:46-53.

Correa-Luna M, Donaghy DJ, Kemp PD, Schutz MM, Lopez-Villalobos N 2019. Effect of genetic merit for milk urea on milk production and efficiency of crude protein utilization of grazing cows with contrasting supplement inclusion. New Zealand Journal of Animal Science and Production 79:112-117

Chapman DF, Dalley DE, Edwards GR, Cameron KC, Malcolm BJ, Clement A, Romera AJ, Pinxterhuis IB, Beukes PC, Di HJ, Bryant RH, Curtis J 2021. Production, profit and nitrogen flows in irrigated dairy systems representing different industry development pathways: the Pastoral 21 experience in Canterbury. New Zealand Journal of Agricultural Research, 64:1, 3-35, DOI: 10.1080/00288233.2020.1807367

Cun GS, Edwards GR, Bryant RH 2018. Milk production does not benefit from mowing previously lax-grazed diverse pastures. New Zealand Journal of Agricultural Research 61: 468-476

Dodd M, Dalley D, Wims C, Elliott D, Griffin A 2019. A comparison of temperate pasture species mixtures selected to increase dairy cow production and reduce urinary nitrogen excretion, New Zealand Journal of Agricultural Research 62: 504-527

Emery RS 1978. Feeding for increased milk protein. Journal of Dairy Science 61:825-828

Handcock RC, Box LA, Glassey C, Garrick DJ 2021 (this proc.). Genetic parameters for urination traits and their relationships with production traits of grazing dairy cattle. New Zealand Journal of Animal Science and Production 81: 153.158.

Hojman D, Kroll O, Adin G, Gips M, Hanochi B, Ezra E 2004. Relationships between milk urea and production, nutrition, and fertility traits in Israeli dairy herds. Journal of Dairy Science 87: 1001-1011

Hoogendoorn CJ, Holmes CW, Chu ACP 1992. Some effects of herbage composition, as influenced by previous grazing management, on milk production by cows grazing on ryegrass–white clover pastures. 2. Milk production in late spring–summer: effects of grazing intensity during the preceding spring period. Grass and Forage Science 47: 316-325.

Jonker JS, Kohn RA, Erdman RA 1998. Using milk urea nitrogen to predict nitrogen excretion and utilization efficiency in lactating dairy cows. Journal of Dairy Science 81:2681-2692.

Marshall CJ, Beck MR, Garrett K, Barrell GK, Al-Marashdeh O, Gregorini P 2020. Grazing dairy cows with low milk urea nitrogen breeding values excrete less urinary urea nitrogen. Science of the Total Environment DOI:10.1016/j.scitotenv.2020.139994

McCarthy KM, McAloon CG, Lynch MB, Pierce KM, Mulligan FJ 2020. Herb species inclusion in grazing swards for dairy cows - a systematic review and meta-analysis. Journal of Dairy Science. 103: 1416-1430.

Nkomboni D, Bryant RH, Edwards GR 2021. Effect of increasing dietary proportion of plantain on milk production and nitrogen use of grazing dairy cows in late lactation. Animal Production Science DOI: 10.1071/AN20440

Spek JW, Dijkstra J, Van Duinkerken G, Bannink A 2013. A review of factors influencing milk urea concentration and its relationship with urinary urea excretion in lactating dairy cattle. Journal of Agricultural Science 151: 407-423.

Sutton JD 1989. Altering milk composition by feeding. Journal of Dairy Science 72:2801-2814

Totty VK, Greenwood SL, Bryant RH, Edwards GR 2013. Nitrogen partitioning and milk production of dairy cows grazing simple and diverse pastures. Journal of Dairy Science 96: 141-149.

AUTHOR INDEX